INSTRUCTOR'S MANUAL TO

LITERATURE
An Introduction to Fiction, Poetry, and Drama

INSTRUCTOR'S MANUAL TO ACCOMPANY

LITERATURE
An Introduction to Fiction, Poetry, and Drama

Fifth Edition

X. J. Kennedy

Dorothy M. Kennedy

HarperCollinsPublishers

Instructor's Manual to accompany *Literature: An Introduction to Fiction, Poetry, and Drama*, Fifth Edition

Copyright © 1990 by X. J. Kennedy and Dorothy M. Kennedy

All rights reserved. Printed in the United States of America. No part of this book may be used or reproduced in any manner whatsoever without written permission, with the following exception: testing materials may be copied for classroom testing. For information, address HarperCollins Publishers Inc., 10 East 53rd Street, New York, NY 10022.

ISBN: 0-673-49882-4

90 91 92 93   9 8 7 6 5 4 3 2 1

Acknowledgments

"Introduction to Poetry" from *The Apple That Astonished Paris*, poems by Billy Collins. Copyright © 1988 by Billy Collins. Reprinted by permission of The University of Arkansas Press. "Buck It" by Jerald Bullis. Reprinted by permission of the author.

Contents

Preface xi

FICTION

1 *Reading a Story* 1
   FABLE AND TALE 1
   THE SHORT STORY 2

2 *Point of View* 4

3 *Character* 9

4 *Setting* 15

5 *Tone and Style* 18
   IRONY 21

6 *Theme* 24

7 *Symbol* 29

8 *Evaluating a Story* 34

9 *Reading Long Stories & Novels* 35

10 *Three Fiction Writers in Depth* 38
   RAYMOND CARVER 38
   ANTON CHEKHOV 40
   FLANNERY O'CONNOR 43

11 *Stories for Further Reading* 48

12 *Criticism: On Fiction* 69

POETRY

Poems Arranged by Subject and Theme 75

Poems for Further Reading, Arranged by Elements 89

13 *Reading a Poem*   99
   LYRIC POETRY   99
   NARRATIVE POETRY   100

14 *Listening to a Voice*   102
   TONE   102
   THE PERSON IN THE POEM   104
   IRONY   106
   FOR REVIEW AND FURTHER STUDY   108

15 *Words*   111
   LITERAL MEANING: WHAT A POEM SAYS FIRST   111
   THE VALUE OF A DICTIONARY   113
   WORD CHOICE AND WORD ORDER   115
   FOR REVIEW AND FURTHER STUDY   116

16 *Saying and Suggesting*   119

17 *Imagery*   122
   ABOUT HAIKU   125
   FOR REVIEW AND FURTHER STUDY   126

18 *Figures of Speech*   130
   WHY SPEAK FIGURATIVELY?   130
   METAPHOR AND SIMILE   130
   OTHER FIGURES   132
   FOR REVIEW AND FURTHER STUDY   134

19 *Song*   137
   SINGING AND SAYING   137
   BALLADS   138
   FOR REVIEW AND FURTHER STUDY   139

20 *Sound*   140
   SOUND AS MEANING   140
   ALLITERATION AND ASSONANCE   141
   RIME   142
   READING AND HEARING POEMS ALOUD   144

21 *Rhythm* 147
    STRESSES AND PAUSES 147
    METER 149

22 *Closed Form, Open Form* 151
    CLOSED FORM: BLANK VERSE, STANZA, SONNET 151
    OPEN FORM 154
    FOR REVIEW AND FURTHER STUDY 157

23 *Poems for the Eye* 160

24 *Symbol* 162

25 *Myth* 166

26 *Alternatives* 169
    THE POET'S REVISIONS 169
    TRANSLATIONS 169
    PARODY 171

27 *Evaluating a Poem* 173
    TELLING GOOD FROM BAD 173
    KNOWING EXCELLENCE 176

28 *What Is Poetry?* 181

29 *Poems for Further Reading* 182

30 *Lives of the Poets* 241

31 *Criticism: On Poetry* 242

DRAMA

32 *Reading a Play* 247
    A PLAY IN ITS ELEMENTS 247
    TRAGEDY AND COMEDY 247

33 *The Theater of Sophocles* 253

34 *The Theater of Shakespeare* 255

35 *The Modern Theater* 258
    EXPERIMENT AND THE ABSURD 260

36 *Evaluating a Play* 264

37 *Plays for Further Reading* 265

38 *Criticism: On Drama* 275

SUPPLEMENT: WRITING

Writing about Literature 281

Writing about a Story 283

Writing a Story 285

Writing about a Poem 286

Writing a Poem 289

Further Notes on Teaching Poetry 293

On Integrating Poetry and Composition 294

Writing about a Play 297

Writing a Play 298

Preface

This manual pretends to be no more than a sheaf of notes to supply you—if you want them—with classroom strategies, comments by critics, and a few homemade interpretations. These last may be wrong, but we set them down in hopes of giving you something clear-cut to disagree with. Wherever it seemed pertinent, facts about a literary work and its background are also supplied.

Every story, poem, and play is dealt with, except for several brief poems given in the text as illustrations.

In the FICTION section of the book itself, the "Stories for Further Reading" have no questions after them. This manual supplies some, and, most of the time, tentative answers (in parentheses). To be sure, other answers are likely and possible.

To help you refer quickly to the parent book, *page numbers are given at the top of each page in this manual*. These correspond to pages in the book itself.

We've always found, before teaching a knotty piece of literature, no preparation more helpful than to sit down and discuss it with a colleague or two. If this manual supplies you with such a colleague or two at inconvenient hours, such as 2:00 A.M. when there's nobody in the faculty coffee room, it will be doing its job.

PLAN OF THE BOOK

There is a plan, but it does not oblige you to follow it. Chapters may be taken up in any sequence: some instructors like to intersperse poetry and plays with stories. Some may wish to teach Chapter 25, on Myth, immediately before teaching *Oedipus the King*. Many find that a useful chapter with which to begin teaching poetry is "Imagery."

If, because you skip around in the book, students meet a term unknown to them, let them look it up in the "Index of Terms." They will be directed to the page where it first occurs and is defined and illustrated.

In POETRY, sections entitled "For Review and Further Study" do not review the whole book up to that moment; they review only the main points of the chapter. Most of these sections contain some poems a little more difficult than those in the body of that chapter.

RENOVATIONS

Besides the major changes in this edition, discussed in the book's Preface, many other renovations have been made. For anyone interested, we will list a few in detail.

Most users of the book seemed to feel no need for a whole chapter on irony in the FICTION section; this material has been kept but now follows the discussion of Tone and Style in Chapter 5. Fresh fiction selections represent Toni Cade Bambara, Nathaniel Hawthorne, Franz Kafka, Katherine Mansfield, and Alice Munro. Stories have been added by the following writers not represented as fiction writers in the last edition: Margaret Atwood, Ambrose Bierce, Jorge Luis Borges, Raymond Carver, Anton Chekhov, Maxine Chernoff, Tess Gallagher, Langston Hughes, Yasunari Kawabata, Jamaica Kincaid, Jack London, Bobbie Ann Mason, Anne Tyler, Kurt Vonnegut, Jr. (this last an old favorite back by demand), and William Carlos Williams.

Once again, the book's representation of minority poets has been strengthened, and so has the number of woman poets, both earlier and contemporary. Now added to this edition are Jimmy Santiago Baca, Aphra Behn, Deborah Digges, Rita Dove, Mary Elizabeth Coleridge, Louise Erdrich, Linda Gregg, Elizabeth Jennings, Lorine Niedecker, Katherine Philips, Grace Schulman, Wole Soyinka, Ruth Stone, Emma Lee Warrior, Nancy Willard, and Elinor Wylie. Several others who were in the last edition are represented now in greater depth. This time the ample *Anthology: Poetry* has been lengthened by only 2 poems, bringing the total to 146. It now includes 44 new selections—traditional, modern, and contemporary.

For classes who have time for it, a knotty problem—how to evaluate an unfamiliar poem—is now treated in detail at the end of Chapter 27, with fresh and almost certainly unfamiliar poems by two excellent, lesser-known poets, Lorine Niedecker and James Hayford. A checklist of questions is supplied to help students make up their own minds. Users of the bad poetry section in this chapter will be happy to see that it now sports major works by William McGonagall, the Bard of Dundee, and Julia A. Moore, The Sweet Singer of Michigan, besides other horrific new examples. (Don't miss those by Francis Saltus Saltus.)

In DRAMA, some will miss *The Tempest*, which scored disappointingly. *Othello* continued to hold its own. The ready availability of Shakespeare's plays in paperback should serve, we trust, anyone who has time to teach a second one. Molière was difficult to drop; only a happy few, it appears, care to teach his work, and for those few, *The Physician in Spite of Himself* did not suffice. We look forward to the day of the tailored-to-order textbook, which will contain the very plays each instructor wishes, and no more. There are still two plays of Sophocles: *Antigone* besides *Oedipus the King*. All in all, it seemed that the book's most valuable new service to DRAMA users (and best use of space) was to include the whole text of August Wilson's major contemporary play *Joe Turner's Come and Gone*. But please let us know if this assumption is mistaken.

TEXTS AND EDITORIAL POLICY

Spelling has been modernized (*rose-lipped* for Campion's *ros-lip'd*) and rendered American, unless to do so would change the sound of a word. But the y has remained in Blake's strange "Tyger," and Whitman has kept his *bloom'd* on the conviction that *bloomed* would no more resemble Whitman than a portrait of him in a starched collar. Untitled poems are identified by their first lines, except for those that have titles assigned by custom ("The Twa Corbies"). Chaucer's poem is given as edited by F. N. Robinson; the poems of Emily Dickinson, as edited by Thomas H. Johnson.

It would have been simpler to gloss no word a student could find in a desk dictionary, on the grounds that rummaging dictionaries is good moral discipline; but it seemed best not to require the student to exchange text for dictionary as many as thirty times in reading a story, poem, or play. Glosses have been provided, therefore, for whatever seemed likely to get in the way of pleasure and understanding.

The spelling *rime* is used instead of *rhyme* on the theory that *rime* is easier to tell apart from *rhythm*.

STATISTICS ON POETRY

This edition of the book includes 440 whole poems, counting translations, haiku, clerihews, and anonymous examples. (Fragments weren't counted.) There are 289 poems in the body of the text, 146 in the *Poems for Further Reading*, and 4 in the *Supplement: Writing*.

In case you wish to teach a poet's work in greater depth than a single poem affords, these 24 poets are most heavily represented (listed by number of poems):

Robert Frost	11
Emily Dickinson	10
Thomas Hardy	10
W. B. Yeats	10
William Shakespeare	9
William Blake	8
Walt Whitman	8
W. C. Williams	8
Wallace Stevens	6
William Wordsworth	6
T. S. Eliot	5
Langston Hughes	5
Gerard M. Hopkins	5
A. E. Housman	5
John Keats	5
Theodore Roethke	5
Alfred, Lord Tennyson	5
Richard Wilbur	5
W. H. Auden	4
Elizabeth Bishop	4
Gwendolyn Brooks	4
E. E. Cummings	4
John Donne	4
George Herbert	4

Besides, there are now 3 poems apiece by James Hayford, Robert Herrick, Ben Jonson, Philip Larkin, Denise Levertov, John Milton, Lorine Niedecker, Sylvia Plath, Alexander Pope, Adrienne Rich, and James Wright; many other poets are represented twice.

THANKS

Many instructors, most of whose names appear in this manual, generously wrote us their suggestions and teaching experiences. We thank them in particular, and also Dan Kennedy, a student at Oberlin College, for supplying some fresh insights.

A CAUTIONARY NOTE ON TEACHING LITERATURE

In the following poem, Billy Collins, who as you might suspect is a professor of English himself—at Lehman College of CUNY—sets forth an experience that may be familiar. (The poem comes from a collection entitled *The Apple That Astonished Paris* [Fayetteville: U of Arkansas P, 1988].)

Introduction to Poetry

I ask them to take a poem
and hold it up to the light
like a color slide

or press an ear against its hive.

I say drop a mouse into a poem
and watch him probe his way out,

or walk inside the poem's room
and feel the walls for a light switch.

I want them to waterski
across the surface of a poem
waving at the author's name on the shore.

But all they want to do
is tie the poem to a chair with rope
and torture a confession out of it.

They begin beating it with a hose
to find out what it really means.

May this manual help you persuade your students to set aside rope and hose, and help turn on a few lights.

INSTRUCTOR'S MANUAL TO ACCOMPANY

LITERATURE
An Introduction to Fiction, Poetry, and Drama

Fiction

1 Reading a Story

For a second illustration of a great detail in a story, a detail that sounds observed instead of invented (besides Defoe's "two shoes, not mates"), you might cite a classic hunk of hokum: H. Rider Haggard's novel of farfetched adventure, *She* (1887). Describing how the Amahagger tribesmen wildly dance by the light of unusual torches—embalmed corpses of the citizens of ancient Kor, left over in quantity—the narrator, Holly, remarks, "So soon as a mummy was consumed to the ankles, which happened in about twenty minutes, the feet were kicked away, and another put in its place." (Pass down another mummy, this one is guttering!) Notice the exact specification "in about twenty minutes" and the unforgettable discarding of the unburned feet, like a candle stub. Such detail, we think, bespeaks a tall-tale-teller of genius. (For this citation, we thank T. J. Binyon's review of *The Private Diaries of Sir Henry Rider Haggard* in *Times Literary Supplement*, 8 Aug. 1980.)

When you introduce students to the *tale* as a literary form, you might point out that even in this age of electronic entertainment, a few tales still circulate from mouth to ear. Ask them whether they have heard any good tales lately (other than dirty jokes). This tale reportedly has been circulated recently in Israel:

> Two Israeli agents, captured in an Arab country, are tied to stakes to be shot. While the firing squad stands in readiness, the Arab commander asks one of the captured men if he has any last request. For answer, the captive spits in the commander's face. From the other captive comes a wail: "For God's sake, Moishe, don't cause *trouble!*"

FABLE AND TALE

W. Somerset Maugham
THE APPOINTMENT IN SAMARRA, page 2

Maugham's retelling of this fable has in common with the Grimm tale "Godfather Death" not only the appearance of Death as a character, but also the moral or lesson that Death cannot be defied. Maugham includes this fable in his play *Sheppey* (1933), but it is probably best known as the epigraph to John O'Hara's novel *The Appointment in Samarra* (New York: Random, 1934).

Students may be asked to recall other fables they know. To jog their memories, famous expressions we owe to Aesop ("sour grapes," "the lion's share," "dog in the manger," and others) may suggest the fables that gave them rise. At least, the fable of the hare and the tortoise should be familiar to any watcher of old Bugs Bunny cartoons.

If you have students write fables (as suggested on page 17), you'll probably find that most will enjoy the task but some will find it difficult. To get them going, you might have them read James Thurber's "The Unicorn in the Garden" at the end of this chapter, or read them other masterworks from Thurber's *Fables for Our Time* (New York: Harper, 1940).

Jakob and Wilhelm Grimm
GODFATHER DEATH, page 4

For all its brevity, "Godfather Death" illustrates typical elements of plot (discussed in the text on pages 6–8). That is the main reason for including it in this chapter (not to mention its intrinsic merits!). It differs from Updike's contemporary "A & P" in its starker characterizations, its summary method of telling, its terser descriptions of setting, and its element of magic and the supernatural. In its world, God, the Devil, and Death walk the highway. If students can be shown these differences, then probably they will be able to distinguish most tales from most short stories.

"Godfather Death" may be useful, too, in a class discussion of point of view. On page 18, you will find some ways in which this tale is stronger for having an omniscient narrator. If you go on to deal with symbolism, you may wish to come back to this tale for a few illustrations of wonderful, suggestive properties: the magical herb, Death's underground cave, and its "thousands upon thousands" of flickering candles.

This is a grim tale even for Grimm: a young man's attentions to a beautiful princess bring about his own destruction. In a fairy tale it is usually dangerous to defy some arbitrary law; and in doing so here the doctor breaks a binding contract. From the opening, we know the contract will be an evil one—by the father's initial foolishness in spurning God. Besides, the doctor is a thirteenth child—an unlucky one.

Possible visual aids: reproductions of the "Dance of Death" woodcuts by Hans Holbein the younger. Have any students seen Ingmar Bergman's film *The Seventh Seal,* and can they recall how Death was personified?

Anne Sexton has a sophisticated retelling of "Godfather Death," in which the doctor's guttering candle is "no bigger than an eyelash," in her *Transformations,* a collection of poems based on Grimm (Boston: Houghton, 1971). "Godfather Death" is seldom included in modern selections of fairy tales for children. The splendid translation by Lore Segal appeared in *The Juniper Tree and Other Tales from Grimm,* selected by Segal, with four tales translated by Randall Jarrell (New York: Farrar, 1973), where a fine drawing by Maurice Sendak accompanies it. Bruno Bettelheim has nothing specific to say about "Godfather Death" but has much of interest to say about fairy tales in his *The Uses of Enchantment* (New York: Knopf, 1976). Though Bettelheim's study is addressed primarily to adults "with children in their care," any college student fascinated by fairy tales would find it stimulating.

THE SHORT STORY

John Updike
A & P, page 9

Within this story, Sammy rises to a kind of heroism. Despite the conventional attitudes he expresses in the first half of the story (his usual male reactions to girls in two-piece bathing suits, his put-down of *all* girls' minds: "a little buzz like a bee in a glass jar"), he comes to feel sympathy for the girls as human beings. He throws over his job to protest against their needless humiliation, and in so doing he asserts his criticisim of supermarket society, a deadly world of "sheep" and "houseslaves" whom dynamite couldn't budge from their dull routines. What harm in a little innocent showing off—in, for once, a display of nonconformity?

Sammy isn't sophisticated. He comes from a family of proletarian beer drinkers and thinks martinis are garnished with mint. His language is sometimes pedestrian: "Really, I thought that was so cute." But he is capable of fresh and accurate observations—his description of the way Queenie walks on her bare feet, his comparison of the "clean bare plane" of her upper chest to "a dented sheet of metal tilted in the light." (Could Sammy be capable of so poetic an observation, or is this Updike's voice?)

Carefully plotted for all its seemingly casual telling, "A & P" illustrates typical elements. The *setting* is clear from Updike's opening paragraph ("I'm in the third checkout slot . . . with my hand on a box of HiHo crackers"). Relatively long for so brief a story, the *exposition* takes up most of the story's first half. Portraying Queenie and the other girls in loving detail, this exposition helps make Sammy's later gesture of heroism understandable. It establishes, also, that Sammy feels at odds with his job and so foreshadows his heroism. He reacts against butcher McMahon's piggishness: "patting his mouth and looking after them sizing up their joints." *Dramatic conflict* arrives with the appearance of Lengel the manager and his confrontation with the girls. Lengel catches Sammy smiling—we can guess the clerk is in for trouble. Crisis and climax are practically one, but if you care to distinguish them, the *crisis* may be found in the paragraph "I thought and said 'No' but it wasn't about that I was thinking," in which Sammy hovers on the brink of his decision. The *climax* is his announcement "I quit"; the *conclusion* is his facing a bleaker future. The last sentence implies not only that Sammy will have trouble getting another job but that if he continues to go through life as an uncompromising man of principle, then life from now on is going to be rough.

In "A & P" and "Godfather Death" the plots are oddly similar. In both, a young man smitten with a young woman's beauty makes a sacrifice in order to defend her against his grim overlord. (Far worse, of course, to have Death for an overlord than to have Lengel.) If this resemblance doesn't seem too abstract, it may be worth briefly pursuing. The stories, to be sure, are more different than similar, but one can show how Updike is relatively abundant in his descriptions of characters and setting and goes more deeply into the central character's motivation—as short-story authors usually do, unlike most writers of tales.

For a fresh explication of this story, see Janet Overmeyer, "Courtly Love in the A & P," in *Notes on Contemporary Literature* for May 1972 (West Georgia College, Carrollton, GA 30117).

James Thurber
THE UNICORN IN THE GARDEN, page 15

More ample than most fables (such as Maugham's or Aesop's spare examples) in its characterizations and its use of dialogue, Thurber's piece still seems briefer and simpler than a short story. Clearly the author is on the side of the husband, who calmly accepts the impossible and takes for granted that a mythic beast belongs in ordinary life. So calm is he that, when the unicorn disappears, he doesn't sound an alarm or call NBC-TV news—he simply goes to sleep. His wife, who can't imagine anything so wonderful, is the crazy one, not he, and well deserves her strait-jacket. More than a shaggy dog story leading to a pun, this "Fable for Our Time" makes a serious (if lighthearted) pitch for the sanctity of miracles and the sanity of people who can imagine.

If you assign suggestion 7 of the "Suggestions for Writing" and students compare this fable with "The Catbird Seat" (in "Stories for Further Reading"), they will probably notice, beyond the differences between a fable and a short story, that in both works the author sides with a mild little man against a hostile and aggressive woman. Is Thurber a sexist pig? Rather, we'd say, a sly champion of meek male underdogs—but the accusation may spark a bit of controversy.

2 Point of View

For other illustrations of the relatively scarce second-person point of view, see Richard Hugo's poetry collection *31 Letters and 13 Dreams* (New York: Norton, 1977): "In Your Fugitive Dream," "In Your War Dream," "In Your Young Dream," and others. These also appear in Hugo's collected poems, *Making Certain It Goes On* (New York: Norton, 1984).

William Faulkner
A ROSE FOR EMILY, page 24

In his opening sentence, the narrator speaks for the whole town of Jefferson, Mississippi, and throughout the story he always says *we*, never *I*, and he is never named. He is an interested bystander, not a participant in the story's main events, some of which presumably took place before he was born. Telling the story from the point of view of such an observer gives Faulkner advantages. The narrator can report events with detachment, and the tone of his story, often matter-of-fact and gossipy ("the Griersons held themselves a little too high for what they really were"), seems earthbound and convincing. It prepares us to accept on trust his later, more surprising, disclosures. Told from, say, the point of view of the central character herself, "A Rose for Emily" would be a radically different story: one without murder or necrophilia (because Miss Emily cannot acknowledge death), without mystery and suspense, and without a final revelation.

Students will want to make sure exactly what happened in the story. From the detail that the strand of hair is *iron-gray*, it appears that Emily lay beside Homer's body recently, long after it was rotten; probably lay beside it many times, for her pillow is clearly indented. Just as she had clung to her conviction that her father and Colonel Sartoris were still alive, she had come to believe that Homer Barron had faithfully married her, and she successfully ignored for forty years all the testimony of her senses. The conclusion of the story is foreshadowed by Emily's refusal to allow her father to be buried, by her purchase of rat poison, by the disappearance of Homer Barron, and by the pervasive smell of decay. In fact, these forshadowings are so evident it is a wonder that, for those reading the story for the first time, the ending is so surprising. Much of the surprise seems due to the narrator's back-and-forth, unchronological method of telling. We aren't told that (1) Emily buys poison, (2) Homer disappears, and (3) there is a mysterious odor—which chain of events might immediately rouse our suspicions. Instead, we hear about (in order) odor, poison, and disappearance. By this arrangement, any connection between these events is made to seem a little less obvious.

Having satisfied their natural interest in the final horror of the story, students can be led to discuss why "A Rose for Emily" isn't a mere thriller. Why (they may be asked) is the story called "A Rose"? No literal rose appears in it, although there may be two serious puns in paragraphs 5 and 6 ("a faint dust rose," "They rose when she entered"), and in the final description of the bedroom there are "curtains of faded rose color" and "rose-shaded lights." Perhaps Emily herself is the white rose of Jefferson (like the heroine of *The White Rose of Memphis*, a novel by Faulkner's scribbling grandfather). But the usual connotations of roses will apply. A rose is a gift to a loved one, and the whole story is the narrator's tribute to Emily.

But, some may object, how can anyone wish to pay tribute to a bloated old poisoner who sleeps with a corpse? The narrator patiently gives us reasons for his sympathy. As a girl, Emily was beautiful, a "slender figure in white," fond of society. But her hopes were

(text pages) 24–31

thwarted by her domineering father, whose horsewhip discouraged suitors from her door. Her strength and pride vanquished all who would invade her house: the new Board of Aldermen who tried to collect her taxes, the Baptist minister sent to lecture her on her morals, the relatives from Atlanta who eventually departed. "It is important," Ray B. West writes, "to realize that during the period of Emily's courtship the town became Emily's allies in a contest between Emily and her Grierson cousins, 'because the two female cousins were even more Grierson than Miss Emily had ever been'" ("Atmosphere and Theme in Faulkner's 'A Rose for Emily,'" *Perspective* Summer 1949: 239–45).

Suggestions abound. Emily's refusal to recognize change is suggested in the symbol of her invisible watch (paragraph 7), with its hint that she lives according to a private, secret time of her own. Her house seems an extension of her person in its "stubborn and coquettish decay" (2). Now it stands amid gasoline pumps, refusing (like its owner) to be part of a new era. The story contains many such images of stasis: when Emily confronts the aldermen, she looks bloated, "like a body long submerged in motionless water"—a foreshadowing, perhaps, of the discovery of Homer's long-guarded dust.

Some have read the story as an allegory: Homer Barron is the crude, commercial North who invades, like a carpetbagger. Emily, with her faithful ex-slave, is the Old South, willing to be violated. In an interview with students at the University of Virginia, Faulkner played down such North-South symbolism. "I don't say that's not valid and not there," he said, "but . . . [I was] simply writing about people." (The whole interview in which Faulkner discusses this story, not very helpfully, is in *Faulkner in the University*, ed. Frederick Gwynn and Joseph Blotner, University Press of Virginia, 1959.)

Still, it is clear that Emily, representing an antebellum first family, receives both Faulkner's admiration and his criticism for resisting change. "The theme of the story," according to C. W. M. Johnson, "can be stated: 'If one resists change, he must love and live with death,' and in this theme it is difficult not to see an implied criticism of the South" (*The Explicator* VI [No. 7] May 1948: item 45). But Faulkner's criticism, Ray B. West, Jr., feels, is leveled at the North as well. West makes much of the passage in which Faulkner discerns two possible views of Time (55). If, for the South, Time is "a huge meadow which no winter ever quite touches," then for the North it is a mere "mathematical progression" and "a diminishing road." West would propose, for a statement of the story's theme: "One must neither resist nor wholly accept change, for to do either is to live as though one were never to die; that is, to live *with* Death without knowing it" (*The Explicator* VII [No. 1] Oct. 1948: item 8).

Studying "A Rose for Emily" may help prepare students for Faulkner's "Barn Burning" (Chapter 5), whose central character, the son of a sharecropper, is Colonel Sartoris Snopes. "A Rose for Emily" clearly is in the tradition of the Gothic story, for it has a crumbling mansion, a mysterious servant, and a hideous secret. For comparison, one might have students read Poe's "The Tell-Tale Heart," if they don't know it already: another story of violence and madness, but told (unlike "A Rose") from the point of view of the violent madman.

For a valuable discussion of the story's point of view see John Daremo, "Insight into Horror: The Narrator in Faulkner's 'A Rose for Emily,'" in Sylvan Barnet, *A Short Guide to Writing about Literature*, 5th ed. (Boston: Little, Brown, 1985). A more superficial view of the story is expressed in this limerick by a celebrated bard, Anonymous:

> Miss Emily, snobbish and cranky,
> Used to horse around town with a Yankee.
>   When she'd wake up in bed
>   With the dust of the dead,
> She would sneeze in her delicate hanky.

When discussion of the story flagged, JoAnna Stephens Mink of Illinois State University, Normal, hit upon a way to rouse her students to fresh attention. She conducted a mock murder trial. Dividing a class of twenty-five into three groups (defense, prosecution, judge-and-jury), she spent half a class period letting each group prepare its evidence (admitting only faithful references to the story), and then devoted the other half to presentation and arguments. It was assumed, of course, that Emily was still alive and able to stand trial. After animated debate, judge and jury found Emily not guilty on grounds of insanity—but other outcomes might be possible. For Professor Mink's full account of this experiment, see "We Brought Emily Grierson to Trial" in *Exercise Exchange*, Spring 1984: 17–19.

Another instructor, after reading about Professor Mink's bright idea in last edition's manual, tried it and reported great success with it. He is Saul Cohen of County College of Morris. He assigned the story to be read and told students to prepare their cases for trial. When the day arrived to bring Emily before the bar of justice, he devoted fifteen minutes to letting the opposing counsel compare notes, determine strategy, and select their spokespersons. After the presentation of briefs, each side was allotted three minutes for rebuttal. The jury then deliberated before the rest of the class so that all could appreciate their discussions and their balloting. He himself served as judge and textual *raisonneur*, summing up positions taken by each side, transcribing on a blackboard summaries of arguments raised, in language acceptable to all, for counsel and jury to consider. The defense argued that reason existed to suspect Emily's faithful, highly protective, fast-disappearing servant of having done the deed and, having raised this uncertainty, persuaded the jury to find the defendant not guilty. For Mr. Cohen's full account of his experience, see *Exercise Exchange* for October 1990.

Katherine Mansfield
MISS BRILL, page 31

For discussion, we present here some possible answers to the questions on page 35. You may not agree with all of them.

1. "Miss Brill" is written with selective omniscience: from the third-person viewpoint of a nonparticipant, but one who sees events through the eyes of the story's protagonist. In paragraph 5, Miss Brill notices that there is "something funny" about her fellow park sitters: "They were odd, silent, nearly all old, and from the way they stared they looked as though they'd just come from dark little rooms or even—even cupboards!" (That she might look the same to them never enters her head—until she is forced to see herself as others see her.) Because we as readers experience the day's happenings from the perspective of Miss Brill, we come to understand and to sympathize with the sweet old dear, even go along with her sudden view of herself as an actress in a play performed every Sunday. To an intense degree we share her dismay and hurt when she overhears herself called a "stupid old thing" and her furpiece ridiculed as "exactly like a fried whiting."

2. In the opening paragraph Mansfield makes clear that Miss Brill is in the Jardins Publiques and, therefore, somewhere in France: in a small town, it seems, since the Sunday band concert is a big feature of local life. In paragraph 6 we learn that the sea is visible from the park. We also know that the season is autumn. Miss Brill has taken her fur out of its box for the first time in a long while. The trees are covered with "yellow leaves down drooping" (6). A chill in the air foreshadows the chill of Miss Brill's dashed spirits at the end of the story.

3. Miss Brill lives frugally, alone in a small room, eking out a meager living by teaching English and reading the newspaper aloud to an invalid gentleman four afternoons a week. She seems happy with her lot even though her daily activities are drab by most standards and her pleasures are small ones. Her well-worn fur delights her, and an almond in the Sunday honey-cake is cause for rejoicing.

4. Because she lives on the fringes of life in her small French town, Miss Brill regards her solitary Sundays in the park as the highlight of her week. Here, watching the people

who come and go and eavesdropping on their conversations, she feels a connection with her fellow human beings. The smallest details about them excite her interest. She comes to feel herself one of them, an actor in life's drama. So caught up does she become in the sudden revelation that all the world's a stage and that she, like everyone else, has a part to play, that she has a mystical experience. In her mind, she merges with the other players, and "it seemed to Miss Brill that in another moment all of them, all the whole company, would begin singing"(10).

5. Miss Brill, who loves her fur as if it were a living pet or companion, is probably capable of thinking it was that "little rogue" she heard crying over the cruelty of the young couple in the park. It's possible, too, for the reader to believe that Miss Brill actually does hear a sound and that the "something crying" is herself. Perhaps she has, as Eudora Welty surmises in "The Reading and Writing of Short Stories," suffered a defeat that "one feels sure . . . is forever" (*The Atlantic Monthly,* Feb.–Mar. 1949).

Toni Cade Bambara
BLUES AIN'T NO MOCKIN BIRD, page 35

This story seems all the stronger for being narrated by the perceptive little girl who, in remarking on what she observes, reveals more than she knows. When the narrator says about her grandmother, "She was making the cakes drunk" (paragraph 1) or "She teaches steady with no letup" (22), the reader experiences the delight of discerning nuances that the child doesn't quite pick up. Cathy is the indispensable sidekick. Somewhat older than the narrator, she supplies useful information that the child storyteller could not possibly know but that is essential to the reader's understanding of the characters and their motives. (That she also skews the details of "Goldilocks and the Three Bears" makes her only a little bit unreliable.)

Bambara's story depicts an understandably proud black family, quick to anger at those who treat them with condescension. It is this pride and anger that impel the family to move from place to place so often and that make the grandmother react as she does to the moviemakers. Their praises, implying as they do that most people who use food stamps are shiftless no-goods, insult rather than compliment her. Their calling her "Auntie" raises her hackles. When she tells the children about the cameramen who filmed a suicide, she reveals another reason for not wanting any moviemakers on her land. They are insensitive clods who make entertainment out of suffering.

That we see the grandfather through the respectful eyes of his little granddaughter gives us enormous respect for him. The details she includes about his days as a train waiter show that people have always looked up to him. He is "like a king." Is he cruel? No, only practical: he kills hawks because they eat chickens and so deprive the family. Wise in the ways of these wild creatures, he knows that by placing the flapping female hawk on a nail he will gain the chance also to kill her mate.

Mockingbirds, favorite songsters, are noted for their ability to mimic other birds. Though environmentalists don't agree, farmers regard hawks as predators to be destroyed. Blues are hawks, not mockingbirds. Bambara seems to suggest that to the grandmother, the moviemakers are like "the people who wouldn't pay her for things, like they said they would. Or Mr. Judson bringin us boxes of old clothes and raggedy magazines. Or Mrs. Cooper comin in our kitchen and touchin everything and sayin how clean it all was" (30). Moviemakers, and untrustworthy or condescending whites, are predators too. Certainly they give Granny the blues. The moviemakers can't be killed like hawks, but the grandfather, who understands his wife well and shares her pride ("This is our own place," he tells the men) can destroy their camera with his hawk-blood-stained hands. He does it with the same quiet skill with which he brought down the hawk. The blues, too, can be dealt with like a predator.

(text pages) 41–46

Toni Cade Bambara speaks of her beginnings, work habits, central themes, and other concerns in a 58-minute audiocassette interview available from American Audio Prose Library, Box 842, Columbia MO 65205. For current price, call (800) 447-2275.

Edgar Allan Poe
THE TELL-TALE HEART, page 41

Poe's story abounds with details that identify his narrator as unreliable, and students will probably enjoy pointing them out. Even in paragraph 1, when he admits to being "very, very dreadfully nervous" and mentions his "disease," the storyteller seems overwrought. By paragraph 2, when he confesses that he made up his mind to kill an old man because of his eye, the narrator's looniness is obvious. Ironically, his madness seems especially clear when he praises his own wisdom, caution, and foresight (paragraph 3).

A rational explanation for the terrifying heartbeat may occur to some students: the madman hears the pounding of his *own* heart. Who knows? All that matters to the story is the narrator's leap to the conclusion that he hears his victim's heart still going strong.

What suggestions do we get from the old man's baleful eye? Is it the all-seeing gaze of the Almighty? Daniel Hoffman has argued that the old man is a father-figure, perhaps even a Father-Figure, and that in striking the old man's eye "the young madman strikes, symbolically, at his sexual power." Perhaps, too, the vulture eye belongs to Time, the narrator's mortal enemy. (See *Poe Poe Poe Poe Poe Poe Poe* [New York: Anchor, 1973] 221–27.)

If Hoffman is right, then perhaps there is a theme in Poe's story; although perhaps it isn't essential to find any. (XJK comments: Once, after a class had read a Poe tale, a wonderful controversy broke out when someone complained, "This story doesn't really say anything." All of us tried to sum up the story's theme, failed miserably, and at last decided that there *is* a place in literature for stories that don't say anything in particular, but supply their readers with memorable nightmares and dreams.)

Fairy tales, too, often do without clearly discernible themes and stir us powerfully. Bruno Bettelheim contends that fairy tales are therapeutic in his study of them, *The Uses of Enchantment* (New York: Knopf, 1976). Bettelheim urges that people who read fairy tales to children not talk about what the tales mean. Is Poe, perhaps, also good for your mental health, and not to be analyzed endlessly?

3 Character

Katherine Anne Porter
THE JILTING OF GRANNY WEATHERALL, page 50

Suggested here are some possible answers to the questions at the end of "The Jilting of Granny Weatherall." Other answers, of equal merit, may occur to you and your students. Other questions may occur as well.

1. That Ellen Weatherall is feisty, accustomed to having her way, and unwilling to be treated like the sick old woman she is, comes through the story from the start.

2. Granny has "weathered all"—unrequited love, marriage, the birth of five children, early widowhood, backbreaking labor, "milk-leg and double pneumonia," the loss of her favorite daughter, and the frustrations of old age. Her victories over adversity have made her scornful of her daughter Cornelia, who seems to her weak and inadequate. Granny is tough and inclined to hold a grudge. She has never forgiven the man who jilted her. So overweening is her pride, in fact, that at the moment of her death, when she wants a sign and fails to perceive one, she decides she will "never forgive."

3. Granny is very ill, her sense of reality distorted by lengthening periods of confusion. The story quite convincingly ushers Granny Weatherall by fits and starts into an altered state of consciousness preceding death. Granny's weakened grasp on reality is again apparent: when the doctor "appeared to float up to the ceiling and out" (paragraph 7); when "the pillow rose and floated under her" and she thought she heard leaves, or newspapers, rustling (8); when she pondered her impossible plans for "tomorrow" (17 and 26); in her belief that she could, if she wanted to, "pack up and move back to her own house" (24); in the distortions that creep into Granny's sense of the passage of time (28, 29, 36, 37, 42, 43, 50, 56, and 57); when she feels the pillow rising again (29); in Granny's sporadic inability to hear (31 and 32); in her belief that there are ants in the bed (40); in Granny's occasional inability to speak clearly enough to be understood (40 and 53); when Granny has hallucinations in which she confronts her daughter Hapsy (41, 50, and 60); in the doctor's "rosy nimbus" (51); in Granny's failure to comprehend that Father Connolly is not tickling her feet but administering the last rites of the Catholic Church (56). (In the anointing of the sick, formerly known as Extreme Unction, the priest makes a sign of the cross with oil upon the eyes, ears, nose, mouth, hands, and feet of the person in danger of death, while praying for his or her soul.)

4. George, the man she was to marry, had failed to show up for the wedding. This blow to Ellen's pride had left permanent scars even though Ellen had subsequently lived a full and useful life. It is hard to say whether her exacting, imperious, unforgiving ways resulted from the jilting—or caused it!

5. The most likely guess about the man who "cursed like a sailor's parrot" is that he was John, the man Ellen eventually married. Presumably John had always loved her and was angry because she had been so deeply hurt. The identity of the man driving the cart is more nebulous. Was he George? John? A confused amalgam of the two? That the reader remains unsure is not a fault in the story. The dreamlike haze surrounding the man's identity beautifully reflects Granny's loosening hold on reality.

6. The story's point of view is one of selective omniscience. The events are reported in the third person by a narrator who can see into granny Weatherall's mind. When Granny

(text pages) 47–57

is lucid, the story proceeds in tidy chronological order. In the story's most interesting passages—especially in paragraphs 17–18, and 24–31—Porter uses stream of consciousness with great skill to present the randomly mingled thoughts and impressions that move through Granny's dying mind. By fragmenting Granny's thoughts, by having her shuttle back and forth between reality and fantasy, by distorting her sense of the passage of time, the author manages to persuade us that the way Granny experiences dying must be nearly universal.

7. Cornelia, evidently the oldest of Granny's children, is a dutiful daughter—a fact resented by her cantankerous mother (10). Granny feels demeaned by her knowledge that Cornelia often humors her to keep peace. And, like mothers everywhere, Granny regards her daughter as far less competent than herself. Cornelia is tenderer, less tough than Granny. She weeps beside the deathbed. The old woman fails to appreciate her daughter's care and compassion because she scorns her own need of them.

8. Hapsy, the youngest child and her mother's favorite, died young. Paragraph 41 suggests that she may have died giving birth to a son.

9. A "flat" character, the doctor is little more than a prop in this account of Granny's dying. There is no reason for him to be more than that, for it is Granny's life and death we are meant to focus on.

10. Before talking about the final paragraph, why not read aloud to your students the parable of the wise and foolish virgins (Matthew 251:1–13).

> Then shall the kingdom of heaven be likened unto ten virgins, which took their lamps, and went forth to meet the bridegroom.
> 2   And five of them were wise, and five were foolish.
> 3   They that were foolish took their lamps, and took no oil with them:
> 4   But the wise took oil in their vessels with their lamps.
> 5   While the bridegroom tarried, they all slumbered and slept.
> 6   And at midnight there was a cry made, Behold, the bridegroom cometh; go ye out to meet him.
> 7   Then all those virgins arose, and trimmed their lamps.
> 8   And the foolish said unto the wise, Give us of your oil; for our lamps are gone out.
> 9   But the wise answered, saying, Not so; lest there be not enough for us and you: but go ye rather to them that sell, and buy for yourselves.
> 10   And while they went to buy, the bridegroom came; and they that were ready went in with him to the marriage: and the door was shut.
> 11   Afterward came also the other virgins, saying, Lord, Lord, open to us.
> 12   But he answered and said, Verily I say unto you, I know you not.
> 13   Watch therefore; for ye know neither the day nor the hour wherein the Son of Man cometh.

Evidently the bridegroom Granny awaits at the end of the story is Christ. When he does not appear, she feels jilted for a second time. Why doesn't he come? Why does Granny not receive the sign she asks for? Is it because her pride is so overweening ("I'll never forgive it") as to keep her from salvation? Is it because of her refusal to stay prepared for death (18)? Or did she receive her sign (the last rites of the Church) and merely fail to perceive it? Students may object to the apparent grimness of the ending. Some of them are likely to insist that Granny gets worse than she deserves, that Porter has allowed her symbolism to run roughshod over her humanity. Divergent opinions may spark a lively discussion.

11. Such a critic will be hard to convince. But perhaps discussion can show that the story is remarkable for its condensation of a long life into a few pages. How does it feel to die? All of us find the question interesting, and Porter answers it. Although Edith Wharton has argued that it is not the nature of a short story to develop a character (in The Writing of Fiction [New York: Scribner's, 1925], "The Jilting of Granny Weatherall" cer-

tainly makes character its central concern. We finish the story persuaded that we know Ellen Weatherall very well indeed—better than she knows herself.

Here is a bleak reading of the story for possible discussion: For sixty years Ellen Weatherall has suppressed the memory of George, the man she loved, who jilted her. She prays not to remember him, lest she follow him down into the pit of Hell (29). If she remembers him, she herself will be damned—yet she remembers him. She longs to see him again (42), imagines him standing by her deathbed (30). In the end she beholds the pit: a darkness that will swallow up light. For the sake of a man, she has lost her soul.

But is the story so grim an account of one woman's damnation? It seems hardly a mortal sin to remember a person and an event so crucial in one's life; damnation seems undeserved. We sense that the author actually admires Granny's defiance in blowing out the light at the end. To say the least, the story is splendidly ambiguous.

"Granny Weatherall" may be a good means to approach Tolstoi's *The Death of Ivan Ilych* (page 218). The leading fact of each story is death, or preparation for it. Ivan Ilych, despite his useless career, overcomes pride and dies in joyous illumination. Ellen Weatherall's life has been fruitful and full; but, proud to the end, she dies in spiritual anguish.

"The Jilting of Granny Weatherall" lent itself effectively to the PBS television film series *The American Short Story*, obtainable on videotape for classroom viewing. Anyone interested in writing for television might care to see Corinne Jacker's excellent script, reprinted in *The American Short Story, Volume 2* (New York: Dell, 1980), together with a revealing interview with the scriptwriter on the problems of adapting "Granny"—such as a dearth of physical action in the present—and Carolyn G. Heilbrun's short critical essay on the Porter story.

Porter's familiarity with illness and the threat of death may have been drawn from memory. As Joan Givner recounts in her biography, *Katherine Anne Porter* (New York: Simon, 1982), Porter had to struggle for years against bronchial troubles and tuberculosis. Defiantly, she endured to age ninety.

Jamaica Kincaid
GIRL, page 58

Without much difficulty, students (who may have heard from their own parents some lists of directives more or less similar in tone) will identify the speaker in "Girl" as a mother and the listener as her daughter. The mother's diatribe is interrupted only twice (in italics) when her daughter protests. From the language, such as the names of foods (*dasheen, doukona*) and the details of daily life, we soon see that the setting is another country: the author's native Antigua, we might guess. The mother's words seem an accumulation of instructions repeated over many years. Her main intent is to teach her daughter to measure up, to grow into a lady and not a slut. Any deviation from her rules seems to her a fatal step toward sluthood. Clearly she sees a woman's role in life as traditionally restricted and restricting. But the mother is not merely delivering negative, binding orders; she tries to impart a whole body of traditional wisdom about the right way to do things. She offers her daughter the secrets of catching fish, avoiding bad luck, curing a cold (or a pregnancy), bullying a man, lovemaking, and making ends meet.

It might be argued that "Girl" is a character sketch (of *two* characters) and not a story. But like many another satisfying work of fiction, "Girl" artfully and sharply delineates its characters, even places them in conflict—the mother anxious for her daughter to be a lady, the girl feeling unjustly blamed and apparently overwhelmed by so many orders.

Margaret Atwood
THE SIN EATER, page 59

Joseph's three wives are flat characters, almost interchangeable. They even look alike. Their function in the story is to throw light on Joseph himself. Joseph is a round character, as is the narrator, who under Joseph's influence grows and changes, acquiring in the process keen insight into both her therapist and herself.

Joseph is richly complex, a study in contradictions. He is both seedy (in appearance) and cultivated (in language and demeanor). Though he implies in paragraph 11 that he regards many of the confidences he hears as "fakery and invention," he is so dedicated to his work that he takes on even patients he knows can never pay him. Compassion is his business, and yet Joseph has never forgiven the little boy who picked his sunflower when he was eight. According to his first wife, Joseph enjoyed utter devotion from many of his patients. Yet he seems lonely (as evidenced by his serial marriages, his characterization of the Sin Eater as a "syphilitic of the soul," the question "Do you like me?" that he asks the narrator early in their sessions). He wants the narrator to prefer being awake to being asleep (86). He has never lost a suicidal patient. And yet the story hints that his own death is a suicide, that he himself prefers sleep after all.

Although Joseph thinks of Sin Eaters as women, it's clear that in the story he tells the narrator he rather ruefully identifies himself with these "destitute old creatures who had no other way of keeping body and soul together" (5). A therapist is, after all, both confessor and healer. Metaphorically, at least, he eats sins.

Some students will perhaps go a step farther and say that the narrator also becomes a Sin Eater. It's true that she refuses in paragraph 74 to hear what the second wife wants to confide about Joseph's unhappiness. But, as Joseph remarks, it's not necessary for a Sin Eater to listen: "Not a blessed word. The sins are transmitted in the food" (14). In her dream the narrator eats the cookies offered to her even though they "look too rich," even though she's afraid they will make her sick (95, 97). In the dream, Joseph actually tells her that the cookies are his sins.

The narrator begins as someone who "wanted things to make sense" (18). Doesn't the dream suggest that through her association with Joseph she has learned that they don't have to make sense and that she doesn't have to be rescued? That she will cope, even with Joseph dead?

Margaret Atwood discusses her craft, her feminism, her Canadian nationalism, and other matters in a 56-minute audiocassette interview available from American Audio Prose Library, Box 842, Columbia MO 65205. For current price, call (800) 447-2275.

Isaac Bashevis Singer
GIMPEL THE FOOL, page 68

Is Gimpel an antihero? Not for Singer:

> I don't think I write in the tradition of the Yiddish writers' "little man," because their little man is actually a *victim*—a man who is a victim of anti-Semitism, the economic situation, and so on. My characters, though they are not big men in the sense that they play a big part in the world, still they are not little, because in their own fashion they are men of character, men of thinking, men of great suffering. . . . Gimpel was not a little man. He was a fool, but he wasn't little (interview in *The Paris Review* 44 [Fall 1968]).

Like Flannery O'Connor's "Revelation," Singer's story grows from a vision—Gimpel's dream of Elka, with her counsel, "Because I was false is everything false too?" (Several critics have pointed out similarities between Singer and Flannery O'Connor, two writers working with a conflict between new and traditional cultures. See Melvin J. Friedman, "Isaac Bashevis Singer: The Appeal of Numbers," in *Critical Views of Isaac Bashevis Singer*, edited by Irving Malin [New York: New York UP, 1969].) Singer is an avid reader of Spinoza, whose concept of the lesser or inferior reality of this world probably influenced the famous statement at the end of "Gimpel the Fool," "No doubt the world is entirely an imaginary world." But, in another sense,

> The world [Singer] recoils from is the world of the market place, of human passions, of vain ambitions, of misguided aspirations, and of all the human relationships which result from them. This is the world of Gimpel the Fool, where the simple and the sensitive are gulled, deprived, humiliated, and despised. It is the world in which the poverty of Frampol distorts the perspective of its people (J.A. Eisenberg, "Isaac Bashevis Singer: Passionate Primitive or Pious Puritan?" in Malin's collection).

Much of Singer's best fiction, including "Gimpel the Fool," takes place in the *shtetl* or Jewish village of Eastern Europe, a place of rutted streets and tumbledown houses—Singer doesn't romanticize it. He writes, according to Ben Siegel, "compassionate fables of Jews who cling tenaciously to their traditions but who otherwise interest no one but God, the devil, and themselves. And God's interest is never certain" ("Sacred and Profane: Isaac Bashevis Singer's Embattled Spirits," *Critique* 5 [Spring 1963]). Phantoms, demons, *dybbuks*, and the devil figure in Jewish folklore and popular stories; in Singer's fiction, a supernatural personage often embodies a character's passions or obsessions. And so in "Gimpel the Fool," when the Spirit of Evil speaks, he voices Gimpel's resentment against Frampol and his momentary distrust of humankind.

An incisive comment on Gimpel's wife Elka comes from Professor Stan Sulkes of Raymond Walters College.

> Do we agree that Gimpel's ability to love his wife emanates from his own goodness rather than hers? (An echo, perhaps, of the prophet Hosea's being commanded to marry a whore so he could appreciate God's love for the errant Israel.) On her deathbed she tells him the truth. (More of her selfishness, I think, since she is anxious over her impending fate rather than his feelings.)

Suggested writing assignment: "A Defense of Elka."

The character of the "simple creature of God," or holy simpleton, ridiculed in this world, rewarded only in afterlife, is a familiar figure in Jewish folk literature. Another famous story with such a character is I.L. Peretz's "Bontsha the Silent." That as a work of storytelling "Gimpel the Fool" is both traditional and original, Sol Gittleman shows in *From Shtetl to Suburbia: The Family in Jewish Literary Imagination* (Boston: Beacon, 1978) 103-07.

Still, some readers have been bothered by Gimpel's passivity, among them Professor Sulkes. His letter deserves further quotation.

> I recall being dismayed when I first read this story. How could a post-holocaust Jewish writer, I wondered, defend the literary convention of the holy fool? Wouldn't someone as naive and accommodating as Gimpel find himself in the Nazi ovens? . . . How could I even be amused by him? Was Singer blinking away the holocaust?
>
> Rereading the story, I realized that it does belong in the context of that horror, though the horror is never mentioned. The holocaust lurks in the background of this tale. As a result I think that Singer is more distanced from Gimpel's simple goodness than he might be if he were a Christian author. He probably *admires* the gullible Gimpel of the first half of the story more than he *approves* of him.
>
> At the end, Gimpel embarks on a journey which reveals to him the evil of the world (events like the holocaust), but he does not allow it to destroy his sweetness and his faith. My class was particularly eager to discuss this point: how does one remain hopeful in light of the atrocities so prevalent in the world? I assume that Singer offers Gimpel's post-journey piety as one response [of] the post-holocaust Jew. But Gimpel's quietism disappoints me, so far removed as it is from the mainstreams of modern Jewry. . . .

13

(text pages) 68–79

Your class might care to discuss this serious objection.

Still another objection might be raised to Gimpel's conclusion, "No doubt the world is entirely an imaginary world." Is, then, the evil in the world only a dream? Isn't this assumption dangerous?

Sol Gittleman, cited above, reads the story to mean that Gimpel, whatever his philosophy, finds truth in his wanderings. Discovering that the world beyond Frampol is less malicious and cruel, Gimpel in the end reaffirms his belief in humanity. Would your students agree or disagree with this paraphrase?

This translation of "Gimpel the Fool" offers the work of one Nobel Prize winner rendered into English by another. First published in *Partisan Review* for May 1953, Saul Bellow's translation was the first appearance in English of any Singer story.

In his speech on accepting the Nobel Prize, Singer said he regarded the award as a recognition of the Yiddish language. The language and its speakers—their conduct, their outlook on life—have been for him always identical.

> There is a quiet humor in Yiddish and a gratitude for every day of life, every crumb of success, each encounter of love. The Yiddish mentality is not haughty. It does not take victory for granted. It does not demand and command but it muddles through, sneaks by, smuggles itself amid the powers of destruction, knowing somewhere that God's plan for Creation is still at the very beginning (*Nobel Lecture* [New York: Farrar, 1978] 8).

Isaac Bashevis Singer reads "Gimpel the Fool" on an audiocassette (SWC 1200) available from American Audio Prose Library, Box 842, Columbia, MO 65205. For current price, call (800) 447-2275.

4 Setting

Kate Chopin
THE STORM, page 83

French names and carefully recorded dialect place Chopin's story in the bayou country of Louisiana. Setting and plot reinforce each other, the storm in nature precipitating and paralleling the storm of passion that engulfs Calixta and her old lover. In Part I we learn that when the storm begins, Calixta's husband and four-year-old son are trapped in the store. Calixta is home alone. Bibi remembers that Sylvie, who apparently helps out with the housework from time to time, is not with Calixta this evening. We realize later that these details matter: Calixta and Alcée will not be interrupted until the storm is over.

What causes Calixta's infidelity? Her husband loves her, in his way. He buys a can of shrimp because he knows she likes it. He stands in awe of her scrupulous housekeeping, but he seems to know nothing of her sensuality. We're told that after he buys the shrimp, he sits "stolidly." It is this stolidness, perhaps, that has prevented his ever having plumbed his wife's passionate nature.

Doesn't Calixta love her husband? She does, apparently: she worries about him and her son when the storm comes up. She seems genuinely glad to be reunited with them when they come home. But passions, Chopin seems to say, cannot be denied. Their force is equal to that of a storm, and the marriage bed is not the likeliest place for release of that force. The author indeed hints that marriage and sexual pleasure are incompatible in Parts IV and V of the story, where Alcée urges his absent wife to stay away another month and we learn that for Clarisse, "their intimate conjugal life was something which she was more than willing to forego for a while." Marriage is dull companionship at best, bondage at worst. Without stating this theme, Chopin clearly suggests it. Do students care to argue with her?

Students might like to ponder whether the story would be as powerful it if took place in Chicago on a sunny fall day. If the setting were changed, would we have any story at all?

Compare another wry view of marriage: Chopin's "The Story of an Hour" in Chapter Eleven, "Stories for Further Reading."

Jack London
TO BUILD A FIRE, page 87

The point of view in "To Build a Fire" is that of an omniscient narrator able to see into the man's mind—even into the dog's. In this powerful story, London lavishes upon the setting the amount of detail usually reserved for a story's characters. Indeed, the Arctic weather functions as if it *were* a character, an unrelenting adversary doing battle with the man. That the man is left nameless seems to emphasize his insignificance in the fight against his mighty enemy.

Skillfully, London builds up almost unendurable suspense. Our hopes, like the protagonist's, are kept alive almost to the end. Detail is piled upon detail as the struggle is gradually lost. The frost on the man's face, the numbing of his extremities and exposed skin, his mental confusion, and, finally, his inability to use his fingers to keep his feeble fire going presage the drowsy calm that signals his end.

Along the way the man makes us aware of the fatal mistakes he has made, the most serious no doubt being his decision to venture alone into the wilderness. Underestimating

(text pages) 87–99

his adversary, he also fails to ascertain the temperature and to cover his face. Finally, suffering from mental confusion, he gets his feet wet and then builds a fire in the wrong place. By that time it is too late to rectify what has gone wrong. The dog, protected by his instincts, survives.

Students with some firsthand experience in snow country will respond heartily to this story. But to respond to it, a firsthand acquaintance with freezing cold isn't necessary. In painful and painstaking detail, which repays close scrutiny, London's account does our imagining for us.

Compare Raymond Carver's story "Where I'm Calling From," in which freezing to death becomes a metaphor at the end. See Carver's meaningful mentions of the alcoholic London and of "To Build a Fire" in his paragraphs 37 and 87–88.

Frivolous writing assignment, perhaps for doing in class: Retell this story in 500 words, from the dog's point of view.

T. Coraghessan Boyle
GREASY LAKE, page 99

It is rebellious adolescence in general that Boyle describes in his opening paragraph; but he's also talking about the late 1960s, when adolescents were in plentiful supply, "bad" behavior was much admired, and "courtesy and winning ways went out of style." In 1967 when the American attack on Khe Sanh (mentioned in paragraph 7) took place, Boyle himself was nineteen years old. We can only guess that's about the year in which "Greasy Lake" is set. Not only the epigraph but also the title of the story come from that bard of a slightly later era, Bruce Springsteen, whose first album appeared in 1973.

Are Digby and Jeff really "bad"? Well, no, and neither is the narrator. They're just engaging in the kind of behavior they think is expected of them (1, 3, 4). When on the night of the story their rebellion backfires, throwing them into a grimmer world than they had bargained for, they feel revulsion. As is clear at the end, they have had enough of being "bad." Like the boy in James Joyce's "Araby," they have grown up painfully. (For other stories of a young man's initiation into maturity, see "A & P," "Araby," and "Barn Burning." Young women who are similarly initiated, or who seem about to be, appear in "The Lover of Horses" and "Where Are You Going, Where Have You Been?")

That the narrator of "Greasy Lake" grows and changes during his adventures is apparent from the two views of "nature" he voices, one in paragraph 2 and one in paragraph 32. Early in the story, "nature" was wanting "to snuff the rich scent of possibility on the breeze, watch a girl take off her clothes and plunge into the festering murk, drink beer, smoke pot, howl at the stars, savor the incongruous full-throated roar of rock and roll against the primeval susurrus of frogs and crickets." By the end of the story, these swinish pleasures have lost their appeal. When at dawn the narrator experiences the beauties of the natural world as if for the first time, he has an epiphany: "This was nature."

Students can have fun demonstrating how Greasy Lake is the perfect setting for Boyle's story. Like the moral view of the narrator (at first), it is "fetid and murky, the mud banks glittering with broken glass and strewn with beer cans and the charred remains of bonfires. There was a single ravaged island a hundred yards from shore, so stripped of vegetation it looked as if the air force had strafed it" (2). The lake is full of "primordial ooze" and "the bad breath of decay" (31). It also hides a waterlogged corpse. Once known for its clear water, the unlucky lake has fallen as far from its ideal state as the people who now frequent its shores have fallen from theirs. (If you teach the chapter on Symbol, hark back to Greasy Lake once more.)

Still, in its way, Greasy Lake is a force for change. Caught trying to rape the girl in the blue car, the narrator and his friends run off into the woods, into the water. Waiting

16

(text pages) 99–107

in the filthy lake, the narrator is grateful to be alive and feels horror at the death of the "bad older character" whose body he meets in the slime. His growth has begun. When at the end of the story two more girls pull into the parking lot, the subdued narrator and his friends are harmless. Cold sober, bone tired, they know they have had a lucky escape from consequences that might have been terrible. Also, the narrator knows, as the girls do not, that Al is dead, his body rotting in the lake. He won't "turn up"—except perhaps in the most grisly way. It is this knowledge and the narrator's new reverence for life that make him think he is going to cry.

Students might enjoy spelling out the change in the narrator's outlook. By what hints does Boyle show us that some time has elapsed since the events of the fateful night? Surely the story displays little admiration for the narrator's early behavior, which he now regards with sarcasm, as when he says, "Digby wore a gold star in his right ear and allowed his father to pay his tuition at Cornell" (3), or when he speaks of "new heights of adventure and daring" (6). Other ironic remarks abound, showing his altered view. The maturity the narrator acquired that night seems to have been permanent.

Critics have cited Boyle as a writer socially and politically disengaged; but satire, he points out, can be corrective. "It can hold up certain attitudes as being fraudulent, and in doing that suggest that the opposite might be an appropriate way to behave. And I hope that if my work is socially redemptive, it is in that way" (interview with David Stanton in *Poets & Writers,* January/February 1990). Surely "Greasy Lake," a story some readers find shocking, is socially redemptive.

5 Tone and Style

Ernest Hemingway
A CLEAN, WELL-LIGHTED PLACE, page 112

This celebrated story is a study in contrasts: between youth and age, belief and doubt, light and darkness. To the younger waiter, the café is only a job; to the older waiter, it is a charitable institution for which he feels personal responsibility. Of course, he himself has need of it: it is his refuge from the night, from solitude, from a sense that the universe is empty and meaningless, expressed in his revised versions of the Hail Mary and the Lord's Prayer. The older waiter feels kinship for the old man, not only because the waiter, too, is alone and growing old, but because both men are apparently atheists. Willing to commit suicide, the old man (unlike his pious daughter) evidently doesn't think he has any immortal soul to fear for. Robert Penn Warren is surely right in calling Hemingway, at least in this story, a religious writer. "The despair beyond plenty of money, the despair that makes a sleeplessness beyond insomnia, is the despair felt by a man who hungers for the sense of order and assurance that men seem to find in religious faith, but who cannot find grounds for his faith" ("Ernest Hemingway," in Warren's *Selected Essays* [New York: Random, 1951] 80–118). What values are left to a man without faith? A love of cleanliness and good light, of companionship, of stoic endurance, and above all, of dignity. (Another attempt to state the theme of the story is at the beginning of Chapter 7.)

At the heart of the story is the symbol of the café, an island of light and order surrounded by night and nothingness. Contrasting images of light and darkness begin in the opening paragraph: the old man, not entirely committed either to death or to life, likes to sit in the shadow of the leaves. Every detail in the story seems meaningful: even, perhaps, as L. Rust Hills points out, "the glint of light on the soldier's collar . . . an attribute of sexual potency" (*Writing in General and the Short Story in Particular* [Boston: Houghton, 1977] 85).

The story has been much admired for Hemingway's handling of point of view. The narrator is a nonparticipant who writes in the third person. He is all-knowing at the beginning of the story: in the opening paragraph we are told how the old man feels, then what the waiters know about him. From then on, until the waiters say "Good night," the narrator remains almost perfectly objective, merely reporting visible details and dialogue. (He editorializes for a moment, though, in observing that the younger waiter employs the syntax of "stupid people.") After the waiters part company, for the rest of the story the narrator limits himself to the thoughts and perceptions of the older waiter, who, we now see, is the central character.

It is clear all along, as we overhear the conversation of the two waiters, that Hemingway sides with the elder's view of the old man. The older waiter reveals himself as wiser and more compassionate. We resent the younger man's abuse of the old man, who cannot hear his "stupid" syntax, his equation of money with happiness. But the older waiter and Hemingway do not see things identically—a point briefly discussed in the text on page 130 in a comment on the story's irony.

A small problem in reading the story is to keep the speakers straight. Evidently it is the younger waiter who has heard of the old man's suicide attempt and who answers the questions at the beginning of the story. Hemingway's device of assigning two successive speeches to the same character without identifying him (paragraphs 20–21 and probably 32–33) has given rise to much confusion among readers—also to twenty or more scholarly articles on the correct reading of the story. David Kerner's "The Foundation of the True Text of 'A Clean, Well-Lighted Place' " settles the question. Demonstrating that the device appears many times in Hemingway's novels and stories, Kerner suggests—with evidence—that Hemingway may have learned it from Turgenev or Joyce (*Fitzgerald/Hemingway Annual* 1979: 279–300). Later, Kerner examined manuscripts of books Hemingway saw through press, and found thirty-eight clear instances of the device ("The Manuscripts Establishing Hemingway's Anti-Metronomic Dialogue," *American Literature* 54 [1982]: 385–96).

How original Hemingway's style once seemed may be less apparent today, after generations of imitators. Ford Madox Ford described the famed style: "Hemingway's words strike you, each one, as if they were pebbles fetched fresh from a brook" (introduction to *A Farewell to Arms* [New York: Modern Library, 1932]). Students may be asked to indicate some of the more prominent pebbles: the repetitions of such words as *night, light, clean, late, shadow, leaves*, and (most obviously) *nada*. The repetitions place emphasis. Students may be asked, too, to demonstrate whether Hemingway's prose seems closer to formal writing or to speech. It might help to have them notice the preponderance of one-syllable words in the opening paragraph of the story and to compare Hemingway's first paragraph with that of Faulkner's "Barn Burning." Faulkner's second sentence is a 117-worder, clearly more recondite in its diction *(dynamic, hermetic)*. Most of Hemingway's story is told in dialogue, and usually we notice that a character *said* (unlike Faulkner's characters, who often *whisper* or *cry* their lines).

Frank O'Connor comments adversely on the Hemingway style:

> As practiced by Hemingway, this literary method, compounded of simplification and repetition, is the opposite of that we learned in our schooldays. We were taught to consider it a fault to repeat a noun and shown how to avoid it by the use of pronouns and synonyms. This led to another fault that Fowler christened "elegant variation." The fault of Hemingway's method might be called "elegant repetition" (*The Lonely Voice* [Cleveland: World, 1963]).

This criticism may be well worth quoting to the class. From "A Clean, Well-Lighted Place," can students see what O'Connor is talking about? Do they agree with him?

In a preface written in 1959 for a selection of his stories that did not materialize as a book, Hemingway congratulated himself for his skill at leaving things out. In his story "The Killer," he had left out Chicago; in "Big Two-Hearted River," the war. "Another time I was leaving out good was in 'A Clean, Well-Lighted Place.' There I really had luck. I left out everything. That is about as far as you can go, so I stood on that one and haven't been drawn to that since" ("The Art of the Short Story," *The Paris Review* 79 [1981]: 100).

Hemingway himself had reason to empathize with the older waiter. "He was plagued all his adult life by insomnia and in sleep by nightmares," notes biographer Carlos Baker (*Ernest Hemingway: A Life Story* [New York: Scribner's, 1969] viii-ix).

William Faulkner
BARN BURNING, page 116

From the opening paragraph, we can tell that the tone of the story will be excited and impassioned—at least in the moments when we see through the eyes of the boy, Colonel Sartoris Snopes. Even the boy's view of canned goods in the general store (where court is being held) is tinged with intense emotion. Fear, despair, and grief sweep over Sarty because his father is on trial as an accused barn burner.

The ten-year-old boy's wonder and dismay are conveyed in a suitably passionate style. Whenever Sarty is most excited, Faulkner's sentences grow longer and more complex and seem to run on like a torrent. The second sentence of the story is a good illustration; or the sentence in which Sarty jumps out of the way of Major de Spain's galloping horse and

(text pages) 116–129

hears the barn going up in flames (at the climax of the story, the long sentence in paragraph 107). A familiar student objection is that Faulkner embodies the boy's feelings in words far beyond a ten-year-old's vocabulary (and Sarty can't even read, we learn from paragraph 1). You might anticipate this objection and get students to realize that a story about a small boy need not be narrated by the boy himself (on page 110 in the remarks on the quotation from *As I Lay Dying*). Evidently "Barn Burning" is told by a nonparticipating narrator who sees things mainly through the eyes of Sarty Snopes, the main character.

Lionel Trilling wrote a good defense of the style in this story: the complexity of Faulkner's rhetoric reflects the muddlement and incompleteness of the boy's perceptions and the boy's emotional stress in moving toward a decision to break away from father and family (*The Experience of Literature* [New York: Holt, 1967] 745–48.)

Now and again, the narrator intrudes his own insights and larger knowledge. At times Faulkner rapidly shifts his perspective from outside Sarty's mind to inside it. Two such shifts take place in a sentence in paragraph 7 ("For a moment . . ."), in which, first, we get the boy's thoughts and then an exterior look at his crouched figure ("small for his age . . . with straight, uncombed, brown hair and eyes gray and wild as storm scud"), and then a return to the boy's perceptions. In the paragraph telling how Ab Snopes's "wolflike independence" impressed strangers (25), Faulkner makes a judgment far beyond the boy's capacity. In paragraph 26, the narrator again separates his view from the boy's to tell us what Sarty would have thought if he were older—"But he did not think this now." The narrator is again clearly in evidence in the story's last two paragraphs.

The effect of these intrusions is, perhaps, to make the reader see that the boy can't quite understand his situation. He is perplexed and innocent, and we realize that somehow he loves his terrible father. In the next-to-last paragraph, Sarty's impressions of his father's war service are set beside the grim truth—which only the father (and the narrator) could know. Wayne C. Booth comments on this passage: "We can say with some confidence that the poignancy of the boy's lonely last-ditch defense of his father is greatly increased by letting us know that even that defense is unjustified" (*The Rhetoric of Fiction* [Chicago: U of Chicago P, 1961] 308).

"Barn Burning" may be taken together with "A Rose for Emily" (Chapter 2) to demonstrate that point of view, in a masterly story, is an essential determinant of the style in which the story is told. "Barn Burning" is the more difficult story, and so you may want to take up "A Rose for Emily" first and to spend a little while getting students into the universe of Faulkner. Any brief discussion of the Civil War and its effects will help prepare them for "Barn Burning." Compared with the latter story, "A Rose" seems more lucid in style and more cool, detached, and wryly humorous in tone—like its narrator, a spokesman for the town, who (unlike the narrator in "Barn Burning") does not enter the leading character's mind and who is somewhat distantly recalling events of years gone by. Though her perch in society is loftier, Miss Emily, like Ab Snopes, is fiercely proud, capable of violent revenge, and she too holds herself above the law. (Faulkner, of course, thinks both Emily and Ab appalling and loves them dearly.)

"A Rose for Emily" also introduces the legend of Colonel Sartoris, war hero, mayor, and first citizen, whose fame and influence linger. Coming to "Barn Burning," students may then appreciate the boy hero's given name. Addressing the boy (in 10), the Justice foreshadows the story's conclusion: "I reckon anybody named for Colonel Sartoris in this country can't help but tell the truth, can he?" Truthfully, Sarty warns Major de Spain that his father is going to burn the major's barn; in so defying Ab, Colonel Sartoris Snopes rises to his namesake's nobility.

Question 1 directs the student to paragraph 107, a crucial passage that repays close attention. From the roar Sarty hears, and from the detail that the night sky is "stained abruptly," it is clear that Ab Snopes and Sarty's older brother succeed in settiing fire to de Spain's barn. We had long assumed that, although a total of three shots ring out, the barn burners get away, for they turn up in a later volume of Faulkner. But Joseph L. Swonk of Rappahannock Community College, North Campus, persuades us that the outcome is grimmer. Sarty cries "Pap! Pap!" —then trips over "something," looks back, and sobs "Father! Father!" "I contend," says Professor Swonk, "that he tripped over the bodies of his father and brother, that he was looking backward at the bodies, and that his shift from 'Pap' to 'Father' was eulogistic. Furthermore, he continues in this manner: 'Father. My Father . . . He was brave! . . . He was! He was in the war!'" Now referring to his father in the past tense, Sarty is delivering a final tribute over Ab's corpse.

"Barn Burning," to place the story in the chronicles of Yoknapatawpha County, is a prelude to the Snopes family history later expanded in Faulkner's trilogy *The Hamlet* (1940), *The Town* (1957), and *The Mansion* (1959). At the conclusion of "Barn Burning," Sarty Snopes turns his back on his father and his clan; so, in the trilogy, the primary figure is Flem, Sarty's brother who remained. Meeting Flem Snopes, whose father Ab is still well known as a barn burner, Jody Varner in *The Hamlet* says dryly, "I hear your father has had a little trouble once or twice with landlords."

"Barn Burning" yields valuable illustrations for any discussion of symbolism: certainly the neat, "shrewd" fire (26) that reflects Ab's cautious nature. Fire is Ab's weapon of revenge, hence to be "regarded with respect and used with discretion." Memorable also are the rug (emblem of the whole social hierarchy that Ab defies and won't make amends to) and Ab's cruddy boots (suggesting his contempt for his employer, his mechanical indifference, his cruddiness).

"Barn Burning" was adapted for television in the PBS-television series *The American Short Story*, and may be available on videotape. A paperback, *The American Short Story, Volume 2* (New York: Dell, 1980), includes short scenes from Horton Foote's television script based on Faulkner's story.

For a stimulating attack on Faulkner's style, see Sean O'Faolain, *The Vanishing Hero* (Boston: Little, Brown, 1957) 101–03, 111–12, 133. Charging that Faulkner can't write plain English because "his psyche is completely out of his control," O'Faolain cites lines of Faulkner that he thinks rely on pure sound instead of sense ("the gasoline-roar of apotheosis"); and complains that Faulkner needlessly uses "second-thought words" ("He did not know why he had been compelled, or anyway needed, to claim it . . . "; "This, anyway, will, shall, must be invulnerable . . . "). In O'Faolain's view, when Faulkner begins a sentence, "I mean," he doesn't know what he means and won't know until he says it. O'Faolain's objections might be entertaining to discuss.

IRONY

James Joyce
ARABY, page 131

Set in the city Joyce called "dear old dirty Dublin," "Araby" reveals a neighborhood so dreary that it seems no wonder a sensitive boy would try to romance his way out of it. In paragraphs 1–3, details stack up tellingly, painting a scene of frustration and decay. We see the dead-end street with an abandoned house at its "blind end"; the boy's house where the priest had died, its room full of "musty air" and "old, useless papers"; dying bushes, rusty bicycle pump; a street of shadows and dark, dripping gardens lit by somber violet light. Still, the description is not unrelievedly sad. Playing in the cold, the boys feel their bodies glow. From "dark, odorous stables" comes the "music" of jingling harnesses. And for the boy, Mangan's sister lends the street enchantment.

Most students won't need help to see that "Araby" is told by its main character. They may need class discussion, though, to realize that the narrator is a man who looks back on his boyhood memories. One indication of the narrator's maturity is his style. In the first paragraph, he remarks, in unboyish language, that the houses, "conscious of decent lives within them, gazed at one another with brown imperturbable faces." Besides, this mature storyteller is about to step back and criticize his younger self: "her name was like a summons to all my foolish blood" (paragraph 4).

Mangan's sister, whose name is never told, seems an ordinary young woman who summons her kid brother to tea. She is a vague figure: the boy glimpses her from afar, sometimes while peering at her from shadow. John J. Brugaletta and Mary H. Hayden suggest that the conversation in which the boy promises to bring her a gift takes place only in his mind. Mangan's sister, they argue, may never have set foot in the room where the priest died (6), into which the boy retreats to have his visionary experience. In this musty shrine, his senses swoon. He clasps his hands in an attitude of prayer, murmurs an incantation over and over (*O love! O love!*), and conjures her face before him—"At last she spoke to me." Even the spikes he sees her clasping are unreal, for they couldn't be there in the dead priest's drawing room. In the end, at the bazaar, the image of Mangan's sister fades before the physical presence of the banal, flirtatious English salesgirl who says, "O, I never said such a thing." Neither did Mangan's sister say a word to the narrator that bound him to his imagined promise ("The Motivation for Anguish in Joyce's 'Araby,'" *Studies in Short Fiction* Winter 1978: 11–17).

The boy's daydreams of Mangan's sister are difficult to take seriously. Amid barrels of pigs' cheeks, he carries her image in his mind as a priest carries a chalice. He regards her with "confused adoration" and feels himself a harp on which she plays (5). From early in the story, the boy has projected a dazzling veil of romance over the commonplace. At the end, he realizes with shock that illusion has had him in thrall. Araby, the enchanted fair, turns out to be merely a drab charity bazaar where gimcracks are peddled, men count money, and a scatterbrained salesclerk makes small talk till the lights go out. The boy's intense anguish seems justified.

Nearly everything Joyce wrote has a thread of allegory, and "Araby" may be no exception. Making much of the identity of Mangan's sister, William York Tindall remarks: "Since [James Clarence] Mangan, one of Joyce's favorite poets, dedicated 'Dark Rosaleen,' his most famous poem, to his country, it seems likely that Mangan's sister is Ireland herself, beckoning and inviting." Tindall thinks the boy's frustrated quest is for Ireland's Church, toward which Joyce, too, felt bitter disillusionment. Rather than pursue Dark Rosaleen, the mature Joyce (and his protagonist Stephen Dedalus) chose exile. "Araby" makes a good introduction to *Portrait of the Artist as a Young Man*. (Tindall discusses the story in *A Reader's Guide to James Joyce* [New York: Noonday, 1959] 20.)

Araby, the bazaar with the "magical name," is paramount. Besides, the apple tree in the unkempt garden (2) hints of the tree in some lost Eden. Dublin, clearly, is a fallen world. Other items also suggest sterility and decay: the "blind" or dead-end street and its "uninhabited house . . . detached from its neightbors" (1); the dead priest's rusty bicycle pump (2). Counting coins on a tray like that used to serve communion, the men in paragraph 25 perform a little act with symbolic overtones. The darkening hall has seemed to the boy "a church after a service," and the two money changers are not driven out of the temple—they drive out the boy.

Elizabeth A. Flynn compares reactions to this story by twenty-six male and twenty-six female college students in "Gender and Reading," *College English* 45 (Mar. 1983): 236–53. Some men felt uncomfortable with the boy's solipsistic infatuation. Recalling similar experiences of their own, they had trouble attaining distance. Several men, Flynn reports, were harsh in their judgment on Mangan's sister. They saw the girl as manipulating the boy for her own ends: "just using him," "playing him along." Most of the women students made better sense of the story. They didn't condemn Mangan's sister, and they understood the ending. They recognized that as the lights of the bazaar go out, the boy passes a painful judgment on himself: he has been a vain fool. Some women saw him gaining from his experience. Freed from his delusion, he can now reenter reality. If your men students have trouble understanding the story, you might have them take a good look at the last line.

(text pages) 131–136

On the popular poem "The Arab's Farewell to His Steed," which the uncle remembers, Matthew A. Fike of the University of Michigan writes, citing an insight by Stanley Friedman. "Joyce's reference to this poem, a work notable for its sentimentality, directs attention to the main significance of 'Araby': the assault on sentimentality and illusion" (Friedman, *The Explicator* 24:5 [Jan. 1966], item 43). There seems more than a little resemblance between the boy's worship of Mangan's sister for "the white curve of her neck" and the Arab's devotion to his horse, with its "glossy neck." Ironically, the Arab's glamorized view of his steed contrasts with the awareness that Joyce's narrator achieves; or as Mr. Fike puts it, "The nomad never parts with his horse, but the boy abandons his illusion." (Thanks to Mr. Fike for prompting XJK to greater precision in his footnote on "The Arab's Farewell.")

Anne Tyler
AVERAGE WAVES IN UNPROTECTED WATERS, page 136

Although the events in Tyler's third-person narrative are seen through Bet's eyes, the point of view is ironic because the author (and therefore the reader) doesn't always interpret scenes and events as Bet does. This is especially true for the hospital scene, where the building that "looked like someone's great, pillared mansion, with square brick buildings all around it," the chicken-wired windows, the corridor full of "fat, ugly women in shapeless gray dresses and ankle socks," the "enormous hallway lined with little white cots" but empty of children, the neatly made bed with the "steely-gray blanket . . . folded across the foot," the clown picture that the nurse fussily straightens, the requirement that the family not visit a new resident for six months all create a chill around our hearts. Bet is concerned with making everyone "see how small and neat [Arnold] was, how somebody cherished him." We know from the clues Bet has inadvertently provided that "cherishing" is probably not to be expected at Parkins State Hospital.

There is irony, too, in our realizing, more sharply than Bet does, how pinched and joyless her life has been. Paragraph 15 makes clear her low self-esteem: "she never could do anything as well as most people." Her early life at home—in "her parents' humped green trailer perched on cinder blocks near a forest of masts in Salt Spray, Maryland"—now seems to her "beautifully free and spacious." Compared with what followed, perhaps it was. She dismisses her memory of "standing staunch" in the breakers with the remark, "As if standing staunch were a virtue, really," seemingly without realizing that it *is* a virtue, and one that has stood her in good stead all her life.

Bright students may be quick to discern a parallel between Bet's early days, when the sea dictated her family's plans, and the present, when her every move depends on Arnold, a boy as moody and unpredictable as the sea. She has to plan her actions around him much as her father had to plan his around the sea, forever aware of "the height of average waves in unprotected waters." When the emotionally disabled Arnold finally grows too big for Bet to manage, she sees no alternative to putting him in a state hospital, a task so heart-wrenching that when she plans the trip, she leaves not a minute unaccounted for. Her anxiety explains why she is so insistent that the taxicab driver be there when she comes out of the hospital. It also explains why the train delay upsets her as much as it does. It obstructs a plan carefully formed to leave her no time for regret or for trying to undo what she has done. Thus she seems to regard the appearance of the workers and the Mayor as a gift from God. They distract her from her misery. "They had come just for her sake, you might think" (77). Indeed, that their ceremony will take twenty minutes, exactly the amount of time Bet wishes to burn, seems an almost supernatural coincidence. But we sense an ironic contrast between a showy ceremony to mark the grand opening of a new wing of the station, and the quiet, profoundly painful closing of a chapter of Bet's life.

6 Theme

Instructors who wish to teach all the stories in the book according to themes (not just the stories in this chapter) will find suggestions at the front of this manual in "Stories Arranged by Theme."

Stephen Crane
THE OPEN BOAT, page 147

It may interest students to know that "The Open Boat" is based on Crane's own experience of shipwreck (as the note on page 147 informs them), but they may need some discussion to realize that factuality does not necessarily make a story excellent. Newspaper accounts may be faithful to the fact, but few are memorable; on the other hand, fine stories often are spun out of imagined experience—as was Crane's novel *The Red Badge of Courage,* a convincing evocation of the Civil War by a writer who had only heard of it.

For what other reasons is "The Open Boat" a superb story? Its characters are sharply drawn and believable; captain, cook, oiler, and correspondent are fully realized portraits, etched with great economy. Crane deeply probes the mind of his primary character, the correspondent, from the point of view of limited omniscience. The author knows all but prefers to confine what he reports to what he sees through the correspondent's eyes.

The story seems written with intense energy; alert students will find many phrases and figures of speech to admire. In a vigorous simile, Crane conveys the motion of the boat and a sense of its precarious balance: "By the very last star of truth, it is easier to steal eggs from under a hen than it was to change seats in the dinghy" (paragraph 31). In the same passage, a man changing places in the boat picks himself up and moves his body as carefully as though he were a piece of delicate china. One way to get students to sense the degree of life in a writer's style is to have them pick out verbs. In the opening sentence of Part II (21): "As the boat *bounced* from the top of each wave the wind *tore* through the hair of the hatless men, and as the craft *plopped* her stern down again the spray *slashed* past them."

Building suspense, Crane brings the men again and again within sight of land and then drives them back to sea. He aligns enemies against them: sharks, the ocean current, the weight of water that sloshes into the boat and threatens to swamp it. The climax of the story—the moment of greatest tension, when the outcome is to be decided—comes in paragraph 204, when the captain decides to make a run through the surf and go for shore.

The situation of having one's nose dragged away before he can "nibble the sacred cheese of life" (70, 143) is a clear instance of irony of fate, or cosmic irony. (These slightly ponderous terms are defined as briefly as possible on page 130.) Crane, who sees Fate as an "old ninnywoman" (70), knows that the rain falls alike on the just and the unjust. There is no one right way to state the theme of this rich story, but here are some attempts by students, each with some truth in it:

> The universe seems blind to human struggles.

> Fate is indifferent, and doesn't always give the brave their just rewards.

> It's an absurd world; only people are reasonable.

This theme (however it is stated) may be seen also in the symbol of the giant tower (204) and Crane's remarks on what it suggests: "The serenity of nature amid the struggles of the individual," a presence not cruel or beneficent or treacherous or wise. Students who like to hunt for symbols sometimes want to see the boat as the universe, in which man is a passenger. A very decent case can be made for reading the story in this way. But everything in the story (ocean, waves, shark, beach, lighthouse), however full of suggestions, is first and foremost a thing concrete and tangible.

Although the secondary theme of comradeship is overtly expressed only in paragraph 43 ("the subtle brotherhood of men that was here established"), it informs the entire story. Until they quit the boat, the men are willing to spell one another at the oars; and even in the water, the captain still thinks of the correspondent's safety. The word heroism remains unspoken but understood: in "The Open Boat" a hero seems to be one who faithfully does what needs to be done. All four men thus qualify for the name.

Remembering the scrap of verse by Victorian bard Caroline Norton (179), the correspondent finds himself drawn into sympathy with any sufferer who, like himself, has to die in a remote place. The sentimental lines give rise to a feeling that, under the circumstances, seems heartfelt and real. In their time, the stormy life and dashing Byronic ballads of Caroline Norton (1808–1877), granddaughter of Richard Brinsley Sheridan, attracted wide notice. Joyce mentions another poem in "Araby," paragraph 23. She signed herself "Hon. Mrs. Norton," having married George Norton, a commissioner of bankruptcy who later went bankrupt himself. Michael R. Turner supplies two of her poems and a biographical note in *Parlour Poetry* (London: Michael Joseph, 1967) 71–72, 228–29.

For its view of people as pawns of nature, Crane's story has been classified as an example of American naturalism. Clearly, "The Open Boat" also has elements in common with recent fiction and drama of the absurd: its notion of Fate as a ninny, its account of Sisyphean struggles in the face of an indifferent universe. For critical comment on these aspects of the story, see Richard P. Adams, "Naturalistic Fiction: 'The Open Boat,'" *Tulane Studies in English* 4 (1954): 137–47; and Peter Buitenhuis, "The Essentials of Life: 'The Open Boat' as Existentialist Fiction," *Modern Fiction Studies* 5 (1959): 243–50.

William Maxwell, novelist and for many years a *New Yorker* fiction editor, was asked in a recent interview, "As an editor, in deciding whether or not to read a story, how much weight do you place on the first sentence?" "A great deal," Maxwell replied. "And if there is nothing promising by the end of the first page there isn't likely to be in what follows. . . . When you get to the last sentence of a [story], you often find that it was implicit in the first sentence, only you didn't know what it was." Asked for his favorite opening lines in fiction, Maxwell's first thought was of "None of them knew the color of the sky," from "The Open Boat" ("The Art of Fiction," *Paris Review* 82 [December 1982]). Students may be asked what this first line reveals. (That the sea compels the survivors' whole attention.) How does the final sentence in the story hark back to the first?

Nathaniel Hawthorne
YOUNG GOODMAN BROWN, page 165

"Young Goodman Brown" is Hawthorne's most frequently reprinted story and probably the most often misunderstood. Some students will take the devil's words for gospel, agree that "Evil is the nature of mankind," and assume that Brown learns the truth about all those hypocritical sinners in the village, including that two-faced Faith.

One likely point of departure is the puns on *Faith* in Brown's speeches:

"I'll cling to her skirts and follow her to heaven" (paragraph 5).

"Faith kept me back awhile" (12).

"Is there any reason why I should quit my dear Faith . . . ?"

"with Heaven above, and Faith below, I will yet stand firm against the devil" (46).

"My Faith is gone!" (50)

What Faith does Hawthorne mean? Surely not Puritanism—in this story, hardly a desirable bedfellow. More likely Brown's Faith is simple faith in the benevolence of God and the essential goodness of humankind. Brown's loss of this natural faith leads him into the principal error of the Salem witch-hangers: suspecting the innocent of being in league with the devil. At first, Brown assumes that his Faith is a pretty little pink-ribboned thing he can depart from and return to whenever he feels like it—he, Brown, the strongman, who will pass a night in the woods and then return to the bosom of his faithful spouse. Of course this is an attitude of blind pride, and it works Brown's ruin.

We realize that this story is often interpreted as highly ambiguous, highly ambivalent in its attitude toward Puritanism and the notion of innate depravity. But we read it as another of those stories in which the Romantic Hawthorne sets out to criticize extreme Puritanism and to chide the folly of looking for evil where there isn't any. In this regard, the story seems much like "Ethan Brand," in which the protagonist sets out to find the unpardonable sin, only to receive God's pardon anyway; and like "The Minister's Black Veil," in which Mr. Hooper makes himself miserable by seeing the world through a dark screen. (The latter is, admittedly, a more ambivalent tale: Hawthorne also finds something to be said in favor of that black veil and its wearer's gloomy view.) Brown's outlook has been tainted by dark illusion, conjured up, it would seem, by the devil's wiles.

Some initial discussion of the story's debt to American history may be helpful—most students can use a brief refresher on the Salem witchcraft trials, in which neighbor suspected neighbor and children recklessly accused innocent old women. The hand of the devil was always nearby, and it was the duty of all to watch for it. From Cotton Mather's *Wonders of the Invisible World* (1693), Hawthorne drew details of his imagined midnight Sabbath. In revealing to Brown the secret wickedness of all the people he knew and trusted, the story seems to illustrate the Puritan doctrine of innate depravity. Humankind was born tarred with the brush of original sin and could not lose the smudge by any simple ritual of baptism. Only the elect—communicants, those who had experienced some spiritual illumination which they had declared in public—could be assured of salvation. Brown's unhappy death at the end of the story seems conventional: Puritans held that how one died indicated his chances in the hereafter. A radiantly serene and happy death was an omen that the victim was Heaven-bound, while a dour death boded ill. The devil's looking like a blood relative may reflect another Puritan assumption. Hawthorne's great-grandfather, the witch trial judge, would have agreed that the devil often appears in disguise. The Salem trials admitted "spectral evidence"—testimony that the devil had been seen in the form of some innocent person. Spectral evidence was part of the case against Goody Cloyse, Goody Cory, and Martha Carrier—all named in Hawthorne's story, all of whom Judge John Hawthorne condemned to death. For more on Puritan doctrines (and how they eroded with time), see Herbert W. Schneider's classic study *The Puritan Mind* (U of Michigan P, 1958). Of Hawthorne's tales and *The Scarlet Letter,* Schneider observes: "[Hawthorne] did not need to believe in Puritanism, for he understood it. . . . He recovered what the Puritans professed but seldom practiced—the spirit of piety, humility, and tragedy in the face of the inscrutable ways of God" (262–63).

Why does Brown go out to the woods? Apparently he has promised the devil he will go meet him, but go no farther. By meeting the devil he has "kept covenant" (15). The initial situation—that Brown has a promise to keep out in the woods—is vague, perhaps deliberately like the beginning of a dream.

And was it all a dream? Hawthorne hints that the devil's revelations to Brown and the midnight Sabbath are all one grand illusion. When Brown staggers against the supposedly flaming rock, it proves cold and damp, and a twig that had been on fire suddenly drips cold dew. (We are indebted here to F. O. Matthiessen's discussion in *American Renais-*

sance [Oxford U P, 1941] 284.) If we read him right, Hawthorne favors the interpretation that Brown dreamed everything ("Be that as it may . . . "). Leave it to the devil to concoct a truly immense deception.

Still, some ambiguity remains. As Hawthorne declared in a letter to a friend in 1854, "I am not quite sure that I entirely comprehend my own meaning in some of these blasted allegories." In the end, Hawthorne leaves the interpretation of the story open; he will not pronounce absolutely. If what Brown saw at the witches' Sabbath really did take place, then his gloom and misery at the end of the story seem understandable. Some have read the story to mean that Brown has grown up to have a true sense of sin, and therefore ends a good Puritan; he has purged himself of his boyish good cheer. (Personally, we find the morose Brown deluded, not admirable, and suspect that Hawthorne does too.)

Students will not have met the term *allegory* in this book unless you have already dealt with the chapter "Symbolism," but some will know it already. Not only Faith can be seen as a figure of allegory, but Young Goodman Brown himself—the Puritan Everyman, subject to the temptation to find evil everywhere. For a class that has already begun symbol-hunting, "Young Goodman Brown" is a fair field. Among the more richly suggestive items are the devil's snaky staff or walking-stick (13), with its suggestions of the Eden snake and the serpentine rods of the Egyptian magicians (Exodus 7:8-12). When the devil laughs, it squirms happily (22). It works like seven-league boots, and its holder enjoys rapid transportation. The devil gives it to Goody Cloyse and then plucks Brown a fresh stick from a maple (38-41); when Brown grasps it in despair (51), it speeds him on to the unholy communion. Other symbolic items (and actions) include the forest—entering it is to be led into temptation, and Brown keeps going deeper and deeper; the withering of the maple branch at the devil's touch (38); and the proferred baptism in blood or fire at the evil meeting (67).

But the richest symbol of all is Faith's lost pink ribbon that flutters to the ground, prompting Brown to conclude that Faith too is a witch and let fall her ribbon while riding on a broomstick to the midnight Sabbath. (Many students, unless you help them, will miss this suggestion.) Sure enough, when Brown arrives, Faith is there too. The ribbon, earlier suggesting youthful beauty and innocence, becomes an ironic sign of monstrous evil and duplicity. This terrible realization causes Brown to decide to follow the devil after all—even though, presumably, the fluttering ribbon was another diabolical trick. When Brown meets the real Faith once more, she is still all beribboned, as if she hasn't lost anything. (A Freudian explanation would find the ribbon a kind of maidenhead, which Faith loses on losing her innocence.)

What are we to make of the resemblance between the devil and Brown's grandfather? Taken literally, perhaps it suggests that evil runs in Brown's family, or in the Puritan line, as the devil asserts (18-19). Or that wickedness lurks within each human heart (as well as good) and that each can recognize it in himself, as if he had looked into a mirror. Of course, donning the family face may be one more trick of the devil: an attempt to ingratiate himself with Brown by appearing as a close relative.

Hawthorne published this story in a magazine in 1835 but, omitting it from *Twice Told Tales* (1837), did not reprint it until eleven years later in *Mosses from an Old Manse* (1846). Did he hesitate to give permanence to this bold blast at the faith of his fathers? The story implies that the Puritan propensity for hunting out hidden sin was inspired by the devil. Perhaps Hawthorne may have had this story in mind when he wrote, in his dedication to a later collection of tales, *The Snow Image* (1851): "In youth, men are apt to write more easily than they really know or feel; and the remainder of life may be not idly spent in realizing and convincing themselves of the wisdom which they uttered long ago."

How to state the theme? Surely not "Evil is the nature of mankind" or "Even the most respected citizens are secretly guilty." That is what the devil would have us believe. A more defensible summing-up might be "Keep your faith in God and humankind" or "He who finds evil where no evil exists makes himself an outcast from humanity."

Compare this story with Shirley Jackson's "The Lottery." There, too, a whole town takes part in a blood-curdling rite—but public, not clandestine.

Kurt Vonnegut, Jr.
HARRISON BERGERON, page 175

Most students, without much trying, can state Vonnegut's theme in one way or another. One way is: Down with mediocrity and conformity. Another is: Hooray for individual excellence.

Although "Harrison Bergeron" isn't a great story—its humor seems elementary and its characters, especially the flat Diana Moon Glampers, seem stick-figures—its message strikes many students with great force. Perhaps, in a dreamlike way, it reminds them of their high school days when, if they showed any spark of mind—even spoke up in class—they would sense the distrust of their peers.

Vonnegut's story objectifies this well-known American distrust of mind and exaggerates it to an insane degree. Vonnegut imagines a society so dedicated to a perverse concept of equality that it condemns absolutely all excellence: good looks, physical grace, imagination. The story may be seen as a fable. At least, there's an apparent moral to it: Don't be afraid to excel. As in most fables, rounded characterization seems unnecessary to its purpose.

This is *anti-utopian fiction*, for, as in Huxley's *Brave New World*, technology serves idiocy. Stanislaw Lem may have a point (in his statement quoted in question 5), but we think Vonnegut's story a playful satire, not an earnest attempt to forecast things to come.

Until Vonnegut in the 1960s began to reach a wide general audience, his following was mainly limited to science-fiction fans—but he has insisted that his work is not confined to science fiction. In an essay, "Science Fiction," for the *New York Times Book Review* (September 5, 1965), he said that until he read the reviews of his first novel, *Player Piano* (New York: Delacorte, 1952), he had not thought of himself as a science-fiction writer: "I supposed that I was writing a novel about life, about things I could not avoid seeing and hearing in Schenectady, a very real town, awkwardly set in the gruesome now. I have been a sore-headed occupant of a file drawer labeled 'science-fiction' ever since, and I would like out."

The Vonnegut Statement edited by Jerome Klinkowitz and John Somer (New York: Delacorte, 1973) is a lively anthology of criticism of Vonnegut's work, by various hands.

Vonnegut reads "Harrison Bergeron" on a tape recording (SWC 1405) available from American Audio Prose Library, Box 842, Columbia MO 65205. For current price, call (800) 447-2275.

7 Symbol

John Steinbeck
THE CHRYSANTHEMUMS, page 185

In the sight of the flower pot and its contents, discarded in the road, Elisa sees the end of her brief interlude of hopes and dreams. Once, the chrysanthemums had embodied fulfillment: living proof of her ability to make things grow and bloom. As the title of the story indicates, they are the primary symbol—and when we state their meaning, we are close to stating the story's theme.

Elisa is a complex study in frustration, a frustration that stems in part from her being a woman in a world of men. That she is a strong, intense woman with far more energy than she can put to use is made clear at once, when we learn that "even her work with the scissors was over-eager, over-powerful" and that her house is "a hard-swept looking little house, with hard-polished windows."

Trapped under a "grey-flannel fog" that encloses the valley like a lid on a pot, Elisa works inside a symbolic barrier: a wire fence "that protected her flower garden from cattle and dogs and chickens" and that also protects her, because she is a woman, from the wider world. Custom denies her and her restless energy the adequate outlets that men enjoy. She cannot buy and sell cattle as Henry does. She cannot drift about the countryside mending utensils as the traveling repairman does, though his gypsy life-style powerfully appeals to her. "That sounds like a nice kind of a way to live," she declares on meeting him. "I wish women could do such things," she tells him later, and adds, "I can sharpen scissors, too. . . . I could show you what a woman might do." But she will never be given such an opportunity.

Besides her frustration at the passive role thrust upon her, Elisa is thwarted because her considerable gifts for nurturing—her "planting hands"—have little value in the wider world. It seems clear that the remarkable chrysanthemums are richly symbolic of her feminine talents. Yet the practical and shortsighted Henry, because the flowers are not a cash crop, says, "I wish you'd work out in the orchard and raise some apples that big." The traveling repairman feigns an interest in the chrysanthemum shoots for his own gain and then throws them on the road and saves the pot. So much for Elisa's creativity. Mordecai Marcus, in his critical comment, sees Elisa's flowers as substitutes for the children she (now thirty-five) was apparently unable to have ("The Lost Dream of Sex and Childbirth in 'The Chrysanthemums,'" *Modern Fiction Studies* 11 [1965]: 54–58). The suggestion, though likely, is not clearly confirmed in the story. "Elisa's need is definitely sexual," according to Elizabeth E. McMahan, "but it does not necessarily have anything to do with a longing for children" ("'The Chrysanthemums': Study of a Woman's Sexuality," *Modern Fiction Studies* 14 [1968]: 453–58).

Given to speaking in impassioned poetry ("Every pointed star gets driven into your body"), Elisa is further thwarted by having to live a life devoid of romance. Although kind enough and considerate enough, Henry is dull. When he tries to turn a compliment, the best he can do is, "Why—why Elisa. You look so nice!"—and when pressed for details, falls into grossness: "You look strong enough to break a calf over your knee, happy enough to eat it like a watermelon."

Sleazy as he is, the gypsy repairman has a touch of the poet about him. He can describe a chrysanthemum as "a quick puff of colored smoke." Elisa's short-lived belief that he values her flowers (and by extension, recognizes her womanliness) releases in her a long-pent eroticism for which the repairman is ill prepared. He changes the subject.

Elizabeth McMahan finds, in the bath-and-pumice scene that follows, "a purification ritual." Elisa "felt shame after her display of passion before the stranger. Now she cleanses herself before returning to her husband, the man to whom she should lawfully reach out in desire." And so Elisa punishes herself with the abrasive pumice until her skin is "scratched and red" in atonement for her imagined infidelity.

In the end, Elisa tries to satisfy her spiritual and erotic cravings by a feeble gesture: by asking Henry if they might order wine with their dinner. "It will be enough if we can have wine. It will be plenty." It isn't enough, of course, and she cries "weakly—like an old woman"—she who had briefly thought herself strong. Her new interest in prize fights, in the spectacle of blood-letting she had formerly rejected, manifests Elisa's momentary wish for revenge on men: her desire to repay them for her injured femininity. At least, this is the interpretation of Marcus and of McMahan.

Elisa's battle with her stifled sexuality is conveyed in detail ("her hand went out toward his legs in the greasy black trousers.... She crouched low like a fawning dog"). A textual critic, William R. Osborne, demonstrates that Steinbeck, in revising the story, heightened Elisa's earthiness and the sexual overtones of her encounter with the repairman (*Modern Fiction Studies* 12 [1966]: 479–84). Of the two extant texts of the story, this book prints the revised version as it appeared in *The Long Valley* (New York: Viking, 1938).

Like his early model, D. H. Lawrence, Steinbeck is fond of portraying people swept by dark forces of the unconscious. A curious book for additional reading: Steinbeck's Lawrencian novel *To a God Unknown* (New York: Viking, 1933). "The Chrysanthemums" invites comparison with Lawrence's "The Blind Man," which also conveys a woman's struggle for intellectual survival while living with a mindless husband on a farm. Henry Allen is no match, though, for Lawrence's impressive Maurice. If the two stories are set side by side for evaluation, Lawrence's may well seem (in our opinion) the deeper and more vivid. Though finely perceptive, "The Chrysanthemums" has something methodical about it, as if the young Steinbeck were deliberately trying to contrive a short-story masterpiece. But this is to dissent from the judgments of Mordecai Marcus, who thinks it indeed "one of the world's great short stories," and of André Gide, who is his *Journals* finds it in a league with the best of Chekhov (Vol. IV [New York: Knopf, 1951] 79).

"The Chrysanthemums" is a rare illustration of an unusual method: the objective point of view, or "the fly on the wall." After some opening authorial remarks about the land, Steinbeck confines himself to reporting external details. Although the reader comes to share Elisa's feelings, we do not enter her mind; we observe her face and her reactions ("Elisa's eyes grew alert and eager"). This cinematic method of storytelling seems a hardship for the author only in paragraph 109, when to communicate Elisa's sadness he has to have her whisper aloud.

Shirley Jackson
THE LOTTERY, page 194

Shirley Jackson's famous story shocks us. By transferring a primitive ritual to a modern American small town and by making clear in passing that the same ritual is being carried out in surrounding towns, the author manages to create in us a growing sense of horror over what is happening. Very early—in paragraphs 2 and 3—she mentions the stones that have been gathered in preparation for the day's events. Not until much later in the story does the importance of the stones begin to dawn.

Students might be asked to sum up the rules of Jackson's lottery, which are simple and straightforward. The head of each household—or, if he is absent, another representative of the family—draws a slip of paper out of a big black box. One householder pulls out a

(text pages) 194–200

piece of paper that has a black circle crudely penciled on it. Each member of his family is then obliged to participate in a second drawing. This time the unlucky recipient of the black circle is stoned to death by the other townspeople, including the members of his or her own family. Whatever justification might ever have existed for the ritual has long since been forgotten. The people simply accept the proceedings as an annual civic duty, the up-to-date version of an ancient fertility ritual ("Lottery in June, corn be heavy soon").

What is spine-chilling in Jackson's story is the matter-of-factness with which the ritual is carried out. Each June the townspeople assemble to murder one of their neighbors. The discrepancy between ordinary, civilized, modern behavior and the calm acceptance of something as primitive as human sacrifice gives "The Lottery" a terrible power. Among the story's many ironies, some of the most notable are:

1. The point of view. An objective narrator tells the story, remaining outside the characters' minds; yet the narrator's detachment contrasts with the attitude of the author, who presumably, like the reader, is horrified. That the day's happenings can be recounted so objectively lends them both credence and force.

2. The setting. The beauty of the June day is out of keeping with the fact that what takes place on the town green is a ritual murder.

3. The misplaced chivalry. Though women can be stoned to death in these yearly proceedings, they are whenever possible protected from having to take part in the general drawing (paragraph 13).

4. The characters. The townspeople are perfectly ordinary types, "surveying their own children, speaking of planting and rain, tractors and taxes" (3). Mr. Summers is in charge because he "had time and energy to devote to civic activities" (4). Old Man Warner is a stickler for tradition. Neighbors chat amiably. Children play. All are grateful that the proceedings will be over in time for them to enjoy their noon meal.

As a matter of course, even the small son of the victim is given some stones to throw at his mother. That is perhaps the most horrifying detail of all.

The story's very outrageousness raises questions about unexamined assumptions in modern society. Do civilized Americans accept and act upon other vestiges of primitive ritual as arbitrary as the one Jackson imagines? Are we shackled by traditions as bizarre and pointless as the lottery in Jackson's story? What determines the line between behavior that is routine and that which is unthinkable? How civilized in fact are we?

In *Private Demons: The Life of Shirley Jackson* (New York: Putnam's, 1988), Judy Oppenheimer gives a good account of the story's genesis. Jackson wrote "The Lottery" in 1948 while pregnant with her third child. She had been reading a book on ancient customs of human sacrifice and had found herself wondering how such a rite might operate in the village of North Bennington, Vermont, where she lived.

Peter Hawkes of East Stroudsburg University finds an obstacle to teaching "The Lottery" in that many students think its central premise totally unrealistic and absurd. How, they assume, can this story have anything to do with me? Hawkes dramatizes the plausibility of the townspeople's unswerving obedience to authority. With a straight face, he announces that the Dean has just decreed that every English teacher give at least one F per class to reduce grade inflation, passes around a wooden box, and tells students to draw for the fatal grade! "While I pass the box around the room, I watch carefully for, and indirectly encourage, the student who will refuse to take a slip of paper. When this happens, I ask the class what should be done. Invariably, someone in the class will say that the person who refused to draw deserves the F. Hearing this, the student almost always draws." See Mr. Hawkes's account in "The Two Lotteries: Teaching Shirley Jackson's 'The Lottery'", *Exercise Exchange* for Fall 1987. We would expect a class to greet this trick with much skepticism! But what if you were to try it on them *before* assigning the story?

(text pages) 194–201

Jackson once remarked that in writing "The Lottery" she had hoped "to shock the story's readers with a graphic demonstration of the pointless violence and general inhumanity in their own lives" (quoted by Lenemaja Friedman, *Shirley Jackson* [New York: Twayne, 1975]). For class discussion: What is the point of Jackson's comment? Is it true? In our own society, what violent behavior is sanctioned? How are we comparable to Jackson's villagers? Don't we too casually accept the unthinkable?

"The Lottery" invites comparison with Hawthorne's "Young Goodman Brown": in each, an entire community is seen to take part in a horrifying rite.

Yet another interpretation is possible. Jackson ran into parental opposition when she announced her intention of marrying fellow Syracuse University student Stanley Edgar Hyman, and some of her housemates warned her of the perils of living with a Jew. Shocked by these early run-ins with anti-Semitism, Jackson once told a friend (according to Judy Oppenheimer) that "The Lottery" was a story about the Holocaust.

There are dangers, of course, in reading more meaning into the story than it will sustain. Jackson herself, in *Come Along with Me* (New York, Viking, 1968), insists that we accept the story at face value. After its debut in *The New Yorker* in 1948, Jackson was surprised by the great amount of mail she received about the story. Some letter writers demanded to know what "The Lottery" meant, others supplied interpretations that they wished confirmed, others angrily abused the author for writing such a story. She even had letters from people asking where these lotteries were held, that they might go and watch the drawings.

Alice Munro
THE FOUND BOAT, page 201

Clearly "The Found Boat" is a story of sexual awakening. The season is spring, the season when the natural world springs to life. Water, an element often associated with life and fecundity, plays an important part in what happens. When the story opens, in the very early spring ("everything bright and cold"), the river's flooding is hard to ignore—as are the first stirrings of puberty in Eva and Carol, Frank, Bud, and Clayton later in the narrative. The story begins with the children in opposite camps, the boys against the girls. Their crude attempts at humor, their slang and profanity, their difficulty in communicating with one another except indirectly through insults, the almost grudging interest each of them develops in the opposite sex hint at their being eleven or twelve years old—old enough to have lost interest in stealing sap (paragraph 48). When first the girls get their feet wet (both literally and metaphorically), they find the water "like blue electric sparks shooting through their veins" (6), evidently both daunting and exhilarating.

The story is told by an all-knowing narrator, a non-participant writing objectively in the third person but sometimes seeing into all of them, as in paragraph 87, sometimes just into Carol and Eva—mostly Eva, perhaps the most aware, the most involved emotionally, the most sensitive to the adventure's sexual overtones. We are privy to her thoughts especially in paragraphs 10, 50–52, and 88.

The boat is satisfyingly real, "painted black, and green inside, with yellow seats, and a strip of yellow all the way around the outside." But it serves also as a luminous symbol, a treasure thrown up from the fecund depths of the water, a gift that the maturing children transform from a ragged joke into something sturdier. As the task gradually unites them into a team, it is clear that Eva is drawn especially to Clayton, whose "concentration, delicacy, absorption" pleasantly intrude upon her dreams (52). He seems equally attracted to her though unable to express his feelings except awkwardly, as when he chooses her to heat the tar (49) and, of course, when he squirts water at her. On the May holiday when the children carry the new-made boat back to the water, all share the load, not just the boys. Riding in it "was not like being on something in the water, but like being in the water itself" (60). They float blissfully downstream with the current, the life force.

Does Eva's name suggest Eve? The naked romp that results from the game of Truth or Dare perhaps recalls Eden before the fall. The run toward the river, idyllic as it is ("They felt as if they were going to jump off a cliff and fly. They felt that something was happening to them different from anything that had happened before, and it had to do with the boat, the water, the sunlight, the dark ruined station, and each other" [87]), can also be read as a joyous leap from innocence toward experience. Then, like Adam's wife, Eva suddenly feels shame (among other emotions) for her nakedness. This happens at the moment when Clayton spits water at her and the two of them mutually realize that the waters they have waded into are more emotionally laden than they can quite handle. Eva immerses herself once more, and this time the current carries her away from Clayton. Symbolically she makes a quick retreat back to childhood, and so do the others. "Anyway. It never was our boat," Eva declares (95). By the end of the story the boys and the girls are once more in opposing camps, the girls convulsed with childish giggles. But, as paragraph 14 reminds us, summer will follow spring. Although the children's awakened sexuality may go under cover for a while, it won't disappear. One season will follow another. The retreat into childhood can only be temporary.

Alice Munro, in an interview available on audiocassette, speaks of influences on her work, the emergence of Canadian literature, and being a woman writer. For current price, call American Audio Prose Library, (800) 447-2275.

8 *Evaluating A Story*

Most students find it a lighter task to place two stories side by side and to decide which is superior than to hold one story before them and evaluate it. This is why we suggest—if you have time for evaluation—assigning pairs of stories in some way comparable. For more suggestions of stories that may be paired, see the two sections at the beginning of this Manual, "Alternative Uses for Stories" and "Stories Arranged by Subject and Theme."

9 Reading Long Stories and Novels

Leo Tolstoi
THE DEATH OF IVAN ILYCH, page 218

Students might respond to the questions like this:

1. "The first thought of each of the gentlemen in that private room was of the changes and promotions [Ivan's death] might occasion among themselves or their acquaintances" (paragraph 5). Then, after these selfish hopes, they thought of their boring duties: to attend the funeral, to visit Ivan's widow (paragraph 18). Tolstoi suggests Ivan had no true friends but only "so-called friends" (18) and "nearest acquaintances" (19).

2. Comic elements include the Church Reader with his imperious tone (paragraph 24), Peter Ivanovich's struggles with the rebellious springs of the couch (33, 37), and his longing to escape to a card game—fulfilled at the end of Chapter I. The widow haggles over the price of a cemetery plot (33–35) and utters the great comic lines (in 40) congratulating herself on enduring Ivan's screams.

3. See paragraph 27, especially the statement that the corpse's expression contains "a reproach and a warning to the living."

4. Tolstoi's arrangement of events seems masterly. We see from the first the selfish, superficial circle of colleagues among whom Ivan lived and his rather obtuse, querulous, self-centered wife. As the story unfolds, we follow Ivan's progress from total absorption in his petty affairs to his final illumination. We begin with an exterior view of the context of Ivan's society and end with one man, all alone, confronting eternity.

5. Ivan seems a usual, sensual man, able to minimize his feelings of guilt (paragraph 59). His life centers in the official world, and he believes it should flow "pleasantly and properly" (82). "The pleasures connected with his work were pleasures of ambition; his social pleasures were those of vanity" (108).

6. Ivan performs his official duties with "exactness and incorruptible honesty" (81); he is decorous and rule-abiding, wields power objectively and does not abuse it (65). As he succeeds in the world, his marriage deteriorates (82), and improves only when Ivan and his wife see little of each other (98).

7. Ivan's excessive concern with every spot on the tablecloth seems part of his fussy overemphasis on material possessions. Tolstoi is suggesting, of course, that Ivan's insistent worldliness goes together with a neglect for the welfare of his soul.

8. Ivan's routine (paragraph 105) omits love, worship, and any profound involvement with his friends and family. His social consciousness is confined to a little chat about politics; his interest in art and science to a little chat about "general topics." Ivan devotes himself only to superficial courtesies and appearances—to maintaining "the semblance of friendly human relations" without any deeper fondness or compassion.

9. The illness might be diagnosed as cancer of the abdomen caused by falling off a ladder while hanging curtains. (The fall is casually described in paragraph 99.) Tolstoi detested physicians and implies (in paragraphs 115–120, 128, 152–153) that Ivan's doctors are know-nothings. In 247, Ivan submits to an examination that he sees as "nonsense and pure deception," like the speeches of certain lawyers to whom he had listened as a judge. The "celebrated specialist" (258–259) seems a quack.

10. See especially paragraphs 308–311. (Tolstoi's story has a theme in common with Thomas Mann's *The Magic Mountain:* The sick and the healthy are races set apart.)

11. From our first meeting with him, Gerasim radiates generosity, warm sympathy, cheerful acceptance of God's will. All men shall come to death, even as Ivan Ilych, he affirms (in 50). Gerasim is the good peasant, faithful and devoted, willing to sit all night holding his master's legs (285).

12. By his reluctance to acknowledge that his life "was not all the right thing" (as he finally admits in paragraph 348), Ivan is hindered in his advance toward enlightenment.

13. Ivan's son, though only briefly sketched, is unforgettable. Early in the story (in 48), Peter Ivanovich finds in his eyes ("tearstained" from genuine grief) "the look . . . of boys of thirteen or fourteen who are not pure-minded." As in all masturbators (according to popular lore), "terribly dark shadows showed under his eyes" (268). Yet it is the schoolboy, alone among Ivan's immediate family, who loves Ivan and feels sorry for him. By kissing his father's hand (349), the boy brings on Ivan's illumination.

14. It is revealed (in paragraph 350) that although Ivan's life has been futile, he still can set it right. "He whose understanding mattered" (351) now understands him. Able to love at last, Ivan feels sorry not for himself but for his wife and son. When at last he is willing to relinquish the life to which he has desperately clung, death holds no fear.

15. Randall Jarrell seems right in his observation that "we terribly identify ourselves with" Ivan Ilych. "Ivan Ilych's life has been a conventional falsehood; *The Death of Ivan Ilych* is the story of how he is tortured into the truth. No matter how alien they may have seemed to him to begin with, in the end the reader can dissociate neither from the falsehood, the torture, nor the truth: he *is* Ivan Ilych" ("Six Russian Short Novels," *The Third Book of Criticism* [New York: Farrar, 1965]).

"Beyond any doubt," said Tolstoi's biographer Henri Troyat, writing of *The Death of Ivan Ilych,* "this double story of the decomposing body and the awakening soul is one of the most powerful works in the literature of the world."

Completed in 1886, when its author was fifty-seven, this short novel stands relatively late in Tolstoi's literary career. *War and Peace* had appeared in 1863–1869; *Anna Karenina,* in 1875–1877. Still to come were *The Kreutzer Sonata* (1889) and *Resurrection* (1899–1900). Tolstoi undertook *The Death of Ivan Ilych* as a diversion from writing his earnest sociological treatise *What Then Must We Do?*—a work that few readers have preferred. Apparently, however, he became deeply involved in the story, for he toiled over it for nearly two years. It was to be (he told a correspondent) an account of "an ordinary man's ordinary death." This irony is stressed in the opening words of Part II: "Ivan Ilych's life had been most simple and most ordinary and therefore most terrible"—a sentence that Jarrell called one of the most frightening in literature.

Tolstoi based details of his story upon actual life: his memories of the agonizing death of his brother Nikolai, whom he dearly loved; and a description of the final illness of one Ivan Ilych Mechnikov, public prosecutor of the Tula district court, which he had heard from Mechnikov's brother.

Embedded in Ivan's story is another story: Praskovya's story, equally convincing as a study of character development. This insight comes to us from Nancy Adams Malone of Mattatuck Community College, who writes—

> Question: How does the girl described in paragraph 68–70 as "attractive, clever" and "sweet, pretty" turn into the appalling woman we meet at the beginning of the story?
>
> To a twentieth-century reader, it's a story as startlingly recognizable as Ivan's own. Praskovya's story begins in paragraph 72 with her first pregnancy. Like any thoughtless, well-meaning, ignorant young couple of our own time, they are not pre-

pared for a change in emotional balance. The successful bride turns into an insecure woman rather suddenly. She's worried about losing her looks, she needs reassurance that her husband still loves her—in short, paragraph 73. Like many another young husband, Ivan is dismayed by the new whiny, tearful, demanding wife, and does the natural (not the necessary) thing: he retreats. This move naturally increases the insecurity and the demands, a predictably vicious circle. . . .

Paragraphs 80-1 show another recognizable pattern: the family moves, and although the husband may be absorbed in his work, the wife has been uprooted from whatever emotional supports she has had earlier, family or friends or both. Money troubles, of course, don't help. And two children died. Even in a century that didn't expect to raise all its children (as we do), this loss couldn't help being an emotional wound. Paragraph 82 shows how Ivan dealt with all these troubles: "His aim was to free himself more and more from unpleasantness . . . by spending less and less time with his family. . . . "

Without making any kind of point of it, Tolstoi shows clearly how Ivan contributes to his wife's emotional and spiritual deterioration. I suppose, in today's climate, I should point out that this is not a feminist accusation that her deterioration is all Ivan's fault.

Jarrell's essay supplies further insights. Ilych is a man whose professional existence—being absurd, parasitic, irrelevant, and (to take a term from Sartre) unauthentic—has swallowed up his humanity and blinded him to the real. A life so misled is terrible, more to be dreaded than "real, serious, all-absorbing death." Revealingly, Jarrell quotes from Tolstoi's autobiographical *Confessions,* in which the author tells how he felt himself, like Ivan Ilych, shaped by society into an artificial mold, encouraged in his "ambition, lust of power, selfishness." Tolstoi himself had experienced a season of grave illness; it had given him time to reflect on his life (before his conversion to Christianity) and to find it meaningless.

In his *Confessions,* Tolstoi also tells of the death of his brother, who "suffered for more than a year, and died an agonizing death without comprehending what he had lived for, and still less why he should die." Some of the details in *Ivan Ilych* are believed to have been taken from Tolstoi's memory of his brother's struggle against disease and futile search for a cure.

For other useful discussions, see Philip Rahv, *Image and Idea* (Norfolk, CT: New Directions, 1949) 111-27; and Irving Halderin, "The Structural Integrity of 'The Death of Ivan Ilych,'" *Slavic and East European Journal* 5 (1961) 334-40.

More clearly, perhaps, than any other modern writer, Tolstoi succeeds in raising an ordinary man to tragic dignity. As a critic observes, writing not of Tolstoi's story but of the nature of tragedy, "To see things plain—that is *anagnorisis,* and it is the ultimate experience we shall have if we have leisure at the point of death. . . . It is what tragedy ultimately is about: the realization of the unthinkable." The statement, from *Tragedy* by Clifford Leech (London: Methuen, 1969), seems closely applicable. Only when Ivan Ilych accepts the unthinkable truth that his life has been lived in vain can he relinquish his tightfisted grip on life, defeat pain, and, on the brink of death, begin to live. Whether or not we accept Tolstoi's religious answer, the story compels us to wonder what in our own lives is valuable—what is authentic beyond petty, selfish, everyday cares.

10 Three Fiction Writers in Depth

Raymond Carver

Background: Shortly after Carver's death, Jay McInerney, who had been his student of fiction writing at Syracuse University in the early 1980s, published a warmly affectionate memoir, "Raymond Carver: A Still, Small Voice," in the *New York Times Book Review* for August 6, 1989. McInerney recalls Carver as a man given to quiet understatement, so shy and soft-spoken that an interviewer once had to put the tape recorder in the author's lap and still found his responses almost inaudible. His harshest criticism to a writing student was, "I think it's good you got that story behind you."

Despite the difficulty of taping Carver's voice, a tape of his reading of three stories ("Nobody Said Anything," "A Serious Talk," and "Fat") and a taped 51-minute interview are available from The American Audio Prose Library, Inc., Box 842, Columbia, MO 65205; phone number for orders: (800) 447-2275. Order numbers 3051 and 3052, respectively, price $10.95 each; or both for $20 (order no. 3053). This information may need to be verified by now; we culled it from a 1989–1990 catalog.

Raymond Carver
CATHEDRAL, page 259

On the surface, "Cathedral" is a simple story told in the flattest possible style by a not-too-bright narrator. His wife has invited a blind friend to stay with them. The encounter is awkward because the narrator has never met a blind person before. His misgivings are spelled out most clearly in paragraphs 1, 5, 9, 17, and 31. As the evening advances, a hearty meal, several drinks, and a few puffs of marijuana put the wife to sleep. The men stay awake and end up drawing a cathedral, the blind man's hand guided by the narrator's.

What does it all mean? Students will have to examine with uncommon thoughtfulness Carter's detailed account of this episode. In the story, one man's preconceptions about blind people are demolished. At the end, he has a revelation in which he experiences what it's like to be blind. How does this new understanding come about? From the start, the wife's behavior toward Robert is accepting and natural. If ever she felt any awkwardness toward blind people, it has long since disappeared. She sees Robert as a valued friend rather than as merely a blind man. In fact, she and Robert seem to know more about one another's thoughts and feelings than she and her husband do. Perhaps sensing this, the narrator exhibits a bit of jealousy mingled with his other negative reactions, especially in paragraph 46.

At first the husband blunders wretchedly, trying to make small-talk by asking the blind man what side of the train he had sat on (25). He notices with some surprise that Robert doesn't wear dark glasses or use a cane (31) and that he smokes (44). Is it just thoughtlessness, or is it a wish to cover an uncomfortable silence that impels him finally to turn on the TV?

The wife grows sleepy. Still wanting to be a good hostess, she suggests that Robert might like to go to bed. When he shows no inclination to do so, she falls asleep in spite of herself. Left alone with Robert, the narrator, too, suggests bed. Robert is, after all, his wife's guest. When Robert expresses a desire to stay up and talk with the husband, a turning point is reached. Perhaps surprised, the husband suddenly realizes that he's lonely, and in a new spirit begins explaining to the blind man what's happening on the TV screen.

(text pages) 270–286

The story ends with a startling revelation. The narrator learns from the blind man not only how to draw a cathedral, but what it's really like to be blind.

Students have phrased the story's theme in different ways: "Barriers tend to break down when people try to communicate with one another," or "Even those not physically blind sometimes need to be taught to see," or "Stereotypes render sighted people blind to the common humanity we all share." (Obviously, the story itself is light years better than any possible statements of its theme.)

For a writing topic, a student might be asked to read D. H. Lawrence's "The Blind Man" (in *Complete Short Stories,* volume II [Penguin Books], or in many anthologies) and compare the characters of the blind man and the sighted man with the similar pair in "Cathedral."

Raymond Carver
A SMALL, GOOD THING, page 270

Though people sometimes act badly, most people are basically good. This simple assumption seems to inform "A Small, Good Thing," in which Carver shows, too, how people can suffer from isolation. When on rare occasions they reach out to one another, sharing their concerns and perceptions, their lives are enriched. Carver hints at much the same thing in "Cathedral."

Carver tells his story in the third person; because the point of view alternates between the mother's and the father's, readers cannot help sharing in the parents' suffering. The flat, numb tone seems to intensify the reader's sense of Howard and Ann's pain. The hospital setting further dramatizes the family's ordeal by cutting them off, as hospitals do, from the rest of the world. When they leave that world, they are subjected to the cruelty of the baker's anonymous phone calls. Thus, except for the mother's brief exchange in the hospital with a family similarly afflicted, the Weisses are completely isolated in their pain. The doctors are false prophets, holding out hope when there is no hope at all.

When Ann finally figures out, in paragraph 160, that it's the baker who has been harassing them with the terrifying phone calls, she and Howard set out to vent their anger upon him. Ironically, in the warmth and fragrance that surround them in the baker's shop, they find solace. The baker is not a menacing madman but a fellow sufferer, alone in the world. Their isolation ends as the baker's does, through the comforting discovery of their common humanity, symbolized by the cinnamon buns, the "small, good thing" the baker offers them along with his explanation for the way he has behaved.

What have students experienced that confirms Carver's observations about people — how they react to suffering and loss, how they seek mutual comfort? Such a class discussion of this story might become too painfully personal, but perhaps the topic might be approached like this: "Has anything like this ever happened to anyone you know?"

Raymond Carver
WHERE I'M CALLING FROM, page 286

"Where I'm Calling From" is a story of recovery. Not permanent, perhaps — alcoholics are never "cured," only "recovering," as Carver, himself a problem drinker, well knew. But the impulse toward good health clearly wins out in "Where I'm Calling From"; and, as is so often true in a Carver story, redemption comes for troubled people when they are able to break out of their isolation, to reach out in some way to their fellow human beings.

Tiny's seizure, coming as it does when he is thought to be on the mend, frightens the other alcoholics. They see clearly what might happen to them. The narrator reacts by distancing himself from Tiny, not daring to talk with him about the incident. He reacts in similar fashion to his girlfriend when she receives bad news from her doctor. He seeks comfort in drink instead of human contact. In a Carver story this is a little death.

"Where I'm Calling From" seems, among other things, an accurate psychological assessment of the kind of person—self-absorbed, lonely, unable to face unpleasantness—who becomes an alcoholic. Ought we to see Freudian implications in J.P.'s rescue from the well or his desire to be a chimney sweep? Probably not. Carver embraces no psychological theories; "who knows why we do what we do?" his narrator asks in paragraph 21. Even love can't always keep an alcoholic from drinking. It's clear that J.P. loves his wife and she loves him. Still he drinks. But at the drying-out facility, as down in the well, for J.P. there is hope, light at the end of the tunnel, if he can just reach out to others.

When J.P. begins to talk and the narrator willingly listens, a kind of healing begins for them both. Then Roxy appears. Strong, loving, forthright, she accelerates the narrator's desire to connect, starts him thinking about a time when he and his wife were in love.

Frank Martin, with his reference from *The Call of the Wild* to the "animal that's half dog and half wolf" (37) is clearly trying to show J.P. and the narrator that they have it in them to be drunks or not, as they choose. They are not ready for his "sermon." "He makes me feel like a bug," J.P. says. When at the end of the story the narrator's thoughts revert to Jack London, it is a different story, "To Build a Fire," that he thinks of. Rejecting the freeze that will kill in favor of the warmth of human contact, he decides to telephone both his wife and his girlfriend. For now at least, he chooses life.

ANTON CHEKHOV

Any student willing to write a long paper on Chekhov might be steered toward the plays as well. In particular, *The Cherry Orchard* invites comparison with "The Darling" and "The Man in a Case." Among the many biographies, Ernest J. Simmons's *Chekhov* (Boston: Atlantic, 1962) remains valuable, readable, compact, and rich in critical insights. For further reading: Chekhov's stories, most of them, have lately been reissued in Constance Garnett's translations by Ecco Press. See also *The Unknown Chekhov*, translated by Avrahm Yarmolinsky (New York: Farrar, 1954), for stories Garnett left out.

Anton Chekhov
THE DARLING, page 297

Chekhov paints a quietly funny, sympathetic portrait of the misfortune-ridden life of Olenka, a woman like litmus paper who takes on the mental coloration of whomever she currently loves. How are we to react to her? With affection, no doubt, for Chekhov makes us appreciate his darling's generous, outgoing nature—but perhaps there's a touch of irony in the title. Personally, we're slightly appalled by her. Olenka's inability to form any ideas or opinions for herself is a terrible affliction, as we learn when the veterinarian leaves town with his regiment and Olenka momentarily finds herself terribly empty and alone (paragraphs 44–48). "How awful it is to have no opinions!" (46). Echoing the views of her love objects, she becomes successively all wrapped up in the theater, the lumber business, veterinary medicine, and (finally) the worship of a child. The story covers a long expanse of time, spanning many years and presumably supplying us with Olenka's nearly-complete biography.

Students need not see Olenka's career as meaningless and meandering. They may enjoy tracing the pattern of her life and perceiving its consistency. Her love-objects are (in chronological order):

Her father, her aunt, and her French teacher (paragraph 7);

Kukin the theater manager (2–23);

Pustovalov the manager of the lumber yard (24–35, 38–40);

Smirnin the veterinarian, who comes and goes (35–37, 41–44, 49–55); and

Sasha the veterinarian's little boy (55–77).

All four males portrayed in the story seem rather surly and ill-favored, not particularly deserving of Olenka's naturally welling love. Kukin is "an emaciated little man with a yellow face and hair combed down over his temples," always bitching about something—the weather or the stupidity of the public—with his face twisted into a look of permanent despair (7). Pustovalov is apparently just a big dull mannerly dolt, his interests confined to the lumber business. Smirnin is hardly a man of wider mind—all he talks about is the infidelity of his hated wife, and cattle diseases. Surly, a sexist pig, he angrily rebukes Olenka for talking about veterinary matters in front of his friends and insists on male intellectual supremacy (42). Though she and the lumber dealer had prayed in church to have children (38), no child is granted Olenka until the unengaging Sasha appears. Although he is a bubbly child full of "joyous laughter," Sasha is a thin, homely runt and something of a brat besides. When Olenka tries to mother him, he snarls, "Oh, leave me alone, please!" In his defense, we don't blame him for being embarrassed when this tall, stout, middle-aged lady insists on convoying him home from school, forcing sweetmeats into his hands (68).

Chekhov's point seems to be that Olenka's loving nature compels her to love whomever she happens to meet, no matter how unlovable. Early in the story, Chekhov gives us a thumbnail rendering of her character: "She was always in love with someone and could not live otherwise" (7). In all her relationships, Olenka seems to give more than she receives.

The title, of course, isn't merely ironic. People call Olenka a "darling" in the beginning (7) and call her that again when she rekindles her life with motherlove (70). Olenka makes a darling out of every one of her love objects, and she calls Sasha a "darling" too (70). Interestingly, Chekhov depicts Olenka as growing less attractive when deprived of love: she becomes "thin and plain," and people in the streets ignore her (45). In love, she brims with an engaging vitality and bubbles over with opinions—even though they are merely received. When the lover is out of sight, she is left utterly empty. Her love of the theater passes away with Kukin. After his death, even exploding skyrockets call up no memories (45). When she begins to love Sasha, his childish concerns and even the opinions he parrots from schoolbooks rush into the vacuum of her mind. She echoes the boy's views of his school, "saying just what Sasha said about them" (72).

Women students may find this story surprisingly pertinent. If they learn anything from Olenka's wistful plight, it might be: Don't let males (or anyone else) manufacture your opinions for you. Discover in your own resources and abilities an independent life of the mind.

Chekhov draws a wonderfully sensitive, balanced portrait, hardly less complex than a figure in life. This story was a favorite of Tolstoi's: the great writer and philanthropist used to read it aloud to his family and friends, calling it "a pearl" and Chekhov "a great, great writer." But he quite misunderstood it, thinks Chekhov's biographer Ernest J. Simmons: he took it for a satire on "the new woman," one who sought equality with men. Chekhov, claimed Tolstoi, started out to satirize Olenka but was unable to fulfill his purpose because his compassionate, poetic soul obliged him to sympathize with her and so overcame his intentions. Simmons believes, rather, that Chekhov intended his compassionate portrait of Olenka all along. "With all his acute insight, Tolstoi never quite perceived the breadth and tolerance of Chekhov's judgment, his tenderness for those who suffered, or his charity in the face of forgivable weakness" (*Chekhov*, [Boston: Atlantic, 1962] 438–439).

In 1901, when Tolstoi lay ailing, Gorky and Chekhov paid him a visit during which (Gorky recalled) Tolstoi rhapsodized over "The Darling." "It is like lace," he insisted through tears of great emotion, "made by a chaste young girl." Obviously embarrassed, Chekhov kept silent for a long time, finally gave a deep sigh, and said in a small, bashful voice, "There are many misprints in it . . . " (Simmons, 547).

Anton Chekhov
THE MAN IN A CASE, page 306

Like "In Exile," included in earlier editions of this book, "The Man in a Case" is told as a story within a story—a story that, like its central character Byelinkoff, seems placed at one remove and packaged in a kind of case. The theme, we conclude, is that Byelinkoff isn't the only man in a case: everyone who lives in the provinces (in the view of Ivan Ivanovich) is a prisoner in a shell of inertia, selfishness, cowardice, and hypocrisy.

What does the character of Mavra, the bailiff's wife, have to do with the story? In paragraph 3, the two men discuss her and her immobility; she herself stages a brief walk-on at the end. Evidently her case suggests to Burkin the story of Byelinkoff: another of those people "who, like hermit crabs or snails, are always trying to withdraw into their shells" (4). We can safely take Mavra for another captive of that inertia, boredom, and moral weakness that the two men attribute to their fellow villagers. As in Chekhov's plays, life in the provinces is depicted as so overwhelmingly dull that people are drawn to commit all sorts of follies just to liven things up. Evidently, trying to make a match between Byelinkoff and Varenka is an entertaining game for the villagers.

This story has also been translated as "The Man in a Shell," a title that somewhat more clearly announces its point. In his biography *Chekhov*, Ernest J. Simmons speculates that the story reflects its author's eleven unhappy early years as a boarding student at the Taganrog School for Boys. In his portrait of the snooping Byelinkoff and his craven colleagues, he may be drawing on memory ([Boston: Atlantic, 1962], 14). Varenka's brother Kovalenko issues a damnation on the whole place: "Oh, friends, how can you live here? Your whole atmosphere here is stifling and nauseating. Are you instructors and teachers? No, you are sycophants, and this isn't a temple of learning; it's a detective office, stinking as sour as a police court" (paragraph 41).

Burkin and Ivan Ivanitch, members of the professional class, see themselves in a position to criticize the passive inertia that informs the lives of their fellow townspeople. Chekhov didn't share Burkin's animosities toward the work of Turgenev and Shedrin; and in this story we don't find him explicitly siding with the veterinarian's views. We suspect that the author sympathizes with the two friends' dour view of provincial life, but he is more concerned in this story to portray it (and the memorable Byelinkoff) than to preach reform. Chekhov himself wouldn't go so far in his condemnation as does Ivan Ivanitch, and neither would the schoolmaster, who finally urges his friend to stop singing grand opera and go to sleep. In the end, Ivan, still sleepless, finds himself isolated, all alone—like yet another man in a case.

Anton Chekhov
VANKA, page 315

Poor Vanka's life is grim; so, apparently, was life in Russia at the time Chekhov lived and wrote. Peasant children were sent away from home to learn a trade at an early age and then often mistreated by those to whom they were sent. (The novels of Dickens paint a similar picture of English life during Victorian times.) Both dogs and children were mercilessly beaten. Life among the "gentlefolk" was pleasanter, to be sure; but, while people like Olga Ignatyevna might be sporadically kind to the peasants who crossed their paths, this kindness did not necessarily alleviate the suffering of the poor for long. Vanka's life after his mother dies is one of grinding poverty and hardship.

Vanka is intelligent (he has learned to read and write), observant, imaginative, and sensitive: traits not likely to make the life he has to lead any easier. His grandfather is skillfully characterized in paragraph 3. He seems a lovable sort, but a drinker. He is too poor—and perhaps too feckless—to provide for his grandson.

Of the dogs, Eel, especially, seems to share the combination of obsequiousness and malice that Chekhov may have observed among the downtrodden peasants. Vanka's fellow apprentices seem to exhibit these traits as well. The last sentence of paragraph 3 might be read as a tribute to the endurance not only of the dog but also of the Russian peasant.

There is irony in the contrast between the beauty of Christmas night and the squalor and misery of the boy. There is irony too in the contrast between the boy's innocent view of Olga, his "favorite," and the reality that Chekhov makes clear to us. She teaches him "to read and write, to count up to a hundred, and even to dance the quadrille" merely for her own amusement (15). When "they" send him first to the servants' quarters and then to Moscow, the bond between Olga and the orphan is easily snapped. He remembers her, but there is no evidence in the story that she ever thinks of him.

The grimmest irony of Chekhov's story has to do with our certainty that Vanka's pathetic letter will never reach his grandfather and the boy's hope of escaping his misery will be dashed.

FLANNERY O'CONNOR

If you think Flannery O'Connor's world, and her theology, are likely to strike your students as remote and unfamiliar, you might begin by telling the class a little about them. Should your library own a copy of Barbara McKenzie's book of photographs, *Flannery O'Connor's Georgia* (Athens, U of Georgia P, 1980), by all means bring it in and show it around. McKenzie recalls John Wesley's remark in "A Good Man Is Hard to Find," "Let's drive through Georgia fast so we won't have to look at it much," and supplies a series of pictures of some "oppressive" landscapes. Besides, there is a stone marker announcing Toomsboro, a town whose name foreshadows the ending of "A Good Man. . . . ", a glimpse of a pig parlor of which Mrs. Turpin might have been proud, and a shot of some urban black people who might be waiting for a bus, like passengers in "Everything That Rises Must Converge."

An essay worth quoting aloud is Alice Walker's sympathetic tribute "Beyond the Peacock: The Reconstruction of Flannery O'Connor," *In Search of Our Mothers' Gardens* (New York: Harcourt, 1983). "She was for me," declares Walker, "the first great modern writer from the South," and she praises O'Connor for not trying to enter the minds of her black characters, not insisting on knowing everything, on being God. "After her great stories of sin, damnation, prophecy, and revelation, the stories one reads casually in the average magazine seem to be about love and roast beef."

O'Connor's *Collected Works* has been accorded the honor of appearing in the Library of America series. This volume, along with three new works of O'Connor criticism, was reviewed by Frederick Crews in the April 26, 1990 issue of *The New York Review of Books*. Crews has interesting things to say about O'Connor's debt not only to New Criticism, but also to Edgar Allan Poe and Nathaniel West. He offers insight as well into what he calls her "stern fanaticism" and her "twisted feelings about segregation."

Flannery O'Connor
EVERYTHING THAT RISES MUST CONVERGE, page 319

Julian's mother, though living in pathetically reduced circumstances, never forgets that she is the granddaughter of a governor and slaveowner, the daughter of a prosperous landowner. Her belief in the innate nobility of her forbears is unshakable. Ancestral values, with their emphasis on the importance of "graciousness," of "knowing who you are," still dictate her often wrong and foolish behavior.

Even so, the mother is a more sympathetic character than is her son. There is an innocence about her that mitigates her faults. Julian, on the other hand, is a hypocrite, for all his pretensions to broadmindedness. (The author's ironic view of him is especially clear in paragraph 62.) Despising his mother for her sacrifices, he yet accepts her money and support. He thinks that sitting down next to a Negro on the bus or trying to strike up a conversation with "the better types, with ones that looked like professors or ministers or lawyers," makes him morally superior to his mother. In truth, such efforts seem undertaken only for the smug sense of virtue he derives from them, and the opportunities they provide for needling his mother. Meanwhile, without ever acknowledging his own aristocratic leanings, he secretly longs for the "threadbare elegance" of the now defunct family mansion, "the house that had been lost for him." He is a snob masquerading as a progressive.

Julian is delighted to find on the bus the black woman wearing a hat identical to his mother's. Apparently he regards the black woman's hat with the same amused contempt. "He could not believe that Fate had thrust upon his mother such a lesson" (82). He expects his mother suddenly to realize that her pride is foolish, that black people are her equals. It seems clear that Julian would like nothing better than a painful shattering of the pathetic illusions that sustain his mother. Only then might she learn to appreciate her son's moral superiority. That never happens, of course. Note the chilling line in paragraph 120, which describes her collapse: "The other [eye] remained fixed on him, raked his face again, *found nothing* [italics ours] and closed." But perhaps her death represents a turning point for Julian. At least he seems suddenly aware that he loves her and needs her still.

We are told in paragraph 90 that Julian's mother directs at Carver's mother "the smile she used when she was being particularly gracious to an inferior." Her offer of a penny to the small boy seems motivated not by bigotry but by the habit of condescension she has cultivated in her fantasy world. For Julian's mother, O'Connor writes, "the gesture would be as normal . . . as breathing" (95) and, we suspect, might just as easily have been directed at a white child. But Carver's belligerent mother apparently finds condescension as hard to take as outright bigotry.

The title of this story evidently contradicts the mother's belief that black people "should rise, yes, but on their own side of the fence" (24). That isn't how it works, O'Connor seems to be saying.

What is the theme of "Everything That Rises Must Converge"? Perhaps that spiritual pride is the worst sin of all. Or, to put it another way, that those who sneer at their supposed moral inferiors may not be so lofty as they think.

Flannery O'Connor
A GOOD MAN IS HARD TO FIND, page 330

Without some attention to its Christian (specifically Catholic) assumptions, this story won't make much sense to students, who might read it for a tale of meaningless violence. Their seeing what O'Connor is driving at depends on their reading with great care the conversation about Jesus between the Misfit and the grandmother, and the account of the old woman's epiphany and death (paragraph 136).

O'Connor loves to pick unlikely, ordinary people and show how, by the sudden and unexpected operation of God's grace, they are transformed into saints: the nit-witted grandmother who dies loving (and presumably forgiving) the Misfit, the awful Mrs. Turpin in "Revelation." In the course of the story, the grandmother grows and changes. At first, she seems a small-minded biddy, selfish or at least self-centered, capable of dumb remarks like "Oh look at the cute little piccaninny!" (18) and "People are certainly not nice like they used to be" (35), capable of blaming Europe for "the way things were now" (44), of regretting that she hadn't married Mr. Teagarden "because he was a gentleman and had bought Coca-Cola stock when it first came out" (26).

In the end when we are told that "her head cleared for an instant," the grandmother becomes newly perceptive. She reveals—and offers to the Misfit—her vast, compassionate heart. We have no reason to doubt the Misfit's shrewd remark, "She would have been a good woman . . . if it had been somebody there to shoot her every minute of her life." In slaying her, the Misfit has done her a tremendous favor: he has made a martyr of her. For one moment just before she dies, the old lady doubts Jesus, or at least feels forsaken: "Maybe he didn't raise the dead" (135). It is an understandable reaction, after the colossal shock she has undergone: she knows that her family has been massacred. But this moment of confusion passes. The grandmother is headed straight to heaven for her marvelous Christ-like act of love a moment before she dies, when she realizes that there is a chance that the Misfit will repent and she reaches out to him lovingly, as though he were a child. By this time, she already knows that the Misfit's gang has murdered her son and daughter-in-law and all the children. The Misfit may be a ruthless murderer, but that doesn't prevent her (in the end) from loving him and hoping for his redemption. Symbolically, in death the old woman's body lies with legs in the form of a cross, a look of childlike sweetness on her face. The Misfit, naturally, is glum, having just declined a chance for salvation.

An insightful and outspoken student might raise this objection: "Isn't it a shame to impose Catholic terms of redemption on all these Protestants?" (That the eight-year-old is named John Wesley suggests that the family is at least nominally Methodist.)

The story may draw other objections. O'Connor's use of familiar racist epithets will outrage some African-American students unless they see that O'Connor is only reporting faithfully how the characters think and talk, not condoning it or thinking that way herself. This will be a problem, too, in teaching "Revelation" (or most Faulkner novels or *Huckleberry Finn*).

Who is the central character? Some will say the Misfit, some the grandmother. This is mainly, we think, the grandmother's story from beginning to end—we take it to be the story of her moral and spiritual growth. She causes things to happen; the Misfit merely reacts to her. She persuades the family to depart from the main road to see the old plantation, she causes the accident by letting the cat out of the basket, she dooms the family when she recognizes the Misfit, and her final gesture incites him to murder her.

The scene at Red Sammy's Barbecue is no mere filler, but an essential. It tightens the suspense and enforces the hint that the much talked-about Misfit is bound to show his face. In his highway signs, Red Sammy boasts of his uniqueness: NONE LIKE FAMOUS RED SAMMY'S. He considers himself a hard-to-find good man. In calling him a good man (37), the grandmother first introduces a theme. The barbecue proprietor agrees with her, even declares, thinking of how many bad characters are on the loose these days, "A good man is hard to find" (43).

In the end, what has the platitudinous title indicated? We are left to think: how true, a good man (a saint) certainly *is* hard to find. We find, at the end, a serenely good woman whose salvation has been achieved only through traumatic suffering and the amazing arrival of grace.

O'Connor herself once commented on the story at a reading of it in 1963; her remarks are included in *Mystery and Manners* (New York: Farrar, 1969) 107–114. Southern students, she had learned, tended to recognize the grandmother as exactly like one of their own relatives, "and they knew, from personal experience, that the old lady lacked comprehension, but that she had a good heart." When her head clears for an instant, the grandmother "realizes, even in her limited way, that she is responsible for the man before her and joined to him by ties of kinship which have their roots deep in the mystery she has merely been prattling about so far. At this point, she does the right thing, she makes the right gesture. I find that students are often puzzled by what she says and does here, but I think myself that if I took out this gesture and what she says with it, I would have no story. What was left would not be worth your attention." She warned against equating the Misfit with the devil and preferred to think that the grandmother's gesture, "like the mustard-seed, will grow to be a great crow-filled tree in the Misfit's heart" and redeem him

(text pages) 330–341

yet. "And in this story you should be on the lookout for such things as the action of grace in the grandmother's soul, and not for the dead bodies."

Still, as Frederick Asals points out in a fruitful critical reading, it is hard to ignore all those dead bodies, either. In Asals's view no other O'Connor story sets up such extreme tension between matters sacred and profane, between comedy and violence. See *Flannery O'Connor: The Imagination of Extremity* (U of Georgia P, 1982) 142–154.

Flannery O'Connor
REVELATION, page 341

Epiphanies, Flannery O'Connor finds, are imminent in the drabbest, most ordinary life. And so she makes the doctor's waiting room the setting for prophecy; a pigpen, the doorstep of beatitude. Mrs. Turpin, bigoted and smugly self-congratulatory, seems an unlikely recipient for a revelation direct from God; and yet even she is capable of accepting such a revelation and by it being transformed and redeemed. Like Malamud's Angel Levine, Mary Grace, a pimpled Wellesley girl with emotional difficulties, seems an unlikely messenger of the Lord. But Mary Grace—an agent of redemption, as her name indicates—announces herself as a kind of biblical prophet: with "churning face," she goes into a trance. Her eyes change color, and Mrs. Turpin knows that the girl is about to utter some profundity meant for Mrs. Turpin alone.

Like the handwriting on the wall, Mary Grace's utterance is baffling and mysterious. Sorely troubled, Mrs. Turpin revolves it over and over in her mind all afternoon. She knows from Whom the message came: "What do you send me a message like that for?" (paragraph 179), and her impulse is to defend herself, to argue back at God. Her irate challenge to the Almighty, "Who do you think you are?" is exactly the question God is asking *her*. God replies immediately: He is Lord of all creation, whose natural world "burned for a moment with a transparent intensity." He is the giver of life and of death, as Mrs. Turpin realizes when she sees Claud's truck, whose driver and passengers at any moment could be destroyed. Then, in the final revelation, God shows her exactly who *she* is: just another sinner, whose pride in her virtues must perish in eternal light. For Mrs. Turpin, the hard road toward sainthood lies ahead.

"Revelation," a beautifully plotted story, builds slowly to its crisis or turning point: Mary Grace's assault on Mrs. Turpin in the doctor's office or, more specifically, the moment when she declares, "Go back to hell where you came from, you old wart hog." The climax we take to be Mrs. Turpin's challenge to God, "Who do you think you are?"; and the conclusion, her shocked acceptance of the final revelation—the vision of the bridge to heaven.

Symbolism pervades the story, although not obviously. The waiting room suggests a microcosm, or human society in capsule form: old man and child, white-trash woman and Wellesley girl. Significantly, the book that strikes Mrs. Turpin and shatters her view of herself is called *Human Development*. The hogs are fat with meanings. They resemble Mrs. Turpin, who is overweight, with eyes small and fierce, and whom Mary Grace calls a wart hog. In her perplexity, Mrs. Turpin is herself like the old sow she blinds with the hose. Her thoughts about pigs resemble her thoughts on the structure of society, "creating a hierarchy of pigs with a pig astronaut at the top"—Josephine Hendin writes in *The World of Flannery O'Connor* (Bloomington: Indiana UP, 1970). Mrs. Turpin gazes into her pig parlor "as if through the very heart of mystery." As darkness draws near—and the moment of ultimate revelation—the pigs are suffused with a red glow. Contemplating them, Mrs. Turpin seems to absorb "some abysmal life-giving knowledge." What is this knowledge? Glowing pigs suggest, perhaps, the resurrection of the body. As Sister Kathleen Feeley says in her excellent discussion of this story, "Natural virtue does as much for fallen men as parlor treatment does for pigs: it does not change their intrinsic nature. Only one thing can change man: his participation in the grace of Redemption" (*Flannery O'Connor: Voice of the Peacock* [New Brunswick, NJ: Rutgers UP, 1972]).

"I like Mrs. Turpin as well as Mary Grace," said O'Connor in a letter to a friend. "You got to be a very big woman to shout at the Lord across a hogpen. She's a country female Jacob. And that vision is purgatorial" (letter to Maryat Lee, May 15, 1964, in *The Habit of Being: Letters of Flannery O'Connor,* ed. Sally Fitzgerald [New York: Farrar, 1979] 577).

One of the last stories Flannery O'Connor finished, "Revelation" appeared in her posthumous collection *Everything That Rises Must Converge* (New York: Farrar, 1965). Walter Sullivan believes that Mrs. Turpin's vision of the bridge of souls, with its rogues, freaks, and lunatics, is the author's vision of humanity and her favorite cast of characters. The story stands as "a kind of final statement, a rounding off of her fiction taken as a whole" ("Flannery O'Connor, Sin, and Grace," in *The Sounder Few: Essays from the Hollins Critic,* ed. R. H. W. Dillard, George Garrett, and John Rees Moore [Athens: U of Georgia P, 1971]).

Joyce Carol Oates, in a comment on "Revelation," finds the story intensely personal. Mary Grace is one of those misfits—"pathetic, overeducated, physically unattractive girls like Joy-Hulga of 'Good Country People'"—of whom the author is especially fond. "That O'Connor identifies with these girls is obvious; it is *she,* through Mary Grace, who throws that textbook on human development at all of us, striking us in the foreheads, hopefully to bring about a change in our lives" ("The Visionary Art of Flannery O'Connor," in *New Heaven, New Earth* [New York: Vanguard, 1974]). In a recent survey of fiction, Josephine Hendin also stresses O'Connor's feelings of kinship for Mary Grace. As the daughter of a genteel family who wrote "distinctly ungenteel books," O'Connor saw herself as an outsider: "She covered her anger with politeness and wrote about people who did the same." In "Revelation," Mary Grace's hurling the book is an act of violence against her mother, for whom Mrs. Turpin serves as a convenient stand-in ("Experimental Fiction," *Harvard Guide to Contemporary Writing* [Cambridge, MA: Harvard UP, 1979]).

At least one African-American student has reacted in anger to the language of this story. Writing from Glassboro (N.J.) State College, she called "Revelation" the "most disgusting story I've ever read" and objected to "the constant repetition of the word *nigger.*" O'Connor never uses the word when speaking in her own voice, but it occurs several times in the thoughts and speeches of Mrs. Turpin (also in one speech of the "white-trash woman," [paragraph 51]). Such a reaction deserves a hearing—if possible, in class. African-American students given to signal reactions are in for some painful shocks unless alerted that they will meet this word in much fiction written twenty years ago or earlier, including *Huckleberry Finn* and Faulkner's stories. In "Revelation," how are we supposed to react to the characters who say or think the word? How is Mrs. Turpin limited and narrow in her views of African-Americans (and of herself)? Where does Flannery O'Connor stand? How does she expect *us* to feel? What do you make of the African-American people in the story? In what ways are they more perceptive than Mrs. Turpin? Should the word be expunged from literature? Or is a writer ever justified in using it to show how certain whites used to think and talk—and unfortunately, still do?

"People talk about the theme of a story," Flannery O'Connor wrote, "as if the theme were like the string that a sack of chicken feed is tied with. They think that if you can pick out the theme, the way you pick the right thread in the chicken-feed sack, you can rip the story open and feed the chickens. But this is not the way meaning works in fiction. . . . A story is a way to say something that can't be said any other way, and it takes every word in the story to say what the meaning is" ("Writing Short Stories," in *Mystery and Manners: Occasional Prose,* selected and edited by Sally and Robert Fitzgerald [New York: Farrar, 1980]).

11 Stories for Further Reading

Ambrose Bierce
AN OCCURRENCE AT OWL CREEK BRIDGE, page 357

QUESTIONS

1. What has brought Peyton Farquhar to the brink of hanging? (Incited by the Federal Scout, he has apparently tried to burn down Owl Creek Bridge, which is in the hands of the Union Army.)

2. One student, after reading Bierce's story, objected to it on the grounds that the actions of the Federal Scout (paragraphs 9–17) are not believable. "Since when do military men dress up like the enemy and incite civilians to sabotage?" was her question. How would you answer it? (One possible answer might be that Farquhar, "a slave owner and like other slave owners a politician," has managed in his own way to engineer damage to the enemy and so has become a marked man. A believer in "at least a part of the frankly villainous dictum that all is fair in love and war" (8), Farquhar is lured into a fatal trap by someone in the enemy camp who assents to the same view.)

3. Are you surprised by the story's conclusion? Can you find hints along the way that Farquhar's escape is an illusion? At what point in the story does the illusion begin? (Bierce lays in enough hints to make most readers accept the surprise ending, though they may want to reread the story just to make sure. In paragraph 5, though Farquhar makes a conscious effort to think about his wife and children, he has already begun to experience an alteration in his sense of time. Recall the old general belief that at the moment of death we see our whole lives pass by again in a flash. As soon as an escape plot forms in Farquhar's mind [6], the sergeant moves off the plank and the noose tightens around the doomed man's neck. Since we know from the ending that the rope does *not* break, what happens to Farquhar after the flashback [8–17] — except for the physical sensations of hanging — has to be a fast-moving hallucination in the mind of a dying man.

Bierce has planted clues that this is so. With "superhuman" effort Farquhar manages to free his bound hands. His senses are "preternaturally keen and alert" [20]. His ability to dodge all the bullets fired at him seems miraculous. The forest through which he walks is unfamiliar and menacing; the stars above him are "grouped in strange constellations"; he hears "whispers in an unknown tongue" [34]. Soon "he could no longer feel the roadway beneath his feet" [35]. His wife greets him with "a smile of ineffable joy" [36] — then he is dead.)

4. From what point of view is "An Occurrence at Owl Creek Bridge" told? (In the third person by a nonparticipating narrator able to see into Farquhar's mind.)

5. At several places in the story Bierce calls attention to Farquhar's heightened sensibility. How would you explain those almost mystical responses to ordinary stimuli? (They are the out-of-body sensations of a dying man. Much has been written about such experiences in the popular press. For some interesting, serious comments on the subject see Elisabeth Kubler-Ross, *On Death and Dying* [New York: Macmillan, 1969]. Dr. Ross quotes the testimony of persons who, after being declared clinically dead, were resuscitated. Some reported finding themselves floating in space, gazing down upon their own bodies.)

6. Where in Bierce's story do you find examples of irony? (There is irony in the contrast between the understated title and the extravagant style of the narrator. Ironic, too, is the mention of Farquhar's belief that "all's fair in love and war" — a belief shared by the Federal

(text pages) 364–369

Scout who lures Farquhar to his capture and death. Verbal irony occurs in "Death is a dignitary who when he comes announced is to be received with formal manifestations of respect, even by those most familiar with him" [paragraph 2] and "The liberal military code makes provision for hanging many kinds of persons, and gentlemen are not excluded" [3]. Students may find other examples as well.)

7. Do you think the story would have been better had Bierce told it in chronological order? Why or why not? (In defense of the way the story is told, we'd argue mainly that Bierce's use of flashback heightens the suspense created by the opening scene on the bridge.)

A topic for writing: Compare this story with either Tolstoi's "The Death of Ivan Ilych" (page 218) or with Borges's "The Secret Miracle" (page 364). All three stories deal with men on the brink of death and their inner experiences.

Jorge Luis Borges
THE SECRET MIRACLE, page 364

QUESTIONS

1. To what extent does the narrator confine his view to Hladik's views? At what moments does he step back from Hladik's consciousness and see the playwright with detachment? Comment, for instance, on the tone of the remark, "Like every writer, he measured the virtues of other writers by their performance, and asked that they measure him by what he had conjectured or planned" (paragraph 4).

2. What do Hladik's two dreams have to do with the rest of the story? The first is the dream of a chess game played for enormous stakes (1); the second, the dream of finding the hiding-place of God (7).

3. What is Hladik's attitude toward his previous writings? Why does completing *The Enemies* mean so much to him?

4. Is there any similarity between the plot of *The Enemies* (summarized in paragraph 5) and the story of Hladik himself?

5. What do you understand to be the central theme (or themes) of the story?

In his *Conversations with Jorge Luis Borges* (New York, 1969), Richard Burgin quotes the distinguished Argentinian writer as saying that, in writing "The Secret Miracle," he was interested in three ideas. These were the idea of "an unassuming miracle . . . wrought for one man only"; the idea of a man's justifying himself to God in a way known only to God; and "the idea of something lasting a very short while on earth and a long time in heaven, or in a man's mind."

Besides, the story is a splendid account of an artist's joy in solitary toil. It also sets forth the theme that a man, by desperate intellectual striving, cannot attain anything so beautiful as the gift of God's grace. This last seems the reason for Hladik's dissatisfaction with his book, *Vindication of Eternity* (a title quite foolish in its pride—as though eternity needs vindicating!). Discovering, by luck, the Word (or Letter) that had eluded the librarian and all his forebears, Hladik is suddenly and gratuitously blessed. (Borges may be giving us an ironic cameo self-portrait in the blind librarian whose hopes are thwarted.)

Throughout the story, Hladik's great enemy is the passage of time. Facing execution, he struggles to find some way to hold fast to the fleeting substance of time. Both his major works, *The Enemies* and *Vindication of Eternity*, set out to demonstrate that time is a fallacy. Although the passage of time is as inexorable as the onslaught of Hitler's armies, Hladik, as it turns out, triumphs over both time and the Gestapo. Essential to this story's tender comedy is our realization that Hladik's great play sounds like an awful dud—a "circular delirium." That it is going to be bad, like all his other works, doesn't matter. It will redeem

49

(text pages) 369–384

Hladik—if only he can bring it to perfection. By so doing he will compensate for a lifetime of writing weak and pretentious books that leave him, knowing their faults, guilty and dissatisfied. Hladik's play brings on stage a "reality" that is all in the mind of the deluded Kubin. Ironically, the secret miracle renders eternal a reality all in Hladik's mind, and in the mind of God. God—"of whose literary preferences he possessed scant knowledge"— may think Hladik a bad playwright, but what matter? The miracle demonstrates that God loves him all the same. And this incredible favor, whose existence Hladik alone perceives, is comparable to the illumination that comes to Tolstoi's Ivan Ilych on the very brink of death, redeeming a wasted life.

For a discussion of the story that relates it to Borges's other work, see Carter Wheelock, *The Mythmaker, a study of motif and symbol in the short stories of Jorge Luis Borges* (Austin, Tex., 1969), pp. 143–147.

Willa Cather
PAUL'S CASE, page 369

QUESTIONS

1. What is it about Paul that so disturbs his teachers? Recall any of his traits that they find irritating, any actions that trouble them. Do they irritate or trouble *you*? (The boy's dandyism, his defiance, and his unconcealed contempt disturb his teachers beyond all reason. They also disturb many readers, who find a discrepancy between their own negative feelings about Paul and the fondness Cather seems to feel toward him. The school principal and Paul's art teacher are able to muster sympathy for the boy [paragraphs 7–8], but they don't claim to understand him.)

2. What do the arts—music, painting, theater—mean to Paul? How does he react when exposed to them? Is he himself an artist? (Oddly, when he hears a concert, it's "not that symphonies, as such, [mean] anything in particular to Paul" [11]. The romance and beauty in the air excite him to hysteria. He reacts in the same way to fine food and wine, good clothes, a lavishly decorated room. For all his sensitivity, he is not an artist, and he doesn't respond to art as an artist would. It's the *idea* of art and beauty, the pleasure they afford his senses, that evoke a response in him. In *Willa: The Life of Willa Cather*, Phyllis C. Robinson points out that Paul "may have had the sensitivity of an artist but, unlike his creator, he was without discipline, without direction and, saddest and most hopeless of all, he was without talent." In that fact lies his private tragedy.)

3. In what different places do we follow Paul throughout the story? How do these settings (and the boy's reactions to them) help us understand him, his attitudes, his personality? (Cather shows us Paul in school, at home, in the art gallery, in the theater, and, finally, in the big city. Everywhere, he functions apart from the ordinary beings who surround him. Isolated and unhappy, Paul regards his esthetic longings as evidence of his superiority to other human beings. Absence of hope and cool disdain for the ordinary ways of attaining wealth and finery and an invitation to New York drive him to one grand, suicidal fling in the beautiful, glamorous world where he imagines he belongs.)

4. Does Cather's brief introduction into the story of the wild Yale freshman from San Francisco (in paragraph 54) serve any purpose? Why do you suppose he and Paul have such a "singularly cool" parting? What is Cather's point? (In New York Paul expects to find wealthy persons of taste and temperament similar to his own. He finds instead the disappointing Yale student who, far from sharing Paul's reverence for beauty, seems to the boy hopelessly crass. Ironically, though he seems to have grown up with all the advantages Paul has lacked, this worldly student disappoints Paul as much as do his family and friends at home. Who

could measure up to Paul's impossible standards? For the rest of his stay, Paul is content to observe his fellow guests from afar.)

5. Is Paul a static character or a dynamic one? If you find him changing and developing in the course of the story, can you indicate what changes him? (The only change in Paul's character during the story seems to be the fairly minor one mentioned in paragraphs 42–43. Once settled in at the Waldorf, he realizes he is no longer "dreading something." Is this change enough to justify our calling him a dynamic character? The change seems to signal in Paul a deepening involvement in an unreal world rather than any growth. We think he's a static character.)

6. Comment on some of the concrete details Cather dwells on. What, for instance, do you make of the portraits of Washington and John Calvin and the motto placed over Paul's bed (paragraph 18)? Of the carnation Paul buries in the snow (paragraph 64)? (Paul's sensibilities recoil from all that is ugly, even ordinary. His room, featuring pictures of those austere and well-disciplined heroes George Washington and John Calvin, is hateful to him. Working-class Cordelia Street, where he lives, induces in him "a shuddering repulsion for the flavorless, colorless mass of every-day existence; a morbid desire for cool things and soft lights and fresh flowers" [19]. By paragraph 64, after Paul has run away from New York, he realizes that fresh flowers do not stay fresh forever. He sees the parallel between "their brave mockery" and his own "revolt against the homilies by which the world is run" and characterizes both as "a losing game in the end." He dies like the flowers, his "one splendid breath" spent.)

7. What implications, if any, does the title "Paul's Case" have for this story? Is Paul mentally ill, or does Cather place the roots of his malaise elsewhere? (Robinson has this to say about Cather's view of Paul:

> As if to underscore her own intuition in regard to the story, Willa gave it the subtitle "A Study in Temperament," and to the modern reader the pathological attributes of Paul's malaise are more persuasive than the romantic aspects. Willa told George Seibel that she drew on two boys who had been in her classes for the character of Paul, but to others she confessed how much of her own hunger and frustration were embodied in the unhappy boy's flight from the drab reality of his daily life and in his instinctive reaching out for beauty.

In other words, we today tend to think Paul seriously disturbed, but apparently Cather didn't see him that way. In conflict with school, neighbors, and family, Paul seems to her an admirable victim. Cather often sticks up for the lonely, sensitive individual pitted against a philistine provincial society—as in her famous story "The Sculptor's Funeral.")

"Paul's Case" may be available for classroom viewing on videotape; it was featured in the PBS Television film series *The American Short Story*. Short scenes from the script of Ron Cowen's adaptation are included in *The American Short Story, Volume 2* (New York: Dell, 1980), and invite comparison with the original.

A detailed plan for teaching "Paul's Case" is offered by Bruce E. Miller in *Teaching the Art of Literature* (Urbana: NCTE, 1980). To get students involved in the story before they read it outside of class, Miller suggests reading passages from the early part of the story (before Paul flees to New York) and summarizing what happens, and then asking students how they'd imagine Cather would continue the story. He advises the instructor to show the class a few photographs of adolescent boys and to ask how closely the pictures reflect students' mental images of Paul. The point is to spark a discussion of Paul's complex character. Is he a cheat? A victim? A hero? Various opinions are likely to emerge, and students may come to see "that the different views of Paul are not necessarily incompatible with each other, and that Cather has accomplished the difficult feat of delineating a complex character who, though flawed, engages the reader's sympathy."

(text pages) 384–386

Maxine Chernoff
THE SPIRIT OF GIVING, page 384

QUESTIONS

1. Describe the tone of "The Spirit of Giving." (Good-humored and humorous, as is Jane, the narrator from whose point of view the story is told.)

2. How would you characterize Jane, her sister Martha, Martha's family, and Ted? (All are yuppies, sophisticated consumers, and one way to look at Chernoff's story is as a good-natured spoof of their values and pursuits. Almost every paragraph contains references to the self-conscious trendiness of the characters' lives: Ted's dismay at receiving earmuffs because he's a translator; the restaurants he and Jane frequent (one featuring pasta and Chianti, another Chinese food); Andy and Martha's desire for both uninhibited children and a white rug. Students will surely find other examples.)

3. How serious is the conflict between Jane and Martha? On what is it based? What is ironic about it? (The sisters merely hold differing views about predation, which Jane regards in a humorous light. The irony has to do with Martha's being a bleeding heart—not what you'd expect of an anthropologist. She manifests her concern about the shedding of innocent blood by hating the Eskimo calendar Jane sends her for Christmas; and, Jane tells us, Martha would object to wearing the slippers in which Marilyn Monroe "thought her last thoughts" [paragraph 14].)

4. What does Chernoff's title have to do with her story? (The story turns traditional ideas about "the spirit of giving" on their ear. Not that Jane doesn't try to please with her gifts. In true yuppie fashion, she just doesn't find pleasing others incompatible with pleasing herself. By gift-giving, she can even threaten to saddle Martha with the care of Mother.)

Kate Chopin
THE STORY OF AN HOUR, page 386

QUESTIONS

1. Is Mrs. Mallard's sorrow at the news of her husband's death merely feigned? Try to account for her sudden changes of feeling. (Undoubtedly her grief is sincere. Caught within a conventional marriage, she has not realized the freedom she has been denied. Notice that her rush of new joy, her looking forward to a whole new life, begins when she observes that spring is bursting forth, too, in the natural world.)

2. Some stories—for instance, "An Occurrence at Owl Creek Bridge"—are noted for their surprise endings. How can it be claimed that "The Story of an Hour" has a *triple* surprise ending? (Surprise ending #1: Mrs. Mallard realizes that her husband's death has left her unexpectedly happy. Surprise ending #2: Her husband is still alive. Surprise ending #3: She dies herself.)

3. "Chopin's little gem gives us a fresh and unconventional view of love and marriage." Discuss, refuting or supporting this opinion by specific references to the story.

Compare this story with Chopin's "The Storm" in Chapter Four—another realistic, unconventional view of love and marriage that, in Chopin's day, undoubtedly helped annul her reputation.

"The Story of an Hour" has been criticized for having a rigged ending that strikes some readers as a cheap trick. A student essay by Jane Betz responds to this charge; it is included in X. J. and Dorothy M. Kennedy, *The Bedford Guide for College Writers*, 2nd ed. (Boston: Bedford Books, 1990).

Tess Gallagher
The Lover of Horses, page 388

Questions

1. What explains the great-grandfather's odd behavior if not drunkenness? (The narrator affirms in her opening paragraph: "I have reason to believe the gypsy in him had more to do with the turn his life took than his drinking.")

2. What is it, exactly, that the narrator has in common with her mother, her father, and her great-grandfather? (She characterizes this family trait as a propensity to be "stolen by things—by mad ambitions, by musical instruments, by otherwise harmless pursuits from mushroom hunting to childbearing" [paragraph 12].)

3. How does this family trait manifest itself in each of the four main characters? How does the mother's way of dealing with it differ from that of the others? (The great-grandfather ran away and joined the circus. The narrator calls him "a man who had been stolen by a horse" [7]. The father is a man stolen "by the more easily recognized and popular obsession with card playing" [12]. The mother "could fall in love with no man but" the feckless card player. Once she has married the charming man who "stole" her, however, she begins to regret her impulsive behavior. From then on she resists being stolen and teaches her daughter to do the same. Once cured of a childhood tendency to whisper, the daughter manages to live pretty much in accordance with her mother's wishes until her epiphany at the end of the story.)

4. What motivates the narrator at the end of the story to sleep on a bed made of boughs stripped from the cedar tree? What happens to her there? (The cedar tree, loved by her father ["The tree whispered, he said," 51], apparently whispers to her, directing her first to smoke cigarettes even though she is a nonsmoker and then to revisit the places her father had loved. As she does so she enters into a trancelike state, acting not of her own volition but as directed by a force beyond her control. It is this force that impels her to make her outdoor bed from cedar boughs and spend the night on it, leaving others to tend her dying father. There, while letting go of her father, she embraces her fate: that she too is a "whisperer.")

5. What does the first-person point of view contribute to Gallagher's story? (Verisimilitude. We are inclined to believe the narrator because she was there.)

6. What if anything does the "whispering" symbolize in "The Lover of Horses"? Do you find other devices in the story that hint at more than their literal meaning? (In Gallagher's story, to be a "whisperer" is apparently more than merely to "possess the power of talking sense into horses" [3]. It is also to be subject to a strong and life-giving force that impels its recipients toward a life often disreputable but also passionately engaged. And this force is contagious [14]. Those under its sway seem mostly to regard it as a precious gift. Though the mother rejects it, the great-grandfather, the father, and the narrator are swept up by it.

Suggestive too are the horses, the circus, the saddle blanket and plumes that the grandfather hangs over the foot of his bed, the cards, and the fishing boat. Students may find others. All have something to do with the joy of being swept by a passion strong enough to plunge its owner into the heart of life.)

7. How would you state the theme of this story? (One way to state it: People who allow themselves to be "stolen" by passion, no matter how disreputable the world into which it draws them, lead richer lives than do those who play it safe.)

You might care to have students compare "The Lover of Horses" with "The Rocking-Horse Winner" (also in "Stories for Further Reading"). What do the two stories have in common, besides horses and betting? (Magic.) How do they differ? (For Lawrence, a major theme is love of money and its sinister consequences.)

(text pages) 397–401

Compare, too, the ending of Gallagher's story with that of Kafka's "The Bucket Rider." Both stories end with the words "lost forever" (or "for ever"). Did Gallagher unconsciously echo Kafka's famous line? (We don't know, but in any case she gives the words a sharply different meaning. For the narrator of "The Lover of Horses," to be lost is positive and desirable.)

Langston Hughes
ON THE ROAD, page 397

QUESTIONS

1. Is "On the Road" a fantasy, or is it a realistic story that contains a dream or vision? Defend your answer with evidence from the story. (Students are likely to express sharply differing views. Some will stoutly maintain that the story is a fantasy in which a man pulls down a church and Christ gets off his cross, takes a walk with the man, then heads for Kansas City—an engaging detail! Others will insist with equal fervor that the white cops, who arrive just as the church door is giving way, club the black man into unconsciousness, and that from this unconscious state arises the dream that begins in paragraph 20. This, they will maintain, explains Sargeant's waking up in jail without knowing how he got there [54–56]. We ourselves are divided on the question, XJK seeing the story as a tale that reveals the marvelous, DMK as a realistic story containing a vision. It is, in any case, a story about racial prejudice.)

2. Where and when is Hughes's story set? (The time is the Great Depression of the thirties, the place a city big enough to have relief shelters for the destitute and far enough north to have snow. Homelessness being the problem it is today, the story seems curiously timely, though hordes of people no longer ride freight trains as they did back in the thirties.)

3. From what point of view is "On the Road" told? (The narrator is omniscient. Mostly he looks at the action through the black man's eyes, even speaks like Sargeant ["He never did see Christ no more," 51], but in paragraph 18 also sees into the thoughts of the people on the street.)

4. To what extent does this story resemble "Godfather Death" and "An Appointment in Samarra" in Chapter One? (All three stories center around a divine or supernatural being. The characters are flat rather than round; there is no appreciable character development. In all three, plot is perhaps the most important element.)

Franz Kafka
THE BUCKET RIDER, page 401

QUESTIONS

1. Like other fantasies in this book ("The Rocking Horse Winner," "On the Road"), this story invites the question, At what point did you become aware that this is not a realistic story? (Students' answers will differ, but the speaker's mode of transportation [paragraph 2] should telegraph the story's being a fantasy.)

2. What elements of "The Bucket Rider" resemble a dream? (Many, but especially the narrator's sense of being light as a feather, of floating on his bucket-steed.)

3. Describe the characters of the coaldealer and his wife. Does the wife see and hear things differently from her husband, or does she merely pretend to? (Hard-heartedly, she pretends that the narrator isn't there.)

4. If the narrator froze to death, as we assume from the final sentence, how can he or she tell us the story? (This question is unanswerable, but go ahead and ask it and see what happens.)

5. What is the tone of this story? What do you understand to be the author's attitude toward the narrator and the events he or she relates? (Sadly funny, as befits this wry comment on the scarcity of human charity.)

"The Bucket Rider" was written in Prague during the severe winter of 1916–17. Max Brod in his biography *Franz Kafka* (New York: Schocken, 1960) calls it "the sole thing of beauty that came out of the coal shortage."

Absurd but not nonsensical, Kafka's story contains both dreamlike logic and some realistic elements. The coaldealer's wife, who prevents her more charitable husband from giving credit, is a villainness of nightmare, but she is also a recognizable businesswoman. The opening metaphor of the sky as a silver shield that protects God from the pleas of mortals prepares us for the story to come, a characteristic Kafkan account (like *The Trial* and *The Castle*) of the impossibility of making oneself heard, or of obtaining mercy. The view of the universe is existentialist, no doubt, but the metaphor also reminds us of Fitz-Gerald's *Rubaiyat*:

> And that inverted Bowl they call the Sky,
> Whereunder crawling cooped we live and die,
>   Lift not your hands to It for help—for It
> As impotently moves as you or I.

In the story, religion lends the narrator no comfort: the commandment "Thou shalt not kill" goes unheeded, and the church bells only conspire to drown his cries.

Students usually enjoy "The Bucket Rider," but some have trouble defining the story's tone of sorrowful fun, as light as the bubble-weight bucket. The tone emerges particularly in what the narrator tells us about himself. Seated upon his fantastic steed, like a child upon a hobbyhorse, he is a powerless pip-squeak, an annoying insect to be waved aside. He expresses the grief of a suffering victim, but his attempted rise to dignity is like that of (says Kafka) a squatting camel; and with the author—if we are right in interpreting his view—we cannot take him seriously.

Yasunari Kawabata
THE MAN WHO DID NOT SMILE, page 403

QUESTIONS

1. Of what importance is the narrator's weariness at the start of Kawabata's story? What do we learn about this narrator as the story goes on? (As sometimes happens with creative people who find themselves in a state of exhaustion, the script writer who narrates Kawabata's story experiences a burst of creativity—the "daydream" that results in a new and artistic ending to the film he has written. It becomes clear as the story progresses that the narrator is a gloomy man [paragraph 9] for whom facing harsh reality is difficult. As a writer—and also, it seems, as a person—he likes to "wrap reality in a beautiful, smiling mask" [7]. He wants even a film set inside an insane asylum to end happily.)

2. What reason does the narrator give for not wanting to try on the masks in front of his wife and children? (The narrator feels "a vague sense of satisfaction" [34] when his children try on the masks, perhaps because there isn't much difference between the smiling

(text pages) 403–406

expressions on the masks and the children's everyday faces. When the children first ask him to try one on [35–40], his refusal seems motivated by some vague, nameless terror. Only after his wife has tried on a mask and has horrified him by doing so does he realize [in 48] what he fears: that putting on a mask will make him "look like an ugly demon" to his wife when he takes it off.)

3. What makes the narrator decide, at the end of the story, to cut the mask scene from the movie? What then do you think motivates him to change his mind? (Horrified initially by the "shadow of a wretched life" that he sees on his wife's face after she removes the mask she has tried on [45], he is further disturbed by the sudden and unpleasant suspicion that perhaps the smile his wife habitually wears [because she knows that's what he wants to see?] is a mask as well [48]. Unwilling to confront such a grim possibility, he decides, "The mask is no good. Art is no good." Then, struck perhaps for the first time by the realization that art ought not to be confused with life, he changes his mind. Masks may be wrong for life, but they're fine for art. They will, after all, provide a tremendously effective ending for his movie.

Another way to regard the script writer's decision to tear up the telegram is as a retreat: Unable to accept the truths he has just discovered, the narrator returns in spirit to the world of pleasant artifice that he has always occupied.)

4. Are the masks in Kawabata's story symbols? If so, what do they suggest? (In art and literature, masks tend to be richly suggestive devices. The ones in this story recall a horrifyingly dramatic skit often performed by the great mime Marcel Marceau. In it, a man dons a grinning mask that he is then unable to remove, despite his increasingly anguished efforts. Kawabata's masks convey similar hints of the suffering that can lie behind smiling faces. [The script writer approvingly regards the brightly lighted hospital windows (23) as masks, too.]

As symbols so often are, the masks in "The Man Who Did Not Smile" are ambiguous and ambivalent. Considered as elements of daydreams, as subtle and artistic props introducing hope into the movie's finale [thus making unpleasant reality easier to bear], they seem benevolent. But they become sinister when they represent real-life efforts to ignore bitter truths. If, like the script writer, we fail to see suffering around us, then how can we ever truly connect with our fellow human beings?)

5. What does "The Man Who Did Not Smile" appear to be saying about the relationship between illusion and reality? (Just as reality influences illusion, illusion can influence reality, even distort it into strange shapes.)

6. What is the theme of this story? (It might be stated in any one of several ways: "Those who look at all of life through rose-colored glasses are likely to miss seeing something important" or "Masks belong in art, not in life" or "To pretend that suffering does not exist is to become alienated from the rest of humankind.")

D. H. Lawrence
THE ROCKING-HORSE WINNER, page 407

QUESTIONS

1. The family members in Lawrence's story harbor a number of secrets. What are they? (The mother's secret is that "at the center of her heart was a hard little place that could not feel love, no, not for anybody" [paragraph 1]. Paul's secret is that, by furiously riding his wooden rocking-horse, he is often able to predict which horses will win races. Bassett's secret, and Uncle Oscar's, is that they profit from Paul's predictions, even while the boy is on his deathbed. Their winnings are kept secret. Paul gives his mother 5,000 pounds, but does it anonymously. The house itself whispers a secret, "There must be more money" [5, 6, 181]. The three children hear the whisper, but no one talks about it.)

(text pages) 407–418

2. What motivates each main character? What sets Paul's quest apart from that of the others? (It is the desire for more money that motivates them all. Perhaps the most blatant evidence of the family's obsession with riches appears in Uncle Oscar's attempt to console his sister after her son's death: "My God, Hester, you're eighty-odd thousand to the good and a poor devil of a son to the bad" [244] — as if he were enumerating her assets and liabilities on an imaginary balance sheet.

Paul's frenzied pursuit of money differs from the greed of the others in that he wants wealth not for himself but for his mother. Clearly he hopes that, by being "luckier" than his father, he will win his mother's love and attention.)

3. Some details in Lawrence's story are implausible. What are they? (Those students who can appreciate Lawrence's particular blend of reality and fantasy will like "The Rocking-Horse Winner." A house that whispers is unusual, but even a hard-headed realist can probably accept it at least as a metaphor. That a boy can learn to predict the winner in a horse race by riding his rocking-horse is perhaps harder to believe.)

4. At what places in his story does Lawrence make use of irony? (Paul, intent upon stopping the whispers in the house, anonymously gives his mother 5,000 pounds as a birthday present. Ironically, his gift has the opposite effect. The whispers grow louder. Given his mother's insatiable greed for money, this result comes as less of a surprise to the reader than to Paul.

There is irony in the story's title. Paul, the rocking-horse winner, loses his life.

Ironic, too, are Paul's final words: "I *am* lucky" [241]. In his mother's definition of luck, in paragraph 18, ["It's what causes you to have money"] Paul *is* lucky, of course. Or was.)

5. In what sense may Paul's periodic rides on the rockinghorse be regarded as symbolic acts? (The single-minded frenzy with which Paul rides his rocking-horse parallels the intensity of the money-lust that dominates Paul's family. There is no joy in his riding, as there is no joy in his house.)

6. What is the theme of "The Rocking-Horse Winner"? (The love of money is destructive of all other love, and even of life itself.)

Bobbie Ann Mason
BIG BERTHA STORIES, page 418

QUESTIONS

1. What problems confront Donald, Jeannette, and Rodney? How does each of them cope? (Donald has an unhealthy preoccupation with the Vietnam War that manifests itself in flashbacks, moodiness, restlessness, inability to hold down a job or to enjoy a normal relationship with his wife and child. All the other problems in the family stem from his. Jeannette has to deal with poverty and loneliness and raising Rodney by herself. Rodney is neurotic: he hides in a closet when his father comes home and suffers from nightmares. He seems to find some release in drawing pictures of the nightmares. His pictures become less disturbed after his father enters the Veterans Hospital and stops coming home.)

2. What is the tone of Mason's story? (Perhaps the flat, disaffected tone is meant to reflect the ordinariness of the characters and the dogged, everyday courage with which they live their lives.)

3. Who or what is Big Bertha? (As Mason explains in paragraph 10, Big Bertha is Donald's name for a colossal strip-mining machine, and he "has Rodney believing that Big Bertha is a female version of Paul Bunyan." The main character in most of the stories Donald tells Rodney, Big Bertha was originally, of course, the name for the giant cannon used by the Germans against the French in World War I.)

4. What motivates Donald to leave home as often as he does? What is it that finally makes him decide to enter the Veterans Hospital? (Donald goes away when he's experiencing dark moods, when "he can't cope with his memories of Vietnam" [14]. Though his absences spare his wife and child, they do not cure his post-traumatic stress. Several painful incidents occur. Finally, he decides to enter the hospital after an attack of uncontrollable trembling that hits him when he's rambling on about his wartime memories [115].)

5. Why is Jeannette seeing a psychologist? How much good do the sessions with him seem to do? (In paragraph 52, Jeannette reminds the therapist that she came to him for advice about Donald. She is frustrated because the therapist keeps trying instead to get her to focus on herself.)

6. How does strip mining function in this story? What seems to be Mason's attitude toward it? (It functions in more than one way. In paragraph 15, Donald sees it as a metaphor for what America did to Vietnam during the war. The difference is that the strip miners have Big Bertha to "put the topsoil back on" [91]. Strip mining takes on more benign associations in paragraph 117, in which Jeannette compares it to what she hopes the hospital treatment will do for Donald: "dig out all those ugly memories.")

Bobbie Anne Mason discusses what she thinks she's doing as a writer on an audiocassette interview available from American Audio Prose Library, Box 842, Columbia MO 65205. For current price, call (800) 447-2275.

Joyce Carol Oates
WHERE ARE YOU GOING, WHERE HAVE YOU BEEN? page 429

QUESTIONS

1. Describe Connie as you see her in the first few paragraphs of the story. In what ways is she appealing? In what respects is she imperceptive and immature? (Connie, Oates implies, is still growing and doesn't know whether to act "childlike" or "languid." Still discovering her identity, she behaves one way at home and another way elsewhere. Her mind is "filled with trashy daydreams," and the first caresses of love seem to her just "the way it was in movies and promised in songs.")

2. Describe the character of Arnold Friend. In what ways is he sinister? What do you make of his strangely detailed knowledge of Connie and her family; of his apparent ability to see what is happening at the barbecue, miles away? Is he a supernatural character? (Perhaps Arnold's knowledge was obtained merely by pumping Connie's friends for information and by keeping close watch on her house, and perhaps his reported vision of the barbecue is merely feigned for Connie's benefit. Much about him seems fakery: his masklike face and "stage voice," his gilded jalopy, his artificially padded boots, his affecting the speech, dress, and music of the youth culture [although he is over thirty]. Still, there are hints that he is a devil or a warlock. Perhaps Ellie Oscar, the "forty-year-old baby," is his imp or familiar; perhaps his bendable boot conceals a cloven hoof. He works a kind of magic: on first spying Connie he draws a sign in the air that marks her for his own. He threatens to possess her very soul: he will enter her "where it's all secret" and then, after the sex act, she will give in to him [paragraph 104]. A charismatic like Charles Manson, he seeks young girls to dominate. According to Oates, the character of Arnold Friend was inspired by the author's reading a newspaper story about a traveling murderer in the Southwest.)

3. Why doesn't Connie succeed in breaking loose from Arnold Friend's spell? (She seems in a trance, like a dazed and terrified bird facing a snake. Everything appears unreal or "only half real" [94]. Like a practitioner of brainwashing, Arnold Friend denies reality: "This place you are now—inside your daddy's house—is nothing but a cardboard box I can

knock down any time" [152]. Arnold suggests that Connie's beating heart isn't real ["Feel that? That feels solid too but we know better"] and soon she thinks her body "wasn't really hers either" [155]. Perhaps, some male student will suggest, Connie really wants to give in to Arnold Friend—after all, she loves to flirt with danger and when Friend first looked at her in the parking lot, she looked back. But Connie's terror seems amply justified; Friend after all has threatened to kill her family unless she submits to him.)

4. What seems ironic in the names of the leading characters? (Friend is no friend; a better name for him would be Fiend. Ellie Oscar, like his first name, seems asexual. His last name suggests a trophy from the Motion Picture Academy—fitting for a media-slave who keeps his ear glued to his transistor radio. His androgynous name perhaps recalls that of another assassin, Lee Oswald. Connie is a connee—one who is conned.)

5. What is the point of view of this story? How is it appropriate? (Limited omniscience, with the author seeing into only one character's mind. Besides, the author perceives more than Connie does and observes Arnold Friend and Ellie more shrewdly than Connie could observe them.)

6. Explain the title. Where is Connie going, where has she been? (She has been living in a world of daydreams, and now, at the end, she is going out into a sunlit field to be raped. In the beginning of the story, we are shown that Connie lives mainly in the present; in the opening paragraph she sees "a shadowy vision of herself as she was right at the moment." She doesn't seem bothered by the ultimate questions asked in the title of the story. Arnold Friend answers the questions in paragraph 152: "The place where you came from ain't there any more, and where you had in mind to go is cancelled out." She has been nowhere, she is going nowhere—for in his view, there is no reality.)

7. What significance, if any, do you find in this story's being dedicated to Bob Dylan? (No doubt some of Dylan's music flows through Connie's mind, and some of Friend's repartee seems a weak-minded imitation of Dylanese: surreal and disconnected, as in his tirade to Ellie [133]: "Don't hem in on me . . . " Arnold Friend bears a faint resemblance to Dylan: he has "a familiar face, somehow," with hawklike nose and hair "crazy as a wig," and he talks with a lilting voice "as if he were reciting the words to a song." His approach is "slightly mocking, kidding, but serious and a little melancholy" and he taps his fists together "in homage to the perpetual music behind him" [77]. Joyce Carol Oates has remarked that Dylan's song "It's All Over Now, Baby Blue" [1965] was an influence on her story. Dylan's lyric addresses a young girl, Baby Blue ["My sweet little blue-eyed girl," Arnold calls Connie], who must make a hasty departure from home across an unreal, shifting landscape. A vagabond raps at her door, and she is told, "Something calls for you / Forget the dead you've left." Oates's title recalls a line from another Dylan song, "Mr. Tambourine Man": "And there is no place I'm goin' to.")

8. Consider this evaluation by a student: "All this story is is a sophisticated cheap thrill. Reading it is sort of like watching a kitten thrown to an octopus."

In teaching this story to her students at the University of Georgia, Professor Anne Williams had "a couple of minor epiphanies." She has written to share them with us:

"Ellie" may be a diminutive of "Beelzebub," lord of the flies. The story is certainly full of references to flies. I also noticed a series of comic allusions to various fairy tales, all of which, according to Bettelheim, concern the difficulties of coming to terms with adult sexuality:

"Snow White" (in reference to the mother's jealousy over Connie's looks, so much like her own faded beauty [1–2]);

"Cinderella" (the pumpkin on Arnold's car [36]);

"Little Red Riding Hood" (here, there seems a fundamental structural parallel—and Arnold is described: "the nose long and hawk-like, sniffing as if she were a treat he was going to gobble up" [46]).

(text pages) 442–448

In spite of *hawk-like*, we agree the description makes Arnold sound distinctly like a wolf. Professor Williams refers, of course, to Bruno Bettelheim's *The Uses of Enchantment* (New York: Knopf, 1976). Perhaps Oates's story might be taken up together with the Grimm tale "Godfather Death."

Joyce Carol Oates reads an excerpt from this story on an audiocassette (order no. AMF-26) available from American Audio Prose Library, Box 842, Columbia MO 65205. For current price, call (800) 447-2275.

A recent film has been based on Oates's story: *Smooth Talk* (1987), directed by Joyce Chopra with screenplay by Tom Cole. In an interview, Oates remarked that although she had nothing to do with making the movie, she respects the "quite remarkable" results. "The story itself is a Hawthornian parable of a kind, 'realistic' in its surface texture but otherwise allegorical" (interview with Barbara C. Millard, *Four Quartets*, Fall 1988). Is it, then, an account of a confrontation with the Devil, comparable with "Young Goodman Brown"?

Frank O'Connor
FIRST CONFESSION, page 442

QUESTIONS

1. Does the narrator of this story seem a boy—an innocent or naive narrator—or a grown-up looking back through seven-year-old eyes? At what moments do you sense an ironic discrepancy between the boy's view of things and Frank O'Connor's? (The latter, as we can tell from the story's knowing selection of details. The small boy's horror, for instance, at the fact that his grandmother drinks porter seems exaggerated; so does his reaction to Mrs. Ryan's yarn about the spirit with burning hands. Why was the priest silent for so long after the boy confesses his murder plot? We assume he was doubled up in silent mirth. Jackie's plan to get rid of the body seems a childish power-fantasy, a solution quite beyond his abilities, as the narrator is evidently well aware.)

2. In what ways is Jackie not entirely naive, but sometimes shrewd and perceptive? (Suggestion: Take a close look at what he says about Mrs. Ryan and about his sister Nora. Most readers feel that the boy's resentment of Nora seems fair and justified. A truly pious girl wouldn't box her brother's ears in church or stick out her tongue at him while he says penance at the altar. Furthermore, the priest sides with him in his opinion. Shrewdly, Jackie notices Mrs. Ryan's overfondness for Hell and her tenacious clinging to her half-crown.)

3. What traits of character do you find in the priest? How does he differ from Mrs. Ryan as a teacher of children? (The easygoing, good-humored priest has a better understanding of child psychology. Unlike Mrs. Ryan, he suggests that wrongdoers are sometimes punished in the here and now. Is he deplorable in trying to implant in Jackie a horror of the gallows? Jackie's reaction is gratitude and relief from his fears of the fiery furnace. Hanging is at least definite and easy to imagine and has an end to it.)

4. Do you believe Frank O'Connor to be saying—as Nora says in the conclusion—that there is no use in trying to lead a virtuous life? (Evidently, Nora's closing speech is meant to be funny: she congratulates herself on living a life of virtue, though in her brother's eyes she is a "raging malicious devil" [paragraph 14] and a hypocrite [17]. O'Connor makes no general statement about the virtuous life but strongly implies that the priest shows the way: by looking at the world with humor, compassion, and forgiveness. In Mrs. Ryan's view, on the other hand, to live a good life is a grim and terrible chore.)

(text pages) 448–454

This appealing story may be useful as a clear illustration of an ironic point of view: the naive boy's impressions as recalled by the grown-up narrator. It is hard to imagine the story told from any other point of view and remaining so effective, or so hilarious.

The boy's intended murder victim is apparently drawn from childhood memory. In his autobiography *An Only Child* (New York: Knopf, 1970), O'Connor relates that his paternal grandmother, an abrasive old soul who swigged porter and ate with her fingers, became a widow and moved in with his family.

For a glimpse of Frank O'Connor's writing habits and his intense revision of "First Confession," see "Writing a Story," page 1487.

Tillie Olsen
I STAND HERE IRONING, page 448

QUESTIONS

1. What is the point of view in "I Stand Here Ironing"? Who is the "you" in Olsen's story? (The story is told in the first person by the mother of a gifted young woman. The author never identifies the "you" in so many words. We can only deduce from internal evidence [in paragraphs 2, 50, 52] that the mother's interior monologue is in response to a call from a counselor or teacher—in college probably, because Emily is nineteen years old—requesting a conference with the mother. The narrator makes clear in paragraph 55 that she has no intention of going in to see the caller.)

2. Briefly describe the circumstances into which Emily was born. (Emily's childhood was one of poverty and deprivation, perhaps best summed up by her mother in paragraph 55. That Emily was a child whose mother loved her fiercely is poignantly evident in paragraph 8: "She was a miracle to me, but when she was eight months old I had to leave her daytimes with the woman downstairs to whom she was no miracle at all. . . .")

3. What are we to make of Emily's telling her mother not to awaken her the next morning even though she has midterms (53)? (By the time she is nineteen, her daughter is apparently over the terrible "goodness" the mother remorsefully recalls as characteristic of Emily when she was a little girl. Emily indicates that she will skip her midterms—not out of lighthearted laziness, however, but out of a sense of futility brought on by the uncertainties of modern life, the conviction that "in a couple of years . . . we'll all be atom-dead. . . ." It is perhaps Emily's neglect of her school work that has occasioned the phone call from her teacher or counselor.)

4. Is there any justification for the mother's determination not to accept the invitation to come in and talk about her daughter? (The mother now realizes that whenever in the past she consulted "experts," their advice caused needless pain. She ignored her own common sense and breastfed her daughter on a rigid schedule because "the books" so decreed [6]. "They" said Emily was ready for nursery school at age two [12]. The mother, rendered helpless by her need to work for a living, therefore sent her daughter to nursery school. "They"—the people at the clinic—persuaded Emily's mother to send her to a convalescent home to recover from the measles [26]. In retrospect the mother realizes her daughter would have been far better off at home. The best advice she ever received came from a man with no special credentials, who kindly entreated her to smile more when she looked at Emily. The happiest moments she remembers are those when, following her own instincts, she kept her children home from school even when they weren't really sick.)

5. What makes "I Stand Here Ironing" more than just a case study of a talented but deprived young woman? What besides the task at hand is suggested by Olsen's use of iron and ironing board in her story? ("I Stand Here Ironing" is enriched by the symbolism of

61

(text pages) 454–465

iron and ironing board. The fervent plea that ends the story is the mother's hope that her daughter will know "that she is more than this dress on the ironing board, helpless before the iron." The iron here suggests fate, heredity, environment—all those forces that limit any individual's aspirations. The mother, who is herself an intelligent, perceptive, sensitive person, has been thwarted by poverty. Like mothers everywhere, she wants a better life for her child. This longing permeates the story. Indeed, her daughter at nineteen leads a very different life from her own at the same age.)

6. One student has called "I Stand Here Ironing" "a very *modern* story. It could have been written yesterday." What elements in the story support that assessment? (Though few people iron anymore and no one rides the streetcar, and in spite of its references to the Depression of the thirties and to World War II, Olsen's story strikes many a contemporary note. Single parents who have to entrust their children to inadequate day-care so that they can hold down jobs are less the exception now than they were when Olsen wrote her story. The author, by demonstrating that the thoughts of a housewife are fit matter for a story, has in a way written a feminist document. There is a sturdy insistence, too, on the part of the narrator-mother, that parents cannot be held eternally accountable for their children's inadequacies. "She has lived for nineteen years. There is all that life that has happened outside of me, beyond me" [3]. Only now has popular psychology begun to concede as much. That sad little Emily, "a child of anxious, not proud, love," should have blossomed into a beautiful, popular comedienne—not troublefree, but at least functioning normally—says something important about the resilience of the human spirit.)

Philip Roth
THE CONVERSION OF THE JEWS, page 454

QUESTIONS

1. What is this story saying? (Ozzie himself probably best states the theme in paragraph 176: "You should never hit anybody about God." "The Conversion of the Jews" is a comic story, but its theme is serious.)

2. What is the story's major conflict? (Friction between Rabbi Binder and the irrepressible Ozzie. By extension, it is a conflict between the rabbi's strong faith and the boy's, which in its unorthodox way is perhaps even stronger. Ozzie is full of questions. The rabbi has answers, but the boy insists that they don't go with the questions he asks. The irony is that the naive child ascribes to God more power than the devout rabbi is able to. The rabbi's explanations arise from religious tradition—a tradition whose truths he has so far been unable to persuade Ozzie to believe. It is surely no coincidence that Roth gave his rabbi the suggestive name of Binder, and the boy, the name of Freedman. "We may also recognize the name," writes Charles Clay Doyle of the University of Georgia, "as an anglicized spelling of Friedman, 'peaceable man'—an appropriate epithet for one who proclaims, 'You should never hit anybody about God.'")

3. What chief differences do you find between the character of Ozzie and that of his friend Itzie? (Itzie is straightforward and practical. He knows how to stay out of trouble. "Itzie preferred to keep *his* mother in the kitchen; he settled for behind-the-back subtleties such as gestures, faces, snarls and other less delicate barnyard noises" [6]. Clearly he regards Ozzie as more than a little crazy for bringing up the Christian belief in God as the father of Jesus. He gets sidetracked from Ozzie's argument by the comedy of Rabbi Binder's using the word *intercourse* in Hebrew School. And when the lively minded Ozzie rhapsodizes about God's having created light, "Itzie's appreciation was honest but unimaginative: it was as though God had just pitched a one-hitter" [21]. Itzie is the perfect foil for Ozzie

Freedman, whose rebelliousness and intellectual curiosity keep getting him into trouble. "What Ozzie wanted to know was always different" [31].)

4. What part do Ozzie's mother and Yakov Blotnik play in Roth's story? (Both exert pressure to conform. Ozzie accuses his mother of calling a plane crash a tragedy only because it had killed eight people with Jewish names [32]. When Ozzie explains to her why she has to come in for yet another conference with Rabbi Binder, thinking she will understand because when she lit candles she looked "like a woman who knew momentarily that God could do anything" [39], she responds by slapping him [40].

As for Yakov, Ozzie suspects him of having "memorized [his] prayers and forgotten all about God [42]. "For Yakov Blotnik life had fractionated itself simply: things were either good-for-the-Jews or no-good-for-the-Jews" [85]. Ozzie gets back at both of them, and at the rabbi, by making them say what to them was blasphemy: first, that God can make a child without intercourse [165] and then that they believe in Jesus Christ [173].)

5. For discussion: Is Ozzie a martyr or a spoiled brat? (Probably most students will see things from the boy's point of view. Certainly the author does. He makes sure that when Ozzie jumps he lands in "the center of the yellow net that glowed in the evening's edge like an overgrown halo" [183].)

In an interview with Italian critic Walter Mauro, Roth comments on this story.

> "The Conversion of the Jews," . . . a story I wrote when I was twenty-three, reveals at its most innocent stage of development a budding concern with the oppressiveness of family feeling and with the binding ideas of religious exclusiveness which I had experienced first-hand in ordinary American-Jewish life. A good boy named Freedman brings to his knees a bad rabbi named Binder (and various other overlords) and then takes wing from the synagogue into the vastness of space. Primitive as this story seems to me now—it might better be called a daydream—it nonetheless evolved out of the same preoccupations that led me, years later, to invent Alexander Portnoy, an older incarnation of claustrophobic little Freedman, who cannot cut loose from what binds and inhibits him quite so magically as the hero I imagined humbling his mother and his rabbi in the "The Conversion of the Jews" ("Writing and the Powers That Be," *Reading Myself and Others* [New York: Penguin, 1985]).

Peter Shaw made a provocative attack on Roth's work (in *Commentary;* quoted in *Contemporary Authors*):

> Roth's insistence that he is a friend to the Jews can only theoretically be squared with the loathing for them that he displays in his work. . . . The real message here, or rather the real aspiration, is not to sweep away anti-Semitism, but to transcend being Jewish. If only you try hard enough, Roth's books tell us, it can be done. This is a message that will not do the Jews any more damage than other specious advice that they have received from time to time, so that one has to agree with Roth that his books are not harmful as charged. But if he has not been bad *for* the Jews, he has decidedly been bad *to* them—and at the expense of his art. . . . Roth has been a positive enemy to his own work, while for the Jews he has been a friend of the proverbial sort that makes enemies unnecessary.

You might care to ask students to consider, in class discussion or in writing, whether they think that Shaw's criticism can be justly applied to "The Conversion of the Jews."

James Thurber
THE CATBIRD SEAT, page 465

Discussion of this story might begin with the question, What is Mr. Martin's true character? To his fellow workers at Fitweiler and Schlosser, Martin is a drab, harmless little drudge of seemingly inhuman efficiency ("Man is fallible but Martin isn't"). To his enemy, Ulgine Barrows, he pretends to be a drinker, smoker, heroin addict, and mad bomber. Neither of these images, of course, reflects the Erwin Martin whom the reader comes to know, a character both rounded and complicated. Early in the story, Thurber indicates that Martin is indeed fussy and meticulous: he remembers the very date when Mrs. Barrows first arrived at the office. Secretly, however, he is crafty and imaginative—a dangerous foe and a dirty fighter.

Martin seems a dynamic rather than static character. Within the story he grows from a somewhat ineffectual ditherer to a purposeful man capable of plotting revenge and boldly carrying it out. At first, his schemes seem merely desperate. In Mrs. Barrows's living room, he casts about helplessly for a murder weapon. Then his "strange and wonderful" Machiavellian idea begins to bloom (paragraph 13). In his eccentric way, Martin is a hero, not a passive antihero. He ends as the triumphant captain of his soul. Robert H. Elias points out in a comment on this story that Martin "saves, indeed, the inner life itself, the life that is not to be harnessed like the atom, streamlined like a spaceship, manipulated, calculated, fed to Univac" ("James Thurber: the Primitive, the Innocent, and the Individual," *American Scholar* 27 [Summer 1958]: 355–63).

Thurber skillfully arouses our sympathy for Martin and his empire of filing cabinets. One of the author's strategies is to make Ulgine Barrows, Martin's opponent, as unattractive as possible. He gives her irritating mannerisms: a "quacking voice and braying laugh," a habit of bombarding people with unanswerable questions: "Are you sitting in the catbird seat?" She is Martin's foil, and students may be asked to point out contrasts between the personalities of these two adversaries. Mrs. Barrows is loud, aggressive, and power hungry, an exponent of radical change. Martin, quiet and retiring, seeks only to keep running his inconspicuous department in his own way; he dreads any change at all.

Humor becomes increasingly broad as the story progresses: from the subtly reasoned trial of Mrs. Barrows that Martin conducts in the courtroom of his head to the final slapstick fracas in which Martin is saved by the adept blocking of a former football player. Beautifully plotted, the story lends itself well to examination of its elements. The crisis occurs when Martin, unable to find a murder weapon, is momentarily at a loss but then switches to his new plan. Among the story's abundant ironies, one of the more delectable occurs when Martin sticks out his tongue and tells his foe he's sitting in the catbird seat (14). Not only does he turn her own words against her, he is suddenly aware (as she isn't) that, indeed, now he is sitting pretty. He transforms her tired cliché into a fresh truth. Actually, the whole story has an ironic situation: the discrepancy between the true Erwin Martin and others' view of him. Martin wins his war by turning this discrepancy to his own advantage.

Edith Wharton
ROMAN FEVER, page 471

QUESTIONS

1. What connotations enrich the title of Wharton's story? (The *Oxford English Dictionary* defines *Roman fever* as "a form of malarial fever prevalent at Rome." It was an illness much feared by tourists and especially, it would seem, by the mothers of marriageable nineteenth-century girls. [The heroine of Henry James's *Daisy Miller* succumbs to it.] Roman fever figures prominently in the anecdote about Mrs. Ansley's great-aunt. But, as Wharton's story tells us, Roman fever was no longer a threat even when Mrs. Slade and Mrs. Ansley were young. Other, less literal fevers, however, consume Wharton's characters during their Roman holiday: Babs's and Jenny's restless efforts to attract eligible young Italian bachelors; the two older women's remembered passions, which had resulted in the youth

ful Grace's indiscretion with Delphin Slade; and, most important, the festering jealousy and hatred with which Mrs. Slade still regards Mrs. Ansley and which, by story's end, she can no longer contain. Like deadly malarial germs, Mrs. Slade's poisonous emotions grow more virulent by moonlight.)

2. How do the two friends, Mrs. Slade and Mrs. Ansley, differ from each other in both appearance and personality? (Mrs. Slade, "fuller, and higher in color, with a small determined nose supported by vigorous black eyebrows," is a forceful, energetic woman with a large share of self-confidence. The aplomb with which she deals with the waiter, her longing for a daughter who needed to be "watched, out-maneuvered, rescued," her pride in her own social gifts all testify to her vigor and her view of herself as a superior being. Mrs. Ansley says to her, "I don't believe you're easily frightened." Grace Ansley seems small, faded, and, especially as seen through Mrs. Slade's eyes, mousy by comparison. "She was evidently far less sure than her companion of herself and of her rights in the world." Though shy and inarticulate, she does have strong opinions about Mrs. Slade. These she keeps to herself.)

3. Analyze Mrs. Slade's feelings toward Mrs. Ansley. Why is she afraid of her? Why does she actually *envy* Mrs. Ansley? (Although Mrs. Slade would like to dismiss Mrs. Ansley as "old-fashioned," a "nullity," a museum specimen of old New York, Mrs. Ansley is a thorn in her side for two reasons. Mrs. Ansley's "vivid" daughter Barbara, apparently on her way to making a brilliant match with a dashing young Italian, is just the kind of girl Mrs. Slade would like as a daughter. Her own Jenny seems bland and tractable, a little boring, by comparison. Even more important, as the story gradually reveals, is Alita's knowledge that when she was engaged to Delphin Slade, her friend Grace was in love with him. Clearly she feared that Delphin Slade was in turn all too appreciative of Grace's "quiet ways" and "sweetness," traits that Alita scorned—and lacked. Perhaps she has sensed through the years the bewildering truth, mentioned twice in the story: meek, respectable, self-effacing Mrs. Ansley feels sorry for her.)

4. Of what importance to the story is the anecdote Mrs. Slade recalls about Grace's great-aunt Harriet and her sister (paragraphs 50–52)? (Its inclusion highlights Wharton's skill as a storyteller. That the "prudent" Grace's great-aunts had behaved so wickedly establishes a family precedent for the young Grace's indiscreet behavior, otherwise a bit hard to believe. Further, the bit of family lore that Grace had shared with Alita when both were in love with Delphin Slade may have served as the inspiration for Alita's cruel plan to lure her own rival into the moonlit Colosseum. Surely we have to accept the truth of her statement that "Of course I never thought you'd die." Wharton has already established that Roman fever was no longer to be feared by the time of Grace's rendezvous with Alita's fiancé. Nevertheless, for her to have sent a girl with "a delicate throat" into the dampness of a Roman evening testifies to a ferocity in Alita that is sobering, to say the least. She seems to feel some guilt about it still.

Ironically, in the anecdote great-aunt Harriet's nefarious plan succeeds with a vengeance, whereas the young Alita's scheme backfires. In the end, it is apparent that she has been more deceived than deceiver.)

5. What additional examples of irony do you find in "Roman Fever"? (The "diffused serenity" [6] with which the two friends contemplate the view before them turns out to be a well-bred disguise, as is the static scene of the story itself. Two women merely sit and talk. What could be pleasanter? But as the author burrows relentlessly through layers of pretense and civility, she gradually uncovers the intense hostility that underlies the relationship between the two women.

There are many other examples of irony, most of them based on the contrast between what Mrs. Ansley knows and what remains hidden from Mrs. Slade—and the reader—until the end of the story. Among the most pointed:

(text pages) 471–480

Paragraphs 9–11, in which Mrs. Slade misinterprets her friend's fond memories as mere random stresses in her speech.

Paragraphs 13 and 15, in which Mrs. Slade indicates her general disapproval of moonlight. In fact, she is only half aware of the moon's important role in her rivalry with her friend.

Paragraph 16, in which Mrs. Ansley's statement, "And perhaps we didn't know much more about each other," contains more truth than either lady realizes.

Paragraph 19, in which Mrs. Slade wonders to herself how Barbara could be so interesting "with those two nullities as parents." In paragraph 36, more politely, though not excessively so, she voices the same thought aloud to her companion: "And I was wondering, ever so respectfully, you understand . . . wondering how two such exemplary characters as you and Horace had managed to produce anything quite so dynamic." She is not prepared for the truth that solves the riddle.

Paragraph 28, in which Mrs. Ansley's statement that the scene before them "is so full of old memories, as you say" conceals more than it reveals.

Paragraphs 30–31, in which Mrs. Slade is probing and Mrs. Ansley defending herself with her knitting. Here Mrs. Slade still believes that her friend had caught nothing but a cold for her long-ago indiscretion.

Paragraphs 41–42, in which Mrs. Slade complacently dismisses the possibility that, for her uninteresting friend, too, "too many memories rose from the lengthening shadows of those august ruins." After all, Mrs. Slade thinks, "There was no one of whom she had less right to think unkindly than of Grace Ansley."

Paragraphs 55–56, in which only Mrs. Ansley knows that her once having been "too happy" sprang from the same source as her friend's happiness.

Paragraph 64, in which Mrs. Ansley's "The most prudent girls aren't always prudent" means more than Mrs. Slade suspects.

Paragraphs 99–100, in which Mrs. Slade smugly assumes that her friend's hasty marriage to Horace Ansley had been an act of mere girlish pique. Here Mrs. Slade is basking in the glow of what she imagines was her victory over her young friend Grace. Actually, by interpreting Grace's marriage as a sign that her friend didn't really care for Delphin Slade, Alita proves how completely she herself has been deceived.)

6. To what extent is the reader prepared for the revelations at the end of "Roman Fever"? (At the start of her beautifully wrought story, Wharton introduces the two matrons as they appear to each other and to the world. They are merely old friends who find themselves in similar circumstances, "each of them the modest appendage of a salient daughter." With great skill, Wharton then begins to introduce causes of friction between them: Mrs. Ansley's daughter outshines Mrs. Slade's. Mrs. Ansley feels sorry for Mrs. Slade. Gradually, what had seemed at first like mere annoyance on the part of Mrs. Slade toward her sweet-tempered friend is revealed as virulent hatred based on the love of both women for one man. As Mrs. Slade, first subtly and then with more force, needles and probes, the scene is set for an explosion. Mrs. Ansley is to be cruelly humiliated at last. That Alita was the writer of the letter of assignation is a revelation foreshadowed in the anecdote about great-aunt Harriet. That Grace went to meet Delphin Slade, Alita had counted on. The biggest surprise, perhaps, is that when Grace went to the Colosseum, Delphin was there. That was the one consequence for which Alita had been unprepared. We, too, are surprised, as the author intended us to be. It is when we go back over the story after its revelations at the end that we see the skill with which Wharton, like any good storyteller, has planted her clues. As a result, the denouement leaves us thinking, "Of course!" And Barbara's sparkle is plausible at last.)

7. What are we to think of Delphin Slade? (Though Wharton tells us no more about him than her plot demands, some classes might like to consider his behavior toward the women in his life. Was he worth fighting over? One student has described him as "an opportunist presented with a great opportunity.")

8. What significance do you find in the fact that, at the end of "Roman Fever," Mrs. Ansley "began to move ahead of Mrs. Slade toward the stairway"? (The battle is over, and Mrs. Ansley is victorious. So convincingly has Wharton punctured Mrs. Slade's pretensions that we as readers rejoice in Mrs. Ansley's stunning victory.)

Edith Wharton was herself childless; but R. W. B. Lewis in his biography *Edith Wharton* (New York: Harper & Row, 1975) sees "Roman Fever," told as it is "from the perspective of a mature and motherly figure," as one of the stories written during the two decades after World War I that arise "from a vital new concern of Edith Wharton's, almost a new compulsion: what might be called the maternal imagination."

"Roman Fever" exhibits superb assurance, verve, and craftsmanship. Lewis writes of the postwar period in Wharton's life as a time when the author "had arrived at a deep harmony with her own life history." He regards "Roman Fever" as a story written with "a serenity that pervades the narrative in a long atmospheric glow."

For another view of "Roman Fever," see Dale M. Bauer's " 'Roman Fever': A Rune of History" in the October 1988 issue of *College English*. Arguing that Wharton was more upset about the rise of Hitler than she seemed, Bauer sees both Alita Slade and Grace Ansley as victims of the old patriarchal order and the story, appropriately set in Rome, a "site of primal violence," as an attack on that order and therefore a condemnation of the past. Wharton deserves credit, Bauer claims, for "articulating the destructive character of a cultural misogyny that led all too quickly to what she saw in 1933," including anti-Semitism.

William Carlos Williams
THE USE OF FORCE, page 481

QUESTIONS

1. To what do you attribute the family's initial wariness toward the doctor? How do you explain the child's refusal to cooperate with him? (Perhaps unaccustomed to doctors, these are new patients unacquainted with the man they have summoned. To be sure, the parents are worried about their daughter. Apparently the child, though she knows she has a sore throat, clamps her mouth shut because she is frightened and suspicious of the doctor.)

2. What is the basic conflict in "The Use of Force"? How is it foreshadowed? How is it resolved? (Foreshadowed in paragraph 4 [where we are made aware of the child's "cold, steady eyes "] and again in paragraphs 6–10, the conflict is between the doctor and the little girl who won't open her mouth. The doctor, suspecting that she has diphtheria, cannot make a true diagnosis without seeing her throat. When all else fails, he learns what he needs to know by brutally forcing her mouth open.)

3. In paragraph 22 the doctor calls the parents "contemptible." To what extent are they to blame for such a startling assessment? What does it reveal about the doctor's state of mind? (The poor worried parents are still trying to be conciliatory while, by paragraph 22, the primitive forces underlying the doctor's psyche have already been engaged. The doctor is determined to fight to the finish. In this battle as in all battles, what count are tenacity and "fury of effort." The doctor increasingly displays these traits as the battle mentality overcomes his professional calm. He sees the little girl as a worthy opponent. Her parents, involved only peripherally in the primal struggle, exist on a separate plane—one that has come to seem irrelevant to the doctor in his altered state.)

4. What makes the narrator persist in his efforts to get the child's mouth open even after it has started bleeding? Is he a sadist, a bad doctor, a potential child molester, or what? (There are two reasons: one rational, the other less so. As the doctor explains in

(text pages) 481–484

paragraphs 31 and 32, he feels compelled to get a diagnosis immediately, before it's too late, and before the child infects others. But something else explains the ferocity of his assault. He has "got beyond reason. . . . It was a pleasure to attack her." The primitive part of his nature has gained control over him.)

5. What is the doctor's attitude about his behavior toward the little girl? Where in the story is it revealed? (Revealing an awareness of his mixed motives, the narrator confesses to "a blind fury, a feeling of adult shame, bred of a longing for muscular release" [32]. Does he also demonstrate a certain pleasure at having won the battle? By story's end, he has not yet had time to reflect sanely upon the whole awful encounter.)

6. What is the theme of "The Use of Force"? (Perhaps that violence lies ready to erupt through the respectable surface of even the most civilized among us.)

12 Criticism: On Fiction

To place at your fingertips several famous (or notorious) observations about fiction is the main purpose of this brief anthology. There is nothing you need to do about it, of course—it is here merely to supply you with resources, if you see any use for them.

Among these critical statements, you will find a few that raise topics for larger, more general discussion. Additional topics for writing will arise from some of these remarks. Certain statements raise points applicable to more than one story in this book, perhaps to all of them. You may find them suggesting topics for longer papers. The process of story-telling—interesting not only to aspiring story writers but to most students—is revealed in a few of these comments. Flaubert tells of his slow and agonizing perfectionism—we think his remarks must reading for every freshman who hacks out a term paper the night before it is due.

Here are some questions to which these statements give rise, perhaps useful for either writing topics or class discussion.

Edgar Allan Poe, THE TALE AND ITS EFFECT, page 485

1. What, according to Poe, does a "short prose narrative" gain by its brevity? (Absorbed at a sitting, it produces one effect.)

2. Paraphrase Poe's notions of how a skilled writer goes about writing a tale (or a short story). What main purpose does such a writer keep in mind? (To start, the writer imagines one effect upon the reader—say, a feeling of growing terror [as in "The Tell-Tale Heart"]. Then the writer carefully invents each incident—indeed, chooses each word—to produce that effect, and not another.)

3. Does Poe strike you as an advocate of "automatic writing," that is, of writing carelessly at top speed, or getting stoned and letting the wild prose freely flow? (On the contrary, he thinks of the storyteller as a conscious, deliberate craftsman, in control of each incident, calculating every word. Poe's declaration of principles may surprise students who have heard that this author of bizarre tales was a drunk and a dope addict. Why not remind them that, like any other good writer, Poe had to work hard. We suspect, however, that Poe's statements of principles [in "The Philosophy of Composition," too] set up ideals to which he aspired and don't record his own working habits. No other writer or critic ever so arrogantly reduced the writing of fiction and poetry to the rigorous application of pure mind.)

Charlotte Brontë, THE WRITER'S PASSIVE WORK, page 486

1. Would Charlotte Brontë agree with Poe's view of the fiction writer as entirely in control of the creative process? (On the contrary, she sees the writer as the helpless instrument of forces greater than she. [Charlotte is trying to absolve Emily of the guilt of having invented Heathcliff by suggesting that Emily *had* to invent Heathcliff; she couldn't help herself.])

2. If you know *Wuthering Heights*, describe Emily Brontë's characterization of Heathcliff. What is he like? (A foundling of obscure origin, perhaps sent "from the Devil," Heathcliff grows up to be a proud, brooding sufferer—a brutal Lord Byron of the Yorkshire moors. He is given to fiery bursts of temper [he throws a knife at old nurse Nelly Dean], to fierce jealousy, and to a love [for Catherine] so passionate and intense it is said to be inhuman—

and, in the end, it presumably outlasts his mortal flesh. Cruel and merciless and unforgiving, he takes years to work revenge.)

Gustave Flaubert, THE LABOR OF STYLE, page 487

1. In Flaubert's methods of writing, what strikes you as remarkable? (Flaubert is a perfectionist, a worrier about "two turns of phrase that are still too much alike," a persistent reviser, a seeker after a nearly unattainable ideal prose. He studies a technical book not only to obtain background information for a story, but to observe its style.)

2. From what source, according to Flaubert, may writers draw strong figures of speech? (From technical treatises, from the language of artisans and craftsmen, of blacksmiths and others who work with their hands.)

Henry James, THE MIRROR OF A CONSCIOUSNESS, page 487

1. Put the gist of James's comment into your own words. (This is a difficult morsel of late Jamesian prose, but the following admittedly partial paraphrase might be reasonably fair to it: "In fiction, just as in Shakespeare's tragedies, it's the sensitive characters we most care about—those who can see what's happening and can respond to it—not the unfeeling dopes.")

2. What, in James's view, have great writers of fiction always done? (They have placed in the midst of their stories some fine-minded character ["a consciousness . . . subject to fine intensification and wide enlargement"] to take in the meaning of what occurs.)

Anton Chekhov, NATURAL DESCRIPTION AND "THE CENTER OF GRAVITY," page 488

1. In Chekhov's own stories, point to some of the "small details" he considers so valuable.

2. If, in writing "The Man in a Case," Chekhov had failed to take his own advice and hadn't "let two people be the center of gravity, he and she," how might he have weakened the story? Does he, or doesn't he, take this particular bit of advice in "The Darling"? In "Vanka"? (In "The Darling," Chekhov pairs Olenka with a series of successive "he's." "Vanka" is a different kind of story.)

James Joyce, EPIPHANIES, page 489

1. An *epiphany* in fiction (as defined in Chapter 1) is "some moment of insight, discovery, or revelation by which a character's life, or view of life, is greatly altered." Is that the same sense in which Joyce's Stephen Dedalus uses the term here? (Stephen's meaning of the word is much more inclusive: any overheard speech, any observed gesture [however trivial] that the young artist takes to be "a sudden spiritual manifestation." Apparently, an epiphany for him can be any perception—or even any thought—that he finds, for any reason, fraught with significance.)

2. What moments in Joyce's "Araby" might Stephen call epiphanies? (For one, compare the overheard conversation between the lady and the gentlemen [paragraphs 26–32] with the trivial conversation Stephen overhears. Why is it an epiphany? Perhaps because the narrator remembers it and thinks it worth recording. In "Araby," the chatters and their English accents suggest [among other things] the controlling British, who sell the Irish dreams.)

Katherine Mansfield, WRITING "MISS BRILL," page 490

1. What do you make of Mansfield's declared attempt to make "the rise and fall of every paragraph" fit Miss Brill? In your own words, explain what you understand the author to be trying. Point to specific passages in the story for examples.

2. Select a paragraph from the story that, in style, strikes you as particularly memorable. Read it aloud to the class and, with the aid of other students, try to define what the writer is doing that makes it so effective.

William Faulkner, "THE HUMAN HEART IN CONFLICT WITH ITSELF," page 490

1. In Faulkner's first paragraph, what point is he making? What has been the effect upon young writers of living in the nuclear age?
2. What, in Faulkner's view, is the duty of the writer today?

Flannery O'Connor, THE SERIOUS WRITER AND THE TIRED READER, page 491

1. What is Flannery O'Connor's opinion of book clubs? Explain her remark on this subject in the second paragraph.
2. Does O'Connor condemn people who want ficton to leave them "lifted up"? What is her attitude toward them? (Bemused contempt.)
3. By what means can writers write the great novels of the future? (By looking into their own hearts, exploring themselves and their native regions, and writing not conventional realistic fiction but a poetic kind of fiction inspired by individual visions.)

Stanley Fish, AN ESKIMO "A ROSE FOR EMILY," page 492

1. How, according to Fish, might this unlikely interpretation of Faulkner's story be justified?
2. What is the writer's point in admitting that such a far-out reading is conceivably possible? (Fish's point is made in his opening sentence.)

Raymond Carver, "COMMONPLACE BUT PRECISE LANGUAGE," page 493

1. What does Raymond Carver's statement tell us about the qualities he prizes in literature? About his personal writing habits?
2. Consider any of Carver's three stories in this book. How does it show that this writer practices what he advocates?

POETRY

Poems Arranged by Subject and Theme

This list sorts out and classifies most of the poems in the entire textbook. Besides subjects or themes, it includes some genres (i.e., elegies, pastoral poems, poems of spring and other seasons).

How to Use This Information. Browse through this list and you will find many poems worth teaching side by side. This list will be particularly helpful to the instructor who wishes to organize a whole poetry course differently from the way the book is structured: to teach poetry not by the elements of poems, but by themes. But, however you prefer to organize your course, you will find this list a ready source of possible writing assignments.

For Writing Topics. You might have students read three or four poems in a group (say, those in the category "Apocalypse," or a few of your choice from "Coming of Age"), then ask them to reply, in a page or two, to the question, "What do these poems have in common?" Or, "How do these poets differ in their expressions of a similar theme?"

What follows is thorough, but not exhaustive. We have left out some categories that sounded unpromising. Would you have cared that the book has four locomotive poems (by Dickinson, Stillman, Walt Whitman, and William Carlos Williams), two (by Bishop and Miller Williams) about gas stations, and three (by Cunningham, Davison, and Tichborne) about persons literally or figuratively decapitated? Not all of the following themes and subjects are central to their poems, but all will be fairly evident.

APOCALYPSE

Frost	Fire and Ice
MacLeish	The End of the World
Yeats	The Second Coming

ART

Auden	Musée des Beaux Arts
Pastan	Ethics
Stevens	Anecdote of the Jar
W. C. Williams	The Dance
Yeats	Long-legged Fly

BEAST AND BIRD

Blake	The Tyger
Clare	Mouse's Nest
Gregg	The Grub
Hall	Names of Horses
Hopkins	The Windhover
Layton	The Bull Calf
Lowell	Skunk Hour
Smart	For I will consider my Cat Jeoffry
Taylor	Riding a One-Eyed Horse
Tennyson	The Eagle

Poems Arranged by Subject and Theme

BELONGING TO A MINORITY (see also BLACK EXPERIENCE, INDIAN LIFE)

Hongo	The Hongo Store
Liu	My Father's Martial Art
Olds	The One Girl at the Boys Party
Sellers	In the Counselor's Waiting Room
Stafford	At the Klamath Berry Festival

BLACK EXPERIENCE (see also BELONGING TO A MINORITY)

Anonymous	Scottsboro
Brooks	The Bean Eaters
Brooks	A Black Wedding Song
Brooks	The Rites for Cousin Vit
Brooks	We Real Cool
Clifton	winnie song
Cullen	For a Lady I Know
Dove	Daystar
Hayden	Those Winter Sundays
Hughes	Dream Boogie
Hughes	Dream Deferred
Hughes	Subway Rush Hour
Kilgore	The White Man Pressed the Locks
Randall	Ballad of Birmingham

BOOKS AND READING (see also LANGUAGE, and WRITING)

Beerbohm	On the imprint of the first English edition . . .
Bradstreet	The Author to Her Book
Keats	On First Looking into Chapman's Homer
Wallace	The Girl Writing Her English Paper

BORING LIFE STYLES

Eliot	The *Boston Evening Transcript*
Stevens	Disillusionment of Ten O'Clock

CARPE DIEM

Herrick	To the Virgins, to Make Much of Time
Housman	Loveliest of trees, the cherry now
Marvell	To His Coy Mistress
Waller	Go, Lovely Rose

CATASTROPHES AND ACCIDENTS

Connellan	Scott Huff
Frost	"Out, Out—"
Hardy	The Convergence of the Twain
McGonagall	The Albion Battleship Calamity
Ruffin	Hotel Fire
Slavitt	Titanic

Poems Arranged by Subject and Theme

CHILDHOOD (see also FATHERS AND CHILDREN, MOTHERS AND CHILDREN)

Bishop	Sestina
Blake	The Chimney Sweeper
Cleghorn	The Golf Links
Cummings	in Just-
Frost	Birches
Hongo	The Hongo Store
Justice	On the Death of Friends in Childhood
Lawrence	Piano
Nemerov	The Snow Globe
Olds	The One Girl at the Boys Party
Roethke	My Papa's Waltz
Thomas	Fern Hill

CITY LIFE

Baudelaire	Recueillement
Blake	London
Brooks	The Rites for Cousin Vit
Brooks	We Real Cool
Eliot	The *Boston Evening Transcript*
Eliot	The Love Song of J. Alfred Prufrock
Eliot	The winter evening settles down
Ginsberg	A Supermarket in California
Hardy	The Ruined Maid
Kilgore	The White Man Pressed the Locks
Miles	Reason
Millay	Recuerdo
Ruffin	Hotel Fire: New Orleans
H. Shapiro	National Cold Storage Company
Simic	Butcher Shop
Swift	A Description of Morning
Wilbur	Transit
Wilde	Symphony in Yellow
W. C. Williams	The Great Figure
Wordsworth	Composed upon Westminster Bridge

CLERICS

R. Browning	Soliloquy of the Spanish Cloister
Cunningham	Friend, on this scaffold Thomas More lies dead
Hardy	In Church (in Four Satires of Circumstance)
Yeats	Crazy Jane Talks with the Bishop

COMING OF AGE

Campion	There is a garden in her face
Housman	When I was one-and-twenty
Pastan	Ethics
Phillips	Running on Empty
K. Shapiro	The Dirty Word
Springsteen	Born to Run

Poems Arranged by Subject and Theme

DEATH (see also ELEGIES)

Anonymous	The Three Ravens
Anonymous	The Twa Corbies
Buson	The piercing chill I feel
Cornford	The Watch
Dickinson	Because I could not stop for Death
Dickinson	I heard a Fly buzz – when I died
Frost	Away!
Frost	Birches
Frost	"Out, Out—"
Hardy	During Wind and Rain
Hardy	In the Cemetery (in Four Satires of Circumstance)
Keats	This living hand, now warm and capable
Knott	Poem (The only response . . .)
Nemerov	The Snow Globe
Owen	Anthem for Doomed Youth
Rossetti	Uphill
Slavitt	Titanic
H. Shapiro	National Cold Storage Company
W. J. Smith	American Primitive
Stevens	The Emperor of Ice Cream
Wordsworth	A Slumber Did My Spirit Seal

ELEGIES (see also DEATH)

Brooks	The Rites for Cousin Vit
Connellan	Scott Huff
Dryden	To the Memory of Mr. Oldham
Gray	Elegy in a Country Churchyard
Herrick	Upon a Child That Died
Housman	For an Athlete Dying Young
Housman	With rue my heart is laden
Jonson	On My First Son
Justice	On the Death of Friends in Childhood
Merwin	Elegy (quoted in Scholes, "How do we make a poem?")
Milton	Lycidas
J. A. Moore	Little Libby
Philips	On Her Son, H.P.
Ransom	Bells for John Whiteside's Daughter
Roethke	Elegy for Jane
Stillman	In Memoriam John Coltrane
Stuart	Crib Death
Tichborne	Elegy
Whitman	O Captain! My Captain!

FAITH AND DOUBT (see also CLERICS, GLORY BE TO GOD)

Arnold	Dover Beach
Ashbery	The Cathedral Is
Cunningham	This *Humanist*, whom no beliefs constrained
Eliot	Journey of the Magi
Hardy	The Oxen

Poems Arranged by Subject and Theme

Hayford	Under the Holy Eye
Hopkins	Thou art indeed just, Lord, if I contend
Milton	When I consider how my light is spent
Ruffin	Hotel Fire: New Orleans
Wordsworth	The World Is Too Much with Us
Yeats	The Magi
Zimmer	The Day Zimmer Lost Religion

FAME

Dickinson	Victory comes late
Gray	Elegy in a Country Churchyard
Guiterman	The Vanity of Earthly Greatness
Keats	When I have fears that I may cease to be
Shelley	Ozymandias

FAMILIES (see also FATHERS AND CHILDREN, MOTHERS AND CHILDREN)

Bishop	Filling Station
Dove	Daystar
Hardy	During Wind and Rain
Hongo	The Hongo Store
Larkin	Home is so Sad
M. Williams	MECANIC ON DUTY AT ALL TIMES
J. Wright	Autumn Begins in Martins Ferry, Ohio

FARM AND COUNTRY

Frost	Birches
Frost	"Out, Out—"
Frost	Stopping by Woods on a Snowy Evening
Gray	Elegy Written in a Country Churchyard
Hall	Names of Horses
Hardy	The Ruined Maid
Hayford	Dry Noon
Larkin	Wedding-Wind
Layton	The Bull Calf
Mew	The Farmer's Bride
Skinner	The Cold Irish Earth
Toomer	Reapers

FATE

Anonymous	The Three Ravens
Anonymous	The Twa Corbies
Hardy	The Convergence of the Twain

FATHERS AND CHILDREN

Allen	Night Driving
Hayden	Those Winter Sundays
Heaney	Digging
Liu	My Father's Martial Art
Jonson	On My First Son

Phillips — Running on Empty
Plath — Daddy
Roethke — My Papa's Waltz
Scott — The Grocer's Children
W. J. Smith — American Primitive
Thomas — Do not go gentle into that good night
Wilbur — The Writer
Winters — At the San Francisco Airport
J. Wright — Autumn Begins in Martins Ferry, Ohio

FRIENDSHIP

Dryden — To the Memory of Mr. Oldham
W. Whitman — I Saw in Louisiana a Live-Oak Growing

GLORY BE TO GOD (see also CLERICS, FAITH AND DOUBT)

Anonymous — I sing of a maiden
Donne — Batter my heart, three-personed God
Herbert — Easter Wings
Herbert — Love
Herbert — Redemption
Hopkins — God's Grandeur
Hopkins — Pied Beauty
Hopkins — The Windhover

GRIEF (see also ELEGIES)

E. B. Browning — Grief
Hughes — Island
Jonson — Slow, slow, fresh fount, keep time with my salt tears
Niedecker — Sorrow moves in wide waves
Tennyson — Dark house, by which once more I stand

HAPPINESS

Anonymous — Carnation Milk
Hunt — Rondeau (Jenny kissed me)
Kenyon — The Suitor
Larkin — Wedding-Wind
Millay — Recuerdo
J. Wright — A Blessing
Yeats — The Lake Isle of Innisfree

HATRED AND INVECTIVE

Anonymous — Edward
Atwood — You fit into me
Betjeman — In Westminster Abbey
Blake — A Poison Tree
R. Browning — Soliloquy of the Spanish Cloister
Burns — The Toad-Eater
Ciardi — In Place of a Curse
Frost — Fire and Ice
H.D. — Helen

Randall   Old Witherington
S. Smith   This Englishwoman
Steele   Epitaph
Stephens   A Glass of Beer

HEROES

Anonymous   Sir Patrick Spence
Carroll   Jabberwocky
Cummings   Buffalo Bill's
Tennyson   Ulysses
Whitman   O Captain! My Captain!

INDIAN LIFE (see also BELONGING TO A MINORITY)

Erdrich   Indian Boarding School: The Runaways
Merwin   Song of Man Chipping an Arrowhead
Stafford   At the Klamath Berry Festival
Warrior   How I Came to Have a Man's Name

INNOCENCE AND EXPERIENCE

Behn   When maidens are young
Blake   The Chimney Sweeper
Hardy   In Church (in Four Satires of Circumstance)
Hopkins   Spring and Fall
Nemerov   The Snow Globe
Pastan   Ethics
K. Shapiro   The Dirty Word

LANGUAGE (see also BOOKS AND READING and WRITING)

Burke   The Habit of Imperfect Rhyming
Carroll   Jabberwocky
Guernsey   Glove
Soyinka   Lost Tribe

LEAVE-TAKING

Baca   Spliced Wire
Davison   The Last Word
Donne   A Valediction: Forbidding Mourning
Drayton   Since there's no help, come let us kiss and part
Hughes   Homecoming
Larkin   Poetry of Departures
Levertov   Leaving Forever
Lovelace   To Lucasta
Shelton   Mexico

LONELINESS

Anonymous   Western Wind
Frost   Desert Places
Ginsberg   A Supermarket in California
Hughes   Homecoming

Lennon & McCartney	Eleanor Rigby
Randall	Old Witherington
P. Williams	When she was here, Li Bo . . .

LOVE

Anonymous	Bonny Barbara Allen
Arnold	Dover Beach
Boyd	Cupid and Venus
Burns	Oh, my love is like a red, red rose
Chaucer	Your ÿen two wol slee me sodenly
Donne	A Valediction: Forbidding Mourning
Drayton	Since there's no help, come let us kiss and part
Frost	The Silken Tent
Guernsey	Glove
Hardy	Neutral Tones
Hayden	Those Winter Sundays
Housman	When I was one-and-twenty
Jennings	Delay
Jonson	To Celia
Larkin	Wedding-Wind
MacDiarmid	Wheesht, Wheesht
Marvell	To His Coy Mistress
Merrill	Laboratory Poem
Nims	Love Poem
Pastan	Jump Cabling
Pound	The River Merchant's Wife: a Letter
Roethke	I Knew a Woman
Ruark	Waiting for You with the Swallows
Schulman	Hemispheres
Sellers	In the Counselor's Waiting Room
Sexton	The Kiss
Shakespeare	My mistress' eyes are nothing like the sun
Shakespeare	Not marble nor the gilded monuments
Shakespeare	Shall I compare thee to a summer's day?
Swift	On Stella's Birthday
Waller	Go, Lovely Rose
Waller	On a Girdle
Whitman	I Saw in Louisiana a Live-Oak Growing
P. Williams	When she was here, Li Bo
Wyatt	With serving still
Yeats	For Anne Gregory

LUST AND DESIRE (see also LOVE)

Anonymous	Western Wind
Chappell	Skin Flick
Donne	The Flea
Graves	Down, Wanton, Down!
MacDiarmid	Wheesht, Wheesht
Rich	The Ninth Symphony of Beethoven Understood at Last as a Sexual Message

Poems Arranged by Subject and Theme

Sexton	The Kiss
Snodgrass	Seeing You Have . . .
Wyatt	They flee from me that sometime did me sekë
Yeats	Crazy Jane Talks with the Bishop

MAGIC AND VISION

Anonymous	The Cruel Mother
Anonymous	The Three Ravens
Anonymous	The Twa Corbies
Bogan	The Dream
S. T. Coleridge	Kubla Khan
de la Mare	The Listeners
Heffernan	Living Room
Hill	Merlin
Yeats	Sailing to Byzantium
Yeats	The Second Coming
Yeats	Who Goes with Fergus?

MARRIAGE AND DIVORCE

Brooks	A Black Wedding Song
M. E. Coleridge	We never said farewell
Donne	A Valediction: Forbidding Mourning
Dove	Daystar
Hardy	In the Room of the Bride-elect and In the Nuptial Chamber (in Four Satires of Circumstance)
Hardy	The Workbox
Heaney	Mother of the Groom
Hecht	The Vow
Larkin	Wedding-Wind
Levertov	Divorcing
Mew	The Farmer's Bride
Pound	The River Merchant's Wife: a Letter
S. Smith	I Remember
Snodgrass	Seeing You Have . . .
Willard	Marriage Amulet

MEDICINE

Frost	"Out, Out—"
Stickney	Sir, say no more
W. C. Williams	Spring and All

MICROCOSMS

Blake	To see a world in a grain of sand
Donne	The Flea
Evans	Wing-Spread
Frost	Design
Tennyson	Flower in the Crannied Wall
Whitman	A Noiseless, Patient Spider

Poems Arranged by Subject and Theme

MOTHERS AND CHILDREN

Anonymous	The Cruel Mother
Anonymous	Edward
Bradstreet	The Author to Her Book
Digges	For *The Daughters of Hannah* Bible Class . . .
Dove	Daystar
Gwynn	Scenes from the Playroom
Hardy	In the Room of the Bride-elect, and In the Cemetery (in Four Satires of Circumstance)
Heaney	Mother of the Groom
Lawrence	Piano
Niedecker	Sorrow moves in wide waves
Olds	The One Girl at the Boys Party
Philips	On Her Son, H.P.
Plath	Metaphors
Plath	Morning Song
Randall	Ballad of Birmingham

MUSIC

Campion	Rose-cheeked Laura, come
Herrick	Upon Julia's Voice
Lorca	La guitarra
Rich	The Ninth Symphony of Beethoven Understood at Last as a Sexual Message
Stevens	Peter Quince at the Clavier
Stillman	In Memoriam John Coltrane

MYTH AND LEGEND (other than poems in Chapter 13, "Myth"; see also SCRIPTURE AND APOCRYPHA)

Anonymous	Silver swan
Carew	Ask me no more where Jove bestows
H.D.	Helen
Hill	Merlin
Tennyson	Ulysses
Yeats	Leda and the Swan
Yeats	Sailing to Byzantium
Yeats	Who Goes with Fergus?

NATURE (see also BEAST AND BIRD, FARM AND COUNTRY, THE SEASONS)

Bishop	The Fish
Blake	To see a world in a grain of sand
Blake	The Tyger
Clare	Mouse's Nest
Dickinson	A Route of Evanescence
Dickinson	The Lightning is a yellow fork
Frost	Design
Frost	Never Again Would Birds' Song Be the Same
Gioia	California Hills in August
Hopkins	Inversnaid

Poems Arranged by Subject and Theme

Hopkins	Pied Beauty
Hopkins	Spring and Fall
Hopkins	The Windhover
Housman	Loveliest of trees, the cherry now
Keats	To Autumn
M. Moore	The Wood-Weasel
Niedecker	I take it slow
Oliver	Rain in Ohio
Roethke	Root Cellar
Snyder	Mid-August at Sourdough Mountain Lookout
Stafford	Traveling through the Dark
Steele	Waiting for the Storm
Stephens	The Wind
Stevens	Anecdote of the Jar
Stevens	Thirteen Ways of Looking at a Blackbird
Tagliabue	Maine vastly covered with much snow
Tennyson	Flower in the Crannied Wall
Wilbur	In the Elegy Season
W. C. Williams	Spring and All
Wordsworth	I Wandered Lonely as a Cloud
J. Wright	A Blessing

OLD AGE

Brooks	The Bean Eaters
Hayford	Dry Noon
Niedecker	Sorrow moves in wide waves
Randall	Old Witherington
Shakespeare	That time of year thou mayst in me behold
Tennyson	Ulysses
R. Whitman	Castoff Skin
W. C. Williams	To Waken an Old Lady
Yeats	The Old Pensioner; The Lamentation of the Old Pensioner
Yeats	Sailing to Byzantium

PRISONS, REFORM SCHOOLS, AND LOCK-UPS

Digges	For *The Daughters of Hannah* Bible Class . . .
Erdrich	Indian Boarding School: The Runaways

POVERTY

Brooks	The Bean Eaters
Niedecker	Popcorn-can Cover
Stallworthy	Sindhi Woman
M. Williams	MECANIC ON DUTY AT ALL TIMES

PROTEST POEMS

Anonymous	Scottsboro
Axelrod	Once in a While a Protest Poem
Blake	London
Cleghorn	The Golf Links

Poems Arranged by Subject and Theme

Clifton	winnie song
Cullen	For a Lady I Know
Hughes	Dream Deferred
Kilgore	The White Man Pressed the Locks
Milton	On the Late Massacre in Piemont
Owen	Dulce et Decorum Est
Randall	Ballad of Birmingham
Rich	Aunt Jennifer's Tigers
Rich	Trying to Talk with a Man
Wordsworth	The World Is Too Much with Us

SCIENCE AND TECHNOLOGY

Ammons	Spring Coming
Eberhart	The Fury of Aerial Bombardment
Frost	Design
Jennings	Delay
Merrill	Laboratory Poem
Reed	Naming of Parts

SCRIPTURE AND APOCRYPHA

Chesterton	The Donkey
Frost	Never Again Would Birds' Song Be the Same
Meredith	Lucifer in Starlight
Stevens	Peter Quince at the Clavier

THE SEASONS

Spring

Ammons	Spring Coming
Housman	Loveliest of trees, the cherry now
Shakespeare	When daisies pied and violets blue
W. C. Williams	Spring and All

Summer

Anonymous	Sumer is icumen in
Gioia	California Hills in August
Lewis	Girl Help
G. Snyder	Mid-August at Sourdough Mountain Lookout
Toomer	Reapers

Autumn

Hopkins	Spring and Fall
Keats	To Autumn
Stephens	The Wind
Stevens	Metamorphosis
Wilbur	In the Elegy Season
J. Wright	Autumn Begins in Martins Ferry, Ohio

Winter

Bly	Driving to Town Late to Mail a Letter
Frost	Stopping by Woods on a Snowy Evening
Haines	Winter News
Hayden	Those Winter Sundays
Keats	Quotation from "The Eve of St. Agnes"
Niedecker	Popcorn-can Cover
Shakespeare	When icicles hang by the wall
Updike	Winter Ocean

SMALL TOWN LIFE

Heffernan	Living Room
Lowell	Skunk Hour
Robinson	Richard Cory
J. Wright	Autumn Begins in Martins Ferry, Ohio

SPORTS

Francis	Catch
Gildner	First Practice
Hongo	The Cadence of Silk
Housman	To an Athlete Dying Young
Liu	My Father's Martial Art
Updike	Ex-Basketball Player
W. Whitman	The Runner
Wright	Autumn Begins in Martins Ferry, Ohio

TIME, THE PASSAGE OF (see also CARPE DIEM, OLD AGE)

Auden	As I Walked Out One Evening
Cummings	anyone lived in a pretty how town
Eliot	The Love Song of J. Alfred Prufrock
Hardy	During Wind and Rain
Housman	Loveliest of trees, the cherry now
Jeffers	To the Stone-cutters
Pastan	Ethics
Shakespeare	Not marble nor the gilded monuments
Shakespeare	That time of year thou mayst in me behold
Shelley	Ozymandias
Wordsworth	Mutability
Yeats	Sailing to Byzantium

WAR

Anonymous	The fortunes of war, I tell you plain
Arnold	Dover Beach
Barth	The Insert
Betjeman	In Westminster Abbey
Eberhart	The Fury of Aerial Bombardment
Hughes	Green Memory
Jarrell	The Death of the Ball Turret Gunner
Lovelace	To Lucasta

Poems Arranged by Subject and Theme

Melville	The Portent
Owen	Anthem for Doomed Youth
Owen	Dulce et Decorum Est
Reed	Naming of Parts
S. Smith	I Remember
W. Whitman	Beat! Beat! Drums!
W. Whitman	Cavalry Crossing a Ford

A WOMAN'S IDENTITY (see also MOTHERS AND CHILDREN)

Brooks	The Rites for Cousin Vit
Dove	Daystar
Olds	The One Girl at the Boys Party
Plath	Daddy
Rich	Aunt Jennifer's Tigers
Warrior	How I Came to Have a Man's Name

WRITING (see also BOOKS AND READING and LANGUAGE)

Barth	Definition
Blake	Her whole life is an epigram
Crawford	My Iambic Pentameter Lines
Heaney	Digging
Herrick	The Bad Season Makes the Poet Sad
Hongo	The Cadence of Silk
Jeffers	To the Stone-cutters
Martial	Readers and listeners praise my books
Pope	True Ease in Writing Comes from Art, Not Chance
Ramsey	A Poet Defended
Shakespeare	My mistress' eyes are nothing like the sun
Shakespeare	Not marble nor the gilded monuments
Wallace	The Girl Writing Her English Paper
Wilbur	The Writer

Poems for Further Reading, Arranged by Elements

Many instructors tell us that they use the poems in the back-of-the-book anthology as an extra reservoir or second fuel tank of illustrations. Others, to be sure, think the book already offers too many examples. If that is your feeling, don't bother with this.

If, however, you would like a few more poems (or some different poems) to illustrate matters taken up in the body of the book, then the following list can help you put your finger on them. It classifies only the "Poems for Further Reading," and it works through the book chapter by chapter.

Obviously, poems, like any living creatures, are made up of many elements. Far from being exclusive, this list merely points to a few poems in which we think a certain element is prominent, visible to students, and in some instances even likely to spark discussion in class. (Are the poems classified under *Irony* to be taken ironically, or straight? Are the poems listed below under "Literal Meaning" all that simple? Can the Child ballads still be sung? Does "The Convergence of the Twain" have bad spots?)

For Writing Topics. After your students have studied a chapter of the book, you can direct them to certain poems in the anthology. Assign a poem or two and a short paper that springs from their reading. (An essay of two or three paragraphs might be enough: at this stage, overlong papers on topics such as figures of speech, rime and meter, stanza form, etc., might be debilitating.) Topics will occur: The Character of the Soliloquist in Browning's "Spanish Cloister" (after studying *The Person in the Poem*); The Effectiveness of the Repeated Booms and Two-steps in Stafford's "At the Klamath Berry Festival" (after studying *Sound* and *Rhythm*), The Attitude of the Daughter in Plath's "Daddy" (*Tone*), The Sex Symbolism of Dickinson's Dog-walk (*Symbol*, referring to "I started Early – Took my Dog"), and more.

For suggesting that this manual could use such a classification of the anthology poems, and for starting to make one of his own, thanks to Prof. Harvey Birenbaum of San Jose State University.

2   Listening to a Voice

TONE

Some poems in which the poet's attitude is especially clear:

Owen	Anthem for Doomed Youth
Philips	On Her Son, H.P.
Plath	Daddy
Tennyson	Dark house, here where once more I stand
Willard	Marriage Amulet
Wordsworth	Composed upon Westminster Bridge

Some poems that express, as Auden says, "mixed feelings":

Baca	Spliced Wire
Bishop	Filling Station
Eliot	The Love Song of J. Alfred Prufrock
Hopkins	Thou art indeed just, Lord, if I contend

Larkin	Poetry of Departures
Lowell	Skunk Hour
Snodgrass	Seeing You Have ...

THE PERSON IN THE POEM

Some poems in which the identity of the speaker is interestingly different from the poet's "I":

R. Browning	Soliloquy of the Spanish Cloister
Chesterton	The Donkey
Levine	Animals Are Passing from Our Lives
Mew	The Farmer's Bride
Pound	The River Merchant's Wife: a Letter

Irony (other kinds besides ironic point of view, as in the five poems just listed):

Chappell	Skin Flick (a discrepancy between X-movie customers and pioneers; verbal irony)
Dickinson	Because I could not stop for Death (a discrepancy between dying and being taken out driving; verbal ironies: "kindness," "Civility," etc.)
Hardy	The Convergence of the Twain (irony of Fate)
Hardy	Four Satires of Circumstance (ironies galore!)
Merrill	Laboratory Poem (a discrepancy between the clinical setting and the tender feelings of the characters)
Reed	Naming of Parts (a discrepancy between the study of a gun and the study of nature, between the voice of the instructor and the view of the soldier; verbal irony in the pun *easing the spring*)
Stafford	At the Klamath Berry Festival (a discrepancy between traditional dances and ways of life today)
Wylie	The Eagle and the Mole (a discrepancy between the course we are directed to follow and the sense that it would be horrible)

3 Words

LITERAL MEANING: WHAT A POEM SAYS FIRST

Poems that can be taken at face value, without looking for symbols, endless suggestions, huge significance (not that they won't repay thought and close reading):

Anonymous	Sumer is icumen in
Gioia	California Hills in August
Hall	Names of Horses
Hongo	The Hongo Store
Larkin	Poetry of Departures
Millay	Recuerdo
Slavitt	Titanic
Shakespeare	When daisies pied and violets blue
Shakespeare	When icicles hang by the wall

Poems for Further Reading, Arranged by Elements

Updike	Ex-Basketball Player
Yeats	The Lake Isle of Innisfree

THE VALUE OF A DICTIONARY

Poems containing two or more brief allusions:

Dryden	To the Memory of Mr. Oldham
Eliot	The Love Song of J. Alfred Prufrock
Hecht	The Vow
M. Moore	The Mind Is an Enchanting Thing
Yeats	Long-legged Fly

Poems with central allusions:

Auden	Musée des Beaux Arts
Boyd	Cupid and Venus
Chesterton	The Donkey
Eliot	Journey of the Magi
H. D.	Helen
Keats	On First Looking into Chapman's Homer (the celebrated blooper in the allusion to Cortez)
Milton	When I consider how my light is spent
Stevens	Peter Quince at the Clavier (allusions to Shakespeare and to the story of Susanna and the Elders)
Tennyson	Ulysses
Yeats	The Magi

WORD CHOICE AND WORD ORDER

Poems in dialect:

Anonymous	Edward
Anonymous	The Twa Corbies
Mew	The Farmer's Bride

Poems in Middle English:

Anonymous	Sumer is icumen in
Anonymous	I sing of a maiden
Chaucer	Your ëyen two wol slee me sodenly

Poems whose diction and syntax depart from those of speech:

Blake	The Tyger
Coleridge	Kubla Khan
Hardy	The Convergence of the Twain
Hopkins	Spring and Fall
Hopkins	Thou art indeed just, Lord, if I contend
Hopkins	The Windhover
Keats	On First Looking into Chapmans' Homer (inverted syntax: "Much have I," "Yet did I never breathe," "Then felt I," etc.)
Keats	To Autumn
Moore	The Mind Is an Enchanting Thing
Thomas	Fern Hill

Poems for Further Reading, Arranged by Elements

Poems containing technical words:

Merrill	Laboratory Poem
Reed	Naming of Parts
Rich	The Ninth Symphony of Beethoven ...

Poems in colloquial diction:

Frost	Birches
Frost	Stopping by Woods on a Snowy Evening
Hughes	Dream Deferred
Levine	Animals Are Passing from Our Lives
Olds	The One Girl at the Boys Party
Updike	Ex-Basketball Player

Poems containing an interesting mix of formal and colloquial diction:

Bishop	Filling Station
Ginsberg	A Supermarket in California
Larkin	Poetry of Departures
Martin	Rough Draft
Sexton	The Kiss
Winters	At the San Francisco Airport

4 Saying and Suggesting

Some poems especially full of words rich in connotations:

Anonymous	The Three Ravens
Anonymous	The Twa Corbies
Coleridge	Kubla Khan
Eliot	The Love Song of J. Alfred Prufrock
Hardy	During Wind and Rain
Keats	To Autumn
Thomas	Fern Hill

5 Imagery

Atwood	All Bread
Bishop	Filling Station
Erdrich	Indian Boarding School: The Runaways
Keats	To Autumn
M. Moore	The Mind is an Enchanting Thing
Shakespeare	When daisies pied and violets blue
Shakespeare	When icicles hang by the wall
Swift	A Description of Morning
Tennyson	Dark house, by which once more I stand
Thomas	Fern Hill
W. C. Williams	Spring and All
W. C. Williams	To Waken an Old Lady

Poems for Further Reading, Arranged by Elements

6 *Figures of Speech*

METAPHOR AND SIMILE

Poems with central metaphors:

Baca	Spliced Wire
Chappell	Skin Flick
Dickinson	My Life had stood – a Loaded Gun
Hopkins	The Windhover
Jennings	Delay
Martin	Rough Draft
Nemerov	The Snow Globe
Phillips	Running on Empty
Rich	Aunt Jennifer's Tigers
W. Whitman	I Saw in Louisiana a Live-Oak Growing
Willard	Marriage Amulet
W. C. Williams	To Waken an Old Lady
Wordsworth	Composed upon Westminster Bridge
Yeats	Long-legged Fly

Other poems with prominent metaphors:

Levertov	Divorcing
Philips	On Her Son, H.P.
Randall	Old Witherington
Sexton	The Kiss
Shakespeare	That time of year thou mayst in me behold
K. Shapiro	The Dirty Word
Simic	Butcher Shop
Wilbur	The Writer
J. Wright	A Blessing

Poems with prominent similes:

Hongo	The Cadence of Silk
Keats	On First Looking into Chapman's Homer ("Then felt I like . . . ")

OTHER FIGURES

Blake	The Sick Rose (apostrophe)
Carew	Ask me no more where Jove bestows (hyperbole)
Chappell	Skin Flick (pun)
Chesterton	The Donkey (personification)
Dickinson	Because I could not stop for Death (personification)
Dickinson	I started Early – Took my Dog (personification)
Donne	A Valediction: Forbidding Mourning (paradox)
Frost	Birches (understatement)
Keats	To Autumn (apostrophe, personification)
Levine	Animals Are Passing from Our Lives (personification)
Marvell	To His Coy Mistress (hyperbole)
Plath	Daddy (hyperbole)
Reed	Naming of Parts (pun)

Poems for Further Reading, Arranged by Elements

Waller	Go, Lovely Rose (apostrophe, personification)
W. C. Williams	Spring and All (personfication)

7 *Song*

SINGING AND SAYING

Some poems originally sung (see also Ballads):

Anonymous	I sing of a maiden
Anonymous	Western Wind
Shakespeare	When daisies pied and violets blue
Shakespeare	When icicles hang by the wall

BALLADS

Anonymous	Edward
Anonymous	The Three Ravens
Anonymous	The Twa Corbies
Mew	The Farmer's Bride

Poems not ballads but balladlike:

Auden	As I Walked Out One Evening
Brodsky	Belfast Tune

8 *Sound*

SOUND AS MEANING

Poems containing onomatopoeia:

Shakespeare	When daisies pied and violets blue
Shakespeare	When icicles hang by the wall
Stafford	At the Klamath Berry Festival
Stevens	Peter Quince at the Clavier
Wilbur	Transit ("click")

ALLITERATION AND ASSONANCE

Blake	The Tyger
Carew	Ask me no more where Jove bestows
Coleridge	Kubla Khan
Hardy	During Wind and Rain
Hopkins	Spring and Fall
Hopkins	The Windhover
Ransom	Bells for John Whiteside's Daughter ("primly propped"!)
Thomas	Fern Hill
Waller	Go, Lovely Rose

RIME

Poems whose rimes may well repay study:

Blake	The Sick Rose
Dickinson	(all poems: masterworks of off-rime)
Eliot	The Love Song of J. Alfred Prufrock

Poems for Further Reading, Arranged by Elements

Lowell	Skunk Hour
Plath	Daddy
Stevens	Peter Quince at the Clavier

9 Rhythm

STRESSES AND PAUSES

In any good metrical poem, rhythms matter, of course, and can't be disentangled from meanings. Here are some poems *in open or syllabic forms* in which rhythms play strong parts:

Hall	Names of Horses
Reed	Naming of Parts
Smart	For I will consider my Cat Jeoffry
Stafford	At the Klamath Berry Festival (Consider the effect of "he took two steps," repeated four times)
Thomas	Fern Hill

METER

Frost	Away!
Wyatt	They flee from me that sometime did me seke (An exercise is suggested in the note after the poem)
Yeats	The Magi (Worth scanning: irregularities battle with regularity for supremacy)

10 Closed Form, Open Form

CLOSED FORM: BLANK VERSE, STANZA, SONNET

Poems in blank verse:

Frost	Birches
Tennyson	Ulysses
Updike	Ex-Basketball Player

Poems in closed (heroic) couplets:

Dryden	To the memory of Mr. Oldham
Jonson	On My First Son
Rich	Aunt Jennifer's Tigers
Swift	A Description of the Morning

Poem in tercets:

Hardy	The Convergence of the Twain

Poems in tightly structured riming stanzas:

Donne	The Flea
Frost	Away!
Hecht	The Vow
Herbert	Love
Keats	To Autumn
Martin	Rough Draft
M. Moore	The Mind is an Enchanting Time

Poems for Further Reading, Arranged by Elements

Poems in syllabic stanzas:

M. Moore	The Mind Is an Enchanting Thing
Thomas	Fern Hill

Sonnets:

Boyd	Cupid and Venus
Brooks	The Rites for Cousin Vit
Herrick	The Bad Season Makes the Poet Sad (a sonnet in couplets)
Hopkins	Thou art indeed just, Lord, if I contend
Hopkins	The Windhover
Keats	On First Looking into Chapman's Homer
Keats	When I have fears that I may cease to be
MacLeish	The End of the World
Meredith	Lucifer in Starlight
Milton	When I consider how my light is spent
Nemerov	The Snow Globe
Owen	Anthem for Doomed Youth
Shakespeare	This time of year thou mayst in me behold
Shakespeare	When, in disgrace with Fortune and men's eyes
Wordsworth	Composed upon Westminster Bridge

A villanelle:

Bishop	One Art

OPEN FORM

A few classics of open form poetry:

Eliot	Journey of the Magi
Ginsberg	A Supermarket in California
H. D.	Helen
Plath	Morning Song
Pound	The River Merchant's Wife: a Letter
Roethke	Elegy for Jane
W. Whitman	I Saw in Louisiana a Live-Oak Growing
W. C. Williams	Spring and All
W. C. Williams	To Waken an Old Lady
J. Wright	A Blessing

A prose poem:

K. Shapiro	The Dirty Word

11 Poems for the Eye

No graphic poems in the anthology section, but if you care to work further with the looks of poems, you might compare the appearances on the page of the open form poems and the tightly structured stanzaic poems listed above. These appearances mean something; at the very least, they announce the name of the game, and the reader knows whether or not to expect rime and meter.

12 Symbol

Ashbery	At North Farm
Atwood	All Bread
Blake	The Sick Rose
Bogan	The Dream
Eliot	The Love Song of J. Alfred Prufrock
Lowell	Skunk Hour
Nemerov	The Snow Globe
Rich	Aunt Jennifer's Tigers
Willard	Marriage Amulet

13 Myth

Carew	Ask me no more where Jove bestows
Eliot	Journey of the Magi
H. D.	Helen
Stafford	At the Klamath Berry festival
Yeats	The Magi

14 Alternatives

TRANSLATIONS

Anonymous	Sumer is icumen in (original and modern version)
Pound	The River-Merchant's Wife: a Letter

15 Evaluating a Poem

TELLING GOOD FROM BAD

Hardy	The Convergence of the Twain (a poem surely bad in parts—e.g., stanza 5—but good in its entirety!)

KNOWING EXCELLENCE

[Your choice.]

13 Reading a Poem

William Butler Yeats, THE LAKE ISLE OF INNISFREE, page 499

As a young man in London in 1887–91, Yeats found himself hating the city and yearning for the west of Ireland. He recalled: "I was going along the Strand and, passing a shop window where there was a little ball kept dancing by a jet of water, I remembered waters about Sligo and was moved to a sudden emotion that shaped itself into 'The Lake Isle of Innisfree'" (*Memoirs* [New York: Macmillan, 1972] 31). In London (he recalled in his *Autobiography*), he sometimes imagined himself "living in imitation of Thoreau on Innisfree, a little island in Lough Gill." The nine bean rows of the poem were evidently inspired by Thoreau's bean patch.

Yeats's lines provide rich rows of sound for the student to hoe: assonance (from *I . . . arise* in the first line through the *o*-sounds in the closing stanza), onomatopoeia *(lapping)*, initial alliteration, internal alliteration *(arise, Innisfree; hear, heart's core)*. Sound images of bees, cricket, linnet, and lake water are predominant. Whatever noises come from roadway or pavement, however, are left unspecified.

Perhaps, in London, Yeats thought himself one of Ireland's prodigal sons. At least, A. Norman Jeffares has noticed in the first line an echo from the parable of the prodigal son (Luke 15:18): "I will arise and go to my father" (*A Commentary on the Collected Poems of W. B. Yeats* [Stanford: Stanford UP, 1968] 35).

In later years, according to John Unterecker, Yeats was shocked that "The Lake Isle" had become his most popular poem. He had taken a dislike to its "Biblical opening lines." But audiences always demanded it of him, and his sonorous reading of the poem is available on a recording (Spoken Arts, 753).

LYRIC POETRY

D. H. Lawrence, PIANO, page 502

About the first question: it's really a quick writing assignment. Ten minutes of class time might be enough to let students write their paraphrases. To be sure, you could let them wing it and paraphrase the poem out loud, but the results may not be so thoughtful or accurate. A few of the students might then be asked to read their efforts aloud, for others to agree or disagree with.

Reader response theory, if crudely applied, might claim that every paraphrase is valid. But we think it greatly helps a class discussion to assume that it is possible to find an interpretation of a poem that all or most will agree comes closest to it.

"Piano" isn't a flawless poem. Lawrence was seldom at ease in rime, and the strained juxtaposition of *clamor* and *glamor* indicates his discomfort. Still, *glamor* is an accurate word in its context: the mature man knows that the child's eyes endowed the past with an illusory beauty. The quality of Lawrence's poem may be seen in the specificity of its detail: "the boom of the tingling strings," "the small, poised feet." Lawrence enters into the child's perspective, while able to criticize it from outside. The speaker is resisting his urge to cry, as the connotations of his words indicate (the song is *insidious*, it *betrays*). But at last he is unable to hold back his tears and, sensibly, yields to them.

How does Lawrence's poem escape bathos? Robert Pinsky has offered an explanation in "Poetry and Pleasure," in *Threepenny Review* (Fall 1983). The subject of "Piano," Pinsky

(text pages) 502–504

finds, is a stock source for poems, "as mothers-in-law or airplanes with ethnically various passengers are stock sources for jokes." Yet the poem strikes us with "something fresh, not stock." Its language is vivid, unconventional; its words *insidious* and *betrays* add a "steely spring"; it sets up an energetic tension between present and past.

Marianne Moore, THE WOOD-WEASEL, page 503

This is a small point, but to avoid any possible confusion you might point out that the title of this poem runs right into the opening line. Marianne Moore is fond of such run-on titles: see also "The Mind Is an Enchanting Thing" (page 851).

This paean to the skunk (or "wood-weasel"—certainly a more sylvan and euphonious name!) cites numerous traits to admire. The animal's motion is dainty, he is beautifully attired (in "regalia," yet, with "glistening goat-fur"). The poet notes how beautifully adapted he is to his surroundings, praises his determination, his dominance over the woods (he is a chieftain, not an outlaw), sweet and playful in personality, noble in character, impervious to insect bites and moth damage. George W. Nitchie has found Moore's skunk "beautiful, independent, and in charge of things . . . playful, but clearly on his own terms and with his own kind" (*Marianne Moore: an introduction to the poetry,* Columbia U P, 1969; 130).

Any attention to the language of the poem might note in particular its rich alliteration (especially in lines 4–5 and 9), the internal rime of *ermined / determination's,* the external rime in lines 9–10. The poem is crammed with minute and specific description—not usual in prose.

Perhaps it's a little early to raise the matter of attitude or tone (dealt with in Chapter 2), but students should realize that the poet isn't writing in deadly earnest.

Marianne Moore loves ingenious and demanding forms that most readers won't even notice. "The Wood-Weasel" is an upside-down acrostic: the first letter of each line spells out the name of the poet's friend Hildegarde Watson, wife of Sibley Watson, copublisher of the literary magazine Moore had edited in the 1920s, *The Dial.* (Nitchie, not we, noticed this ingenuity.) For another acrostic poem, see Dabney Stuart's "Crib Death" in Chapter 15 (page 751).

Moore also loves naming things. The tale is told of the occasion in 1955 when the Ford Motor Company engaged her to come up with a name for a new car. She made several gorgeous suggestions—the Ford Silver Sword, the Arc-en-ciel, the Resilient Bullet, the Anticipator, the Thunder Crester, the Turquoise Turbo-turtle—but in the end, the name the company settled on was the Edsel. (Moore's exchange of letters with a Ford executive is reproduced in *A Marianne Moore Reader* [New York: Viking, 1961].)

NARRATIVE POETRY

Anonymous, SIR PATRICK SPENCE, page 504

On the questions in the book: We really don't think the king's motives can be known for sure from this bare portrait of him; we ask this question mainly to prompt students to pay attention to what they find on the page and be wary of deep extrapolations. As far as we see him in the poem, the king sits around drinking wine, leading a life of ease, and (with a deliberate official gesture) sends his best sea-captain and a loyal contingent of naval officers to their doom. Although the poet takes a sour view of the comfortable life at court, he feels for the Scots nobles, and we too are moved by his spare sketch of the bereaved ladies, futilely waiting for their men, who will never return. The great stanza about the new and old moons, apparently an ill omen, serves further to heighten the tension of the story and foreshadow its conclusion.

Here are two more questions:

(text pages) 504–508

Comment on Sir Patrick's character. What do you make of his abrupt transition from laughter to tears (lines 13–16)? (He is not only brave and loyal to obey the king's order; he is a passionate man with quick, open, unconcealed feelings.)

In what lines do you notice a wry comment on the soft life that the nobles led at court? What does this attitude suggest about this anonymous poet? (29–30: The nobles are loath to get their fine shoes wet. Probably the poet wasn't a noble, but a sarcastic commoner.)

Robert Frost, "Out, Out–," page 506

Like Sir Patrick Spence, the boy's initial reaction to his terrible realization is to laugh—then, almost at once, dismay sets in. And like the folk ballad, "Out, Out—" tells a story of sudden, meaningless death, and does so with spare economy.

Perhaps the "they"—the doctor and the hospital staff?—who turn to their own affairs are not merciless. The "watcher at his pulse" grows frightened when the pulse fails; no one wants to believe the boy will die. Radcliffe Squires finds no one to blame for the "faceless accident." In his view, "Simultaneously, one sees the human watchers touched by normal griefs and fears. And yet life must turn to a more important task finally, that of continuing. . . . Only the grand composer could hold together in one poem the two severe and mutually accusing ideas that one must be moved to pity and compassion and that one must coldly and sternly pursue the duty of endurance and survival" *(The Major Themes of Robert Frost,* [U of Michigan P, 1963] 46). Frost's poem offers no comfort, but seems a realistic view of what happens in an emergency ward. Any student interested in a career in medicine might be asked for a response to this poem.

Frost's allusion to *Macbeth* is part of the meaning of the poem, and students may be asked to think about it. Perhaps Frost suggests that the snarling buzz-saw full of sound and fury, reaching out its friendly handshake, just doesn't make sense. This, as Stanley Burnshaw has noticed, is one among several of Frost's poems that seem to question the existence of a benevolent order in the universe. Others include "A Servant to Servants," "The Housekeeper," "Home Burial," and (we would add) "Design" *(Robert Frost Himself,* [New York: Braziller, 1986] 298).

Frost based his poem on an actual incident: an accident that had happened in 1910 to a sixteen-year-old boy he knew in Bethlehem, New Hampshire; five years went by before the poem took form (in 1915–16). See Lawrance Thompson, *Robert Frost: The Early Years* (New York: Holt, 1966) 566–567.

14 Listening to a Voice

TONE

Theodore Roethke, MY PAPA'S WALTZ, *page 510*

Fred Roux of Shippensburg University of Pennsylvania reports that his students' interpretations of "My Papa's Waltz" have differed according to sex. Young men almost unanimously respond to the poem as a happy childhood memory of a loving father's exuberant horseplay. A few young women react negatively. For them, "I hung on like death" and "You beat time on my head," as well as "battered" and "scraped," suggest that the speaker's recollection is unhappy. They also assume that a man with whiskey on his breath must be drunk. None has perceived an ironic parallel between their responses and that of the speaker's frowning mother. "From this," adds Professor Roux, "it would appear that student response to 'My Papa's Waltz' is, to a degree, the result of a difference in socializing experiences during early childhood." Haven't any young women had boisterous fathers? We'd like to hear other classroom experiences.

As Alan Seager discerns in his biography of Roethke, *The Glass House* (New York: McGraw, 1968) 23, the mature Roethke seems to have felt a certain guilty resentment against his father, a sense of how (as an awkward, chubby, bookish, and sensitive child) the young poet had failed to make the old man proud of him.

"My Papa's Waltz" may have had its genesis in a wish-fulfilling dream. After his father's death, Roethke wrote a memoir (calling himself "John"): "Sometimes he dreamed about Papa. Once it seemed Papa came in and danced around with him. John put his feet on top of Papa's and they'd waltz. Hei-dee-dei-dei. Rump-tee-tump. Only babies expected dreams to come true" (qtd. in Seager, 24).

Countee Cullen, FOR A LADY I KNOW, *page 511*

From Cullen's first book, *Color* (1925), this is one of a series of twenty-nine epitaphs. Compare it with another brief poem that makes a biting social comment: Sarah N. Cleghorn's "The Golf Links" (page 526). Cleghorn's poem seems angrier; the tone of Cullen's poem seems to be wry amusement at stupidity.

Cullen's early biography is sparsely documented. Raised by his grandmother until he was eleven, he was then adopted by the Reverend Frederick A. Cullen, pastor of a Methodist church in Harlem, who gave the future poet not only a name but a new life of books and conversation. Famed as the leading poet of the Harlem Renaissance, Cullen suffered a decline in reputation when militant black critics of the 1960s reevaluated his work and found it wanting in anger and social consciousness. But his wit can bite, as it does in "For a Lady I Know"; and Houston A. Baker has rightly called much of his work an "ironical protest . . . against economic oppression" in his short study of Cullen, *A Many-colored Coat of Dreams* (Detroit: Broadside, 1974).

Anne Bradstreet, THE AUTHOR TO HER BOOK, *page 511*

The "rags" (line 5) worn by this bastard brat of a book may have been the first edition's abundance of typographical errors. Although Bradstreet patiently revised her work, she did not live to see her "brat" appear in better dress. This poem prefaced the Boston edition published in 1678, six years after the poet's death.

(text pages) 511–514

Robert Hutchinson, in the introduction to his edition of *Poems of Anne Bradstreet* (New York: Dover, 1969), gives a concise account of the book's publication. Evidently the author's family, proud of her poetry, felt that it deserved more notice than New England could then give. The Reverend John Woodbridge, Bradstreet's brother-in-law, took with him to England the manuscript of the collection. London at the time had sixty printers; New England, one—and so it must have been difficult, even then, to print poetry in America. "The fact," notes Hutchinson, "that Herrick's *Hesperides* had just appeared in England while the latest venture of Samuel Green, the Cambridge, Massachusetts, printer, was a revision of *The Bay Psalm Book* to rid it of its crudities, gives an indication of the intellectual distance between the two countries."

Walt Whitman, TO A LOCOMOTIVE IN WINTER, page 512

Emily Dickinson, I LIKE TO SEE IT LAP THE MILES, page 513

Though both of these great nineteenth-century Americans take almost the same subject, in tone and in form the two poems differ as sharply as opera differs from chamber music. (Some students might argue that the mutual subject isn't a moving locomotive, but the poets' praise of it. While seeing a real similarity, they would be missing the distinction between subject and tone.) Whitman addresses his machine in awe and exultation. In lines 14–19 he practically prays to it (almost like Henry Adams on bended knees before the dynamo in *Education*). Dickinson is evidently more playful in her affectionate view of the locomotive as a great beast. It is horselike in that it neighs and has a stable, but it isn't quite a horse: it crawls and hoots. Both poets, incidentally, see not only a locomotive, but a whole train. Dickinson's seeing it "chase itself" suggests cars trying to catch their locomotive as they roll downhill. Dickinson's allusion to Boanerges means no more, I think, than that the locomotive is a servant, and is thunderous.

Whitman's poem is full of diction from music: *recitative, beat, ringing bell, notes, chant, harp, piano, trills.* The locomotive embodies poetry, too, in its *metrical* pant and roar, and in its ability to serve the Muse. The word *recitative* indicates the form the poem will be cast in. In Italian opera, to which Whitman was devoted, Rossini had introduced the use of the full orchestra to accompany the recitative, the passage of half-sung, half-spoken declamation; and it may be that, as Robert D. Faner has argued, such recitative was a basic model for Whitman's poetry. "The recitative, highly rhythmic and emotional, punctuated by instrumental accompaniment with thrilling effect, and in its chanted delivery giving the impression of the rhythms of speech, he found well adapted to the bulk of his work, which he thought of as a sort of bardic chant" (*Walt Whitman & Opera* [Carbondale: Southern Illinois UP, 1951] 234).

Langston Hughes, HOMECOMING, page 514

The last lines make the speaker's feeling unmistakable: he no longer has a lover, all he has now is vacant space. The title is ironic: *homecoming* usually suggests a warm welcome, not a cold and empty bed. The poem is remarkable for what it leaves unsaid, for the economy with which Hughes portrays utter loneliness.

John Milton, ON THE LATE MASSACRE IN PIEMONT, page 514

Sorrow, wrath, and hatred seem mingled in this powerful sonnet. Milton sees the Waldenses as true primitive Christians. They had broken with Rome in the twelfth century, refusing to accept rituals and dogmas which, they thought, had been too recently decreed. Protestant Europe was outraged at the massacre. As Cromwell's secretary, Milton was instructed to write letters of protest to the heads of state. His own indignant prayer to a wrathful God has, as William Riley Parker has said, "the awesome sound of a great wave pounding against a wall."

The Person in the Poem

Trumbull Stickney, Sir, say no more, page 515

The death at age 30 of the brilliant Stickney, scholar of Greek and recipient of the only doctorate in letters that the University of Paris had granted to an Anglo-Saxon, silenced one of the finest lyric voices in American poetry. One of his sonnets, "Near Helikon," ends:

> To me my troubled life doth now appear
> Like scarce distinguishable summits hung
> Around the blue horizon: places where
> Nor even a traveller purposeth to steer,—
> Whereof a migrant bird in passing sung
> And the girl closed her window not to hear.

Indispensable to anyone who wishes to read further is *The Poems of Trumbull Stickney,* ed. Amberys R. Whittle, with a foreword by Edmund Wilson (New York: Farrar, 1972).

Philip Larkin, Wedding-Wind, page 516

Larkin is usually regarded as a poet of exclusively glum, bleak sentiments; this warm, loving, and joyous poem reveals a different side of him. (For the better-known side of Larkin, see "Home Is So Sad" and "Poetry of Departures" in the Anthology: Poetry.)

On the questions in the text: The speaker is apparently a poor farmer's wife. She didn't honeymoon in Bermuda; the day after the wedding finds her home feeding the chickens. But the language sounds nothing like that of any simple country woman: the metaphysical metaphor (joy runs through her actions like a thread that carries beads), the Psalmlike diction of the last line, the closing rhetorical questions. In this a fault of the poem? On the contrary, it's a wonderful poem. Maybe Larkin is trying to convey complex feelings, and to do so has to put an adequate language into the speaker's mouth—much as William Faulkner fills the mouth of young Sarty Snopes with his own glorious purple prose in the story "Barn Burning." Although in each case the speaker wouldn't say these words, they express what he or she feels.

Although we wouldn't belabor this point—students could get bored or skeptical—the rain that pours down on the wedding night invites a contrast with the missing rain in "Western Wind" (in the Anthology). Withheld and yearned for, that rain goes together with a dearth of love. Such symbolism, which must be as old as poetry, is central in Eliot's *The Waste Land* and present in "The Love Song of J. Alfred Prufrock."

This poem might be compared with another monologue by a woman in love, also by a male poet: Ezra Pound's version from Li Po, "The River-Merchant's Wife" (also in the Anthology).

William Wordsworth, I Wandered Lonely as a Cloud, page 517

To point out the distance between art and reporting, it may be helpful to read Wordsworth's poem aloud—at least parts of it. In their rhythm, lines such as "fluttering and dancing in the breeze" and "Tossing their heads in sprightly dance" make the motion of the daffodils come alive. By comparison, Dorothy Wordsworth's record of the incident ("the rest tossed and reeled and danced") seems merely excellent prose.

Actually, Wordsworth's sister was a distinguished poet in her own right, as Hyman Eigerman demonstrates in *The Poetry of Dorothy Wordsworth* (New York: Columbia

UP, 1940), an anthology of passages from her journals arranged into formally open verse.

James Stephens, A GLASS OF BEER, page 519

The high regard of the Irish for the magical powers of speech has given them a long and glorious tradition of poetic cursing. In the ancient tales of the Ulster saga, we read of kings who wouldn't go to battle without an accompanying druid: a poet-priest charged with pronouncing magnificent metrical curses upon the enemy. Who knows?—in the pubs of Stephen's native Dublin, curses like the one in "A Glass of Beer" may well have seemed ordinary, even mild.

Although the speaker—some frustrated drinker hard up for cash—is in a towering rage at the barmaid who denied him, the tone of the poem is not anger but high amusement. There is irony, too, in the obvious contrast between the speaker's stupendous hyperboles and the puny occasion for them. Save this poem, if you like, for teaching figures of speech.

There is hardly a better modern poem, however, for reminding students that the feelings expressed in poetry aren't always positive. A poem may be written in rage or chagrin, as well as in love or joy. This seems an essential truth, and one that XJK has tried to demonstrate at some length in *Tygers of Wrath: Poems of Hate, Anger, and Invective* (Athens: U of Georgia P, 1981), an annotated anthology showing the tradition of dark emotion in British, Irish, and American poetry from the Middle Ages to the present. Naturally, in this tradition, "A Glass of Beer" holds an honored place.

"A Glass of Beer" is a free translation from the Irish of Dáibhí Ó Bruadair (c. 1625–98). The original with a translation by Thomas Kinsella ("A Shrewish, Barren, Bony, Nosey Servant") is given by Seán Ó Tuama and Kinsella in *An Duanaire: An Irish Anthology* (Philadelphia: U of Pennsylvania P, 1981) 116–17.

Mary Elizabeth Coleridge, WE NEVER SAID FAREWELL, page 519

The speaker, we assume, is the wife in a marriage that has become cold and loveless. In truth, we have no way to know it isn't the husband speaking, other than the fact that a woman wrote the poem. The speaker can't be sure of the precise moment when the pair soured on each other permanently ("broke the level line"), but they still dwell side by side, together and yet apart. The poet clearly perceives the paradox in the speaker's situation. "Our places fixed for life upon the chart" recalls an era when marriages were harder to undo. The second stanza, with its metaphor of the two islands and the rime throwing weight upon that final *apart,* closes the poem powerfully. A flat statement of theme: "What a grim marriage we're trapped in!"

Mary Coleridge was distantly related to Samuel Taylor Coleridge: a great-great niece. A well-known English lyric poet in her own right, championed in her day by Robert Bridges, she is now remembered mainly for a few anthology set-pieces. High time her work received a serious revaluing.

Paul Zimmer, THE DAY ZIMMER LOST RELIGION, page 520

Paul Zimmer writes usually comic, sometimes touching poems featuring a central character who bears his own last name. If Zimmer in this poem is the poet and not a fictitious character, then the speaker may be the mature Zimmer, looking back through his own younger eyes. The "old days" mentioned in line 8 would seem an even earlier time when, as a schoolboy, Zimmer assisted at Mass. Now (as an adolescent?) he has come to doubt—but he still takes a boyish view, expecting Christ, "like the playground bully," to punish him. The last two lines seem the mature Zimmer's view. Only the grown-up are ready for Christ. Without knowing the actual Zimmer's present convictions, we can assume that the mature poet speaks either as a believer or as one with a deepened respect for belief.

(text pages) 520–524

For more Zimmer poems, see *The Zimmer Poems* (Washington: Dryad, 1976). The world in which the character Zimmer moves often seems dreamlike.

William Carlos Williams, THE RED WHEELBARROW, page 521

Evidently many readers have found it easy to admire this poem without feeling a need to know the circumstances in which it was written. For a fairly recent appreciation, see Louis Untermeyer, *The Pursuit of Poetry* (New York: Simon, 1969) 25. Untermeyer views the poem as a kind of haiku that makes us aware of glories in commonplaces. A more sharply critical estimate is that of Roy Harvey Pearce in his fine essay "Williams and the 'New Mode'" in *The Continuity of American Poetry* (Princeton: Princeton UP, 1961) 335–48. Pearce charges the poem with sentimentality: "At its worst this is togetherness in a chicken-yard." However, in Pearce's view, the poem also has a better aspect: what "depends" is the poet's vocation as a poet. He needs common objects in order to write poems, and the objects in turn need him to imagine them into poetry.

If the librarian is right about the situation in which the poem was written, "The Red Wheelbarrow" seems a better poem than we had realized: a kind of prayer, a work of compassion. However, that the poem fails to give us an intimation of the reasons for the poet's feelings (and of why we ought to share them) does expose it to Pearce's accusation that it is sentimental. Whatever the instructor's opinion, students may be invited to debate the merits and demerits of the poem.

IRONY

Robert Creeley, OH NO, page 521

"What interests me about 'Oh No' is its tone," Cynthia Edelberg remarks in an interview with the poet. "How would you describe it?" Creeley replies that he sees it as wry irony, the poem being "self-parody," a comment on his feelings at the time. "As Joel Oppenheimer said, that would qualify me to be a Jew. He really liked that poem. It's that kind of humor" (Edelberg's *Robert Creeley's Poetry: A Critical Introduction* [Albuquerque: U of Mexico P, 1978] 168).

W. H. Auden, THE UNKNOWN CITIZEN, page 523

For making students better aware of irony, Auden's familiar satire remains as dependable as any poem we know. Little seems to have dated in it, other than the praise of the citizen for adding five children to the population. Students are usually good at seeing that, unlike the unknown soldier, the citizen is all too thoroughly identified; and that, nevertheless, his true nature and inmost wants remain unknown. Meaty questions for discussion naturally arise: What are the premises of such a society? It seems dedicated to the proposition that to conform to a norm is the highest virtue—any individual traits, of course, being an annoyance to statisticians. What is a "Modern Man?" One with animal needs, but no aspirations. The epitaph, often overlooked, is worth dwelling on: it tells us at once that the unknown citizen is only a number, and that bureaucrats keep track of him—and, incidentally, like the rest of the poem, the epitaph is in rime.

"The Unknown Citizen" is one of six poems in this chapter in which we hear a voice obviously not the poet's. (The others are the ones by Jarrell, Robinson, Betjeman, Stephens, and Blake.)

Herbert Scott, THE GROCER'S CHILDREN, page 524

An unusually sharp reader might detect irony in the first two lines. That the grocer, that dealer in good food, feeds his own children what is stale, moldy, spoiled, and insect-infested, becomes unmistakably clearer as the poem goes on, adding to its inventory of outrages. We suspect the speaker feels sorry for the poor kids and resents their father's economies. At least, he troubles to report all this, and he builds up a case for the children, against the father, that practically turns our stomachs.

Scott, who now teaches literature and creative writing at Western Michigan University, spent years working in the grocery business before switching careers. "The Grocer's Children" and other poems reflecting this experience appear in his second collection, *Groceries* (U Pittsburgh P, 1976).

John Betjeman, IN WESTMINSTER ABBEY, page 524

Cadogan Square was an especially fashionable London address around the turn of the century, and the fact that the speaker owns stocks (line 30) also indicates her style of life. Her mind, however, is ordinary: her ideals seem bounded by drugstore novels and by plumbing that works properly.

Students usually have a fine time picking out the easy contradictions in the lady's beliefs: that the Lord may allow bombs to hit German women, but not English women; that He protects whites more dutifully than blacks; that it is all very well for the "gallant blacks" to die, but let the Empire remain united; that democracy and class distinction go hand in hand.

The speaker's attitude seems to be: "Let God wait upon my convenience." To call His word a "treat" reduces Scripture to the importance of candy. That Betjeman first printed this ironic blast at smug, hate-mongering chauvinism in the midst of World War II strikes us as a brave and large-minded plea for genuine Christian charity.

Sarah N. Cleghorn, THE GOLF LINKS, page 526

What a great epigram!—no verbal irony in it, just matter-of-fact notation of a social condition that seems ironic in the extreme. As Robert Frost said, in his introduction to Cleghorn's autobiography, *Threescore* (1936), "There is more high explosive for righteousness in the least little line of Sarah Cleghorn's poem about the children working in the mill . . . than in all the prose of our radical-bound-boys pressed together under a weight of several atmospheres of revolution." (The conservative Frost didn't like Marxists, but he called Cleghorn "a saint and a reformer" anyway.) For a more recent tribute, see Irving Dilliard, "Four Short Lines," *The Nation* 222 (10 Apr. 1976): 444–45.

Stanley Kunitz and Howard Hayward's *Twentieth Century Authors* (New York: Wilson, 1942), in an article on Cleghorn that she apparently helped write, explains the twenty-year hiatus between her early books and her later ones. "This was caused by the fact that her socialism and pacifism made editors and publishers reluctant to use her later writing, and partly by the fact that in middle age she became a teacher." Among her other works is a novel, *The Spinster* (1916), and a last collection, *Poems of Peace and Freedom* (1945).

Thomas Hardy, THE WORKBOX, page 526

Dramatic irony is present in the discrepancy between the carpenter's limited knowledge and the reader's growing conviction that the wife knew John much better than she cares to admit. Her phrase "mere accidental things" contains verbal irony, and in general the whole speech in lines 25–28 is a verbal irony. Cosmic irony may be operating too (and one is sure that it is, knowing Hardy) in the Fate or chance that caused the carpenter to select a piece of poor John's coffin out of all pieces of wood in the world.

To us, the situation in the poem had seemed like that in James Joyce's "The Dead": the wife, by remembering a young man who died of love for her, has a bleak realization that she might have known a joyous life had she married him instead. However, Albert Furtwangler and his students at Mount Allison University found other possible levels of irony, as he kindly wrote to report. For Professor Furtwangler, "The Workbox" is marred by an excess of irony that runs too deep: "it remains fascinating in the long run more as a puzzle than as a clear disclosure of character." Among other readings he considered the two following, which he thinks overingenious and yet consistent with the poem.

The husband, aware of his wife's past, has contrived his present as a cunning torture for her. "He seems to offer it in love, but takes pleasure in drawing out his wife's confused replies . . . thus trapping her in her own hypocrisy."

The husband knows his wife's history; and she knows that he knows it. "But they coexist uneasily with each other by exercising an elaborate fiction of ignorance."

What will you and your students decide?

J. O. Bailey sees in this poem the "ballad-like theme of the lover who died of grief when his beloved married another." Like traditional English and Scottish ballads, the poem has a question-and-answer structure and ends in a surprise. (See *The Poetry of Thomas Hardy* [Chapel Hill: U of North Carolina P, 1970].) Compare "The Workbox" in these respects with "Bonny Barbara Allan" (page 612) and "Edward" (page 774).

Robert Burns, THE TOAD-EATER, page 527

One of Burns's best (and nastiest) epigrams, this blast at a toady, or fawning sycophant, implies that its object is a louse.

For comparison, it might be worth reading to a class Burns's "To a Louse, On Seeing One on a Lady's Bonnet at Church," or at least quoting its famous prayer, "O wad some Power the giftie gie us / To see oursels as others see us!"

FOR REVIEW AND FURTHER STUDY

EXERCISE: Telling Tone, page 527

Richard Lovelace, TO LUCASTA, page 528

Wilfred Owen, DULCE ET DECORUM EST, page 528

"To Lucasta" may refer to an actual parting. During the Puritan Revolution of 1642–45, Lovelace fought in the service of Charles I. Students will readily see the poet's theme that Honor (duty to God and King) takes priority over duty to Lucasta; the tone of the poem may give them greater difficulty. The closing line makes a serious affirmation: Honor for Lovelace is not an "old Lie," but a creed. Neither grim nor smug, the poem also has wit and loving tenderness. The witty second stanza seems almost comic in its figures of speech: having renounced Lucasta's nunlike chastity and calm, the speaker will now go whet his sword upon the body of someone wilder.

Owen's theme is apparent: death in battle is hideous, no matter what certain ignorant poets say about it. For us, there seems irony in the fact that Owen himself was to be killed in action in France. Although in a wartime letter he called himself "a conscientious objector with a very seared conscience," Owen in this poem does not question that to die for one's country may be necessary. His attitude is overpowering disgust—with the butchery of war, with those who idealize it.

Bettie Sellers, IN THE COUNSELOR'S WAITING ROOM, page 529

Evidently this student has been referred to the college psychologist. "Home soil" implies that the mothers have been plowed and planted and made to bear, a condition their daughters appear to be resisting.

Still, this "terra cotta" girl is the product of her soil. Unconsciously, her toes furrow the rug as if she is plowing it. Like the mothers, she seems to be one with the earth; but unlike them, she now (in the mothers' view) holds no promise of procreation. In this discrepancy lies the poem's main irony. (Other mildly ironic discrepancies: that a farm girl from a Bible-reading background now studies existentialism; the girl's guilt feelings juxtaposed with existential anguish over the nature of the universe.)

Because the flat-footed girl is the subject, because we witness her unease (toes furrowing the rug as she clutches the comfortless book on existentialism), certainly she receives the larger share of the poet's—and our own—sympathy. She may be an awkward rustic, but she has found love and can't help loving. Sellers regards both student and mothers with (we think) affectionate humor: the girl with her "big flat farm feet," the mothers so eager to see their daughters reproduce that they weep over the prospect of a crop failure.

Yet clearly she understands the feelings of both generations. Without taking sides, she sets forth their conflict with wonderful brevity. Students who see her poem as an editorial for or against gay/lesbian rights will be reading into it.

Jonathan Swift, ON STELLA'S BIRTHDAY, page 530

Swift's playfully tender birthday gift kids Stella about her size, but artfully turns a dig into a compliment. Imagining her split in two, he declares that even half of her would surpass any other whole woman. For most students, the only difficult lines will be 7–8. Swift's argument seems to be that, while Stella has lost much of the slender beauty she had at sixteen, she hasn't greatly declined in total worth, for an increase of her mental gifts (among them wit) has amply compensated.

Robert Flanagan, REPLY TO AN EVICTION NOTICE, page 530

In few lines, this poem makes clear a tone of withering resentment. The poet sees clearly and reports honestly: rich landlord and ousted tenant both lead lives advantaged *and* disadvantaged.

The poem is roughly a sonnet: its fourteen lines, various in length, end in near-rimes and some exact rimes.

Although Robert Flanagan, who teaches at Ohio Wesleyan University, has written much poetry, he is best known as a writer of fiction: for *Maggot*, a best-selling novel about Marine Corps basic training (1971; reissued in a Warner paperback, 1988), and for *Naked to Naked Goes*, a collection of stories (Scribners, 1986).

John Ciardi, IN PLACE OF A CURSE, page 531

The opening line would be hard to top for outrageously swaggering egotism, if we didn't sense that the speaker is well aware that his chances of being elected are slight. (Shades of a line John Updike once put into the mouth of a Greek-letter fraternity lad, "At seventeen, I was elected Zeus.") We say "the speaker," but we believe this to be John Ciardi speaking. The poem reflects his personal attitude—he favored the pugnacious—and it sounds like his speaking voice.

Ciardi, it might be pointed out, doesn't berate the meek whom Jesus blessed (Matthew 5:5), but those who practice meekness as a trade (9), who employ meekness in their calculations (14).

For the very model of hypocritical meekness, recall Uriah Heep in *David Copperfield*.

Since his death, Ciardi's various work as poet, teacher, essayist, lexicographer, critic, translator of Dante, writer for children, lecturer, and broadcaster has been receiving fresh attention and reconsideration. University of Arkansas Press, publisher of Ciardi's *Selected Poems* (1984) and his last collection, *The Birds of Pompeii* (1985), has also brought out Vince Clemente's compilation of critical essays and friendly memoirs, *John Ciardi: Measure of the Man* (1987) and Ciardi's *Saipan: the War* (1988) and *Echoes: Poems Left Behind* (1989). Edward M. Cifelli of County College of Morris (N.J.), is working on an edition of the letters and a critical biography.

William Blake, THE CHIMNEY SWEEPER, page 532

Set next to Cleghorn's "Golf Links" (page 526), Blake's song will seem larger and more strange; and yet both poets seem comparable in their hatred of adults who enslave children. Though Blake is not a child, he obviously shares Tom Dacre's wish that the chimney sweepers be freed from their coffinlike chimneys, washed clean, and restored to childhood joys. The punning cry " 'weep! 'weep! 'weep!" is the street cry of the sweepers, sent through London to advertise their services. Compare the tone of this poem to that of Blake's "London" (page 562); the anger is similar, but in "The Chimney Sweeper," a poem also touching and compassionate, anger is not stated outright, but only implied.

Music to "The Chimney Sweeper" has been supplied by Allen Ginsberg, who sings the resulting song on *Songs of Innocence and Experience* (MGM recording FTS 3083), assisted by Peter Orlovsky.

15 Words

Literal Meaning: What a Poem Says First

Why a whole section on literal meaning? The need first occurred to XJK in a conversation with Robert Reiter and David Anderson of Boston College. Professor Reiter, who had been using the book in a previous edition, averred that, while it was well to encourage students to read poetry for its suggestions, his students tended to go too far in that direction, and sometimes needed to have their attentions bolted down to the denotations of words on a page. Early in a poetry course, the problem seemed especially large—"I try not to let them look for any symbols until after Thanksgiving!" Mr. Anderson had felt the same difficulty. In teaching Donne's "Batter my heart" sonnet, he had had to argue with students who couldn't see how, in a poem of spiritual aspiration, Donne possibly could be referring to anything so grossly physical as rape. They needed to see the plain, literal basis of Donne's tremendous metaphor, that they might then go on to understand the poet's conception of sanctifying grace.

With these comments in mind, the publishers sent a questionnaire to more than one hundred instructors who had used the book, asking them (among other questions) whether they felt the need for more emphasis on denotation. All who replied said that they would welcome such an emphasis (in addition to the old emphasis on connotation)—all, that is, except for one instructor (God help him) who reported that he couldn't persuade his students ever to rise *above* the level of the literal, if indeed he could get them to rise that far.

Most instructors like to discuss imagery fairly early. They will find nothing to hinder them from taking the chapter on imagery ahead of this one. Another procedure would be to defer "Imagery" until after having discussed both denotation and connotation—taking in sequence the present chapter, "Words," and Chapter 16, "Saying and Suggesting."

William Carlos Williams, This Is Just to Say, page 535

Williams once recalled that this poem was an actual note he had written to his wife—"and she replied very beautifully. Unfortunately, I've lost it. I think what she wrote was quite as good as this" (conversation with John W. Gerber and Emily M. Wallace in *Interviews with William Carlos Williams*, ed. Linda Welshimer Wagner [New York: New Directions, 1976]).

For parodies of this famous poem, see Kenneth Koch's "Variations on a Theme by William Carlos Williams" in *Contemporary American Poetry*, ed. A Poulin (Boston: Houghton, 1980) and other anthologies.

Knute Skinner, The Cold Irish Earth, page 537

Skinner is an American poet who lives most of each year in Killaspuglonane (pronounced the way it looks, with the accent on POOG), a village of about 200, near Liscannon Bay in the west of Ireland. This poem is one of several recollections of Irish country life in his book *A Close Sky over Killaspuglonane* (Dublin: Dolmen, 1968; 2nd ed., St. Louis: Burton, 1975). The Hag's face is a rock formation in the Cliffs of Moher (or Mohee), in whose crevices members of the IRA once hid from the British.

The familiar phrase, of course, is "to become wet to the bone." Every image in the poem indicates that it is to be taken literally.

(text pages) 537–540

Henry Taylor, RIDING A ONE-EYED HORSE, *page 537*

Henry Taylor comments in a letter: "I like the question about the one-eyed horse; the answer is of course, both."

The poem was first published in *Practical Horseman*.

Robert Graves, DOWN, WANTON, DOWN!, *page 538*

This poem can be an astonisher, especially if students haven't read it in advance. One freshman group XJK sprang it on provided a beautiful gamut of reactions: from stunned surprise to hilarity. At first, most didn't know quite what to make of the poem, but they soon saw that its puns and metaphors point to details of male and female anatomy; and, in catching these, they found themselves looking to literal meanings. After further discussion, they decided that the poem, however witty, makes a serious point about the blindness of lust. To get at this point, students may be asked to sum up the contrast Graves is drawing between Love and Beauty and the wanton's approach to them.

The title (and opening line) echo a phrase from Shakespeare in a passage about eels being rolled into a pie (*King Lear,* II, iv, 118–23):

> Lear: O me, my heart, my rising heart! But down!
> Fool: Cry to it, nuncle, as the cockney did to the eels when she put 'em i' th' paste alive. She knapped 'em o' th' coxcombs with a stick and cried, "Down, wantons, down!" 'Twas her brother that, in pure kindness to his horse, buttered his hay.

One instructor at a community college in New Jersey has reported an embarrassing experience. One morning, not having had time to prepare for class, he introduced this poem without having read it first. "What's it about?" he queried, and someone in the class replied, "An erection." "WHAT?" he exploded. "Come on, now, let's look at it *closely*. . . . " But as he stared at the poem before him, a chill stole over him. Luckily, he was saved by the bell.

Peter Davison, THE LAST WORD, *page 539*

The tangible side of Davison's central metaphor—that to part with a lover is to chop off her head—is plainly enforced by lines 9–13: painful physical actions that end with the *k*-sound of an abrupt crack in *nick . . . creak . . . block*.

Bruce Guernsey, GLOVE, *page 539*

Unless students pay close attention to the words on the page, this little poem—a charmer—will cruise right by them. The first stanza plays with the distinction between words and things. The word *glove*, like the thing it stands for, shelters something inside it: the word *love* like a hand made of bone and blood. So "the form that warms your lovely hand" can refer either to a *glove* or to the word *glove*. The second stanza might be read in at least two ways: (1) I accept your love but don't possess your body; therefore, my love is empty; or (2), in a way that makes more sense to us, if I were to take your love but ignore your body, which I am not about to do, that would be meaningless, for what good is the word without the thing? The theme, in either case, seems the same: Love depends on the flesh. Compare Yeats's "Crazy Jane Talks with the Bishop" (page 901).

Like a glove, this poem has a definite form (iambic dimeter).

Bruce Guernsey, who teaches at Eastern Illinois University, has published a collection, *January Thaw* (U of Pittsburgh P, 1982) and more recently a chapbook, *The Invention of the Telephone* (Stormline Press, Box 593, Urbana, IL 61801; 1987), from which we take "Glove."

John Donne, BATTER MY HEART, THREE-PERSONED GOD, FOR YOU, page 540

On Donne's last line: the literature of mysticism is full of accounts of spiritual experience seen in physical terms; and any students who wish to pursue the matter might be directed, for instance, to the poems of St. John of the Cross (which have been splendidly translated by John Frederick Nims).

John E. Parish has shown that Donne's poem incorporates two metaphors, both worn and familiar: the traditional Christian comparison of the soul to a maiden and Christ to a bridegroom, and the Petrarchan conceit of the reluctant woman as a castle and her lover as an invading army. Donne brilliantly combined the two into a new whole. In lines 1–4, the sinner's heart is like a walled town, fallen to Satan, the enemy. Now God the rightful King approaches and knocks for entrance. But merely to knock won't do—the King must break open the gates with a battering ram. The verbs in these lines all suggest the act of storming a citadel, "and even *blowe* may be intended to suggest the use of gunpowder to blow up the fortress" ("No. 14 of Donne's *Holy Sonnets*," *College English* 24 [Jan. 1963]: 299–302).

"The paradox of death and rebirth, the central paradox of Christianity" is (according to A. L. Clements in another comment) the organizing principle of the poem. To illustrate the paradox of destroying in order to revive, Donne employs two sorts of figurative language: one, military and destructive; the other, marital and uniting ("Donne's 'Holy Sonnet XIV,'" *Modern Language Notes* 76 [June 1961]: 484–89).

Both the Clements and the Parish articles are reprinted, together with the four other discussions of the poem, in *John Donne's Poetry*, edited by Clements (New York: Norton, 1966).

It is hard to talk for long about rhythm in poetry without citing the opening lines of "Batter my heart." Both in meter and in meaning, they must be among the most powerful lines in English poetry.

THE VALUE OF A DICTIONARY

Richard Wilbur, IN THE ELEGY SEASON, page 542

Rich with imagery, this early Wilbur poem makes a revealing companion piece to Keats's "To Autumn" (page 838). But unlike Keats, the speaker in Wilbur's poem accepts the season neither with mind (which gazes backward, remembering summer) nor with body (which strains ahead, longing for spring). The poem is also wealthy in allusions. Perhaps the "boundless backward of the eyes" echoes Shakespeare's *Tempest*: "the dark backward and abysm of time." The goddess heard climbing the stair from the underworld is Persephone. We are indebted to Donald Hill's reading of the poem in his study *Richard Wilbur* (New York: Twayne, 1967).

A brief glossary of etymologies:

potpourri: rotten pot (denotation: an incongruous mixture)

revenance: a return (denotation: the return of a spirit after death)

circumstance: condition that surrounds

inspiration: a breathing in

conceptual: taking in (denotations: perceived, apprehended, imagined)

commotion: co-motion, moving together (a wonderful word for what a bird's wings do!)

cordial: pertaining to the heart (*cor* in Latin) (denotations: friendly, stimulating)

azure: lapis lazuli

(text pages) 543–544

Exercise: *Catching Allusions*, page 543

J. V. Cunningham, Friend, on this scaffold Thomas More lies dead, page 543

Herman Melville, The Portent, page 543

Lucille Clifton, winnie song, page 544

Laurence Perrine, Janus, page 544

Cunningham's epigram states a metaphor: it likens two famous separations decreed by Henry VIII. Separation of the Body (the Church of England) from the Head (the Pope) is like the decapitation of More, who had opposed it. A possible original for Cunningham's epigram, a Latin epigram by John Owen (1606), has been discovered by Charles Clay Doyle:

> Abscindi passus caput est a corpore Morus;
>   Abscindi corpus noluit a capite.

In 1659 Thomas Pecke rendered it into English:

> What though Head was from Body severed?
> *More* would not let Body be cut from Head.

Doyle remarks that in fact More played down the role of the Pope as "head" of the Church, preferring the allegorical view (derived from Paul) of Christ as head upon the Church's body ("The Hair and Beard of Thomas More," *Moreana* 18, 71–72 [Nov. 1981]: 5–14).

Melville's symbolic poem also refers to an execution: the hanging of John Brown on 2 December 1859, for seizing the arsenal at Harpers Ferry, where the Shenandoah River meets the Potomac. Captured by Robert E. Lee, Brown suffered a wounded scalp, concealed by a cap he wore to the gallows. Melville, seeing the swinging corpse with its long beard as a comet with a streaming tail, refers to the ancient belief that comets and meteors are omens of war or catastrophe—as students may recall from Shakespeare's *Julius Caesar*. Melville's recognition of the portent proved right, of course: Union troops were soon to go into battle singing "John Brown's Body."

Lucille Clifton alludes, of course, to Winnie Mandela, wife of long-imprisoned South African black leader Nelson Mandela, freed in 1990, who withstood persecution and harassment during her husband's incarceration—including a fire that leveled her home. Line 10 puns on Mandela/Mandala and so alludes to a Hindu or Buddhist graphic symbol for the universe.

In "Janus," this two-faced husband is, to be sure, named for the Roman god who looks in opposite directions at once. Perrine, poet and textbook author, taught English at Southern Methodist University from 1946 until his retirement in 1981. This epigram appeared in *Poetry* for June 1984.

John Clare, Mouse's Nest, page 544

The connection between the final couplet and the rest of the poem is one of metaphor. Small trickles of water that "scarce could run" are newborn mice; "broad old cesspools," their mother.

Milton Klonsky has praised the poem in his anthology of graphic and pictorial poetry, *Speaking Pictures* (New York: Harmony, 1975). He admires "the cinematic flow of Clare's imagery, with each picture flashing by to be replaced by the next before its own afterimage has completely faded." This comment might be discussed—do students agree that Clare's poem seems cinematic and contemporary?

A few facts of Clare's heartbreaking life might interest students. Born into grinding poverty, the son of a field laborer in Northamptonshire, Clare enjoyed brief fame for his

(text pages) 544–551

Poems Descriptive of Rural Life (1820). Lionized by Coleridge and other London literati as an untutored genius, he was then forgotten. The latter half of his life was spent in lunatic asylums, where he wrote some remarkable lyrics and (under the delusion that he was Lord Byron) a continuation of *Don Juan*. Theodore Roethke, whose work shows a similar delight in close-up views of living creatures, has paid tribute (in his poem "Heard in a Violent Ward") to "that sweet man, John Clare."

WORD CHOICE AND WORD ORDER

An exercise to make a class more aware of *le mot juste* is suggested by W. Jackson Bate and David Perkins in *British & American Poets* (San Diego: Harcourt, 1986). Print out several lines of a poem, with an admirably chosen word or words left out. Let students suggest ways to fill in the blank, and debate their choices. Then the instructor whips out a trump card: the way the poet filled in the blank—if you're lucky, to "a collective sigh of appreciation."

Josephine Miles, REASON, page 547

Only the reference to Gary Cooper's horse at all dates this concise story-poem, a thing of lasting freshness. "The real characters in this story are the fragments of slang, not the speakers," Lawrence R. Smith has noticed. "Their absence is emphasized by the absence of personal pronouns at the beginning of each line. In place of the pronouns, we simply have 'Said.' This is a poem of pure language" ("Josephine Miles: Metaphysician of the Irrational," *A Book of Rereadings*, ed. Greg Kuzma [Lincoln, Pebble and Best Cellar, 1979] 29).

Hugh MacDiarmid, WHEESHT, WHEESHT, page 548

MacDiarmid, the most eminent twentieth-century poet to write in Scots, is often compared with Burns; and for another brief love lyric at least partly in dialect, students might be asked to see "John Anderson my jo, John."

Emma Lee Warrior, HOW I CAME TO HAVE A MAN'S NAME, page 550

No wonder the family didn't change her name, after Yellow Dust's wonderful prayer, after all they endured to reach the hospital. Even if the student struggles over them, the Blackfoot words lend the poem some splendid sounds, not to mention a ring of authenticity.

Why is the poet's name Emma Lee Warrior? Why isn't it Ipisowahs Warrior, or Emma Lee Ipisowahs? It seems she has a Blackfoot name (Ipisowahs—a name for a warrior) and uses her English name as an alternate.

A Peigan Indian born in Brocket, Alberta, Emma Lee Warrior went on to earn her master's degree in English; she has been working on the Blackfoot Reserve in Alberta developing curricula in the Blackfoot language. This poem was first published in *Harper's Anthology of 20th Century Native American Poetry*, ed. Duane Niatum (New York: Harper & Row, 1988).

Thomas Hardy, THE RUINED MAID, page 551

In a London street, an innocent girl from Dorset encounters a friend who has run away from life on the farm. Now a well-paid prostitute, 'Melia calls herself *ruined* with cheerful irony. That this maid has been made, it would seem, has been the making of her. Hardy, of course, is probably less stricken with awe before 'Melia's glamorous clothes than is the first speaker. As the *ain't* in the last line indicates, 'Melia's citified polish doesn't go deep.

For a sequel to "The Ruined Maid," see "A Daughter Returns" in Hardy's last collection of poetry, *Winter Words*. With "Dainty-cut raiment" and "earrings of pearl," a

115

(text pages) 551–554

runaway daughter returns to her country home, only to be spurned by her father for having lost her innocence.

Richard Eberhart, THE FURY OF AERIAL BOMBARDMENT, page 552

Dr. Johnson said that technical language is inadmissible to poetry, but in the case of Eberhart's poem, it is hard to agree. We do not need to know the referents of "belt feed lover" and "belt holding pawl" in order to catch the poet's meaning. Indeed, he evidently chooses these terms as specimens of a jargon barely comprehensible to the unlucky gunnery students to who failed to master it. At a reading of his poems in public, Eberhart once remarked that he had added the last stanza as an afterthought. The tone (it seems to us) remains troubled and sorrowful but shifts from loftiness and grandeur to matter-of-fact. This shift takes place in diction as well: from the generality of "infinite spaces," "multitudinous will," "eternal truth," and "the Beast" in man's soul down to "Names on a list," "lever," and "pawl." The poem is a wonderful instance of a poet's writing himself into a fix—getting snarled in unanswerable questions—and then triumphantly saving the day (and his poem) by suddenly returning with a bump to the ordinary, particular world.

Wole Soyinka, LOST TRIBE, page 553

Soyinka finds us taking our language from merchants, advertisers, and television. Thus the stress on imperatives: "Enjoy your meal" is supposed to project firm decision and sincerity, while to make a wish—"I hope you'll enjoy your meal"—seems to us wishy-washy. No wonder our trite phrases are counterfeit coins, he suggests in lines 7–8: we let private enterprise mint them for us.

Americans, in his view, wander about "in search of lost community"—like the Lost Tribes of Israel. In the Biblical account, there were thirteen Hebrew tribes. The tribes of Judah, Benjamin, and Levi became the southern kingdom of Judah; the other ten tribes formed a northern kingdom of Israel. In the eighth century B.C. these ten were conquered and their people transported to Assyria. Since known as the Lost Tribes of Israel, their subsequent fate has incited never-ending speculation. Various peoples have been identified with the Lost Tribes; sarcastically, Soyinka nominates Americans to be among them.

Novelist, poet, playwright, and critic, Wole Soyinka was born in a small town in western Nigeria. Educated at government schools and at the University of Leeds, he writes in English. In 1986 he was awarded the Nobel Prize for Literature. Currently, Soyinka is Professor of Comparative Literature at the University of Ife, Nigeria. This poem comes from a series, "New York, U.S.A.", from *Mandela's Earth and Other Poems* (New York: Random House, 1988).

FOR REVIEW AND FURTHER STUDY

David B. Axelrod, ONCE IN A WHILE A PROTEST POEM, page 553

The crucial word in this disturbing poem is *silicone*, a substance injected into flat bosoms to make them buxom, in order to put up a false front—like those sympathies we merely pretend to feel. Carefully cropped by someone in an advertising agency, the breast of the starving woman (like a breast treated with silicone) is artificially changed and becomes abstract. Although the photograph is supposed to rouse our sympathies (and our contributions), it often has the opposite effect of making us callous. To the poet, it seems meant to "toughen us" (as silicone toughens the bosoms of Playboy bunnies?); it seems meant to "teach us to ignore."

Lewis Carroll, JABBERWOCKY, page 554

"Jabberwocky" has to be heard aloud: you might ask a student to read it, alerting him or her in advance to prepare it, and offering tips on pronunciation. ("The *i* in *slithy* is like the *i* in *slime*; the *a* in *wabe*, like the *a* in *wave*.")

Although Carroll added *chortled* to the dictionary, not all his odd words are invented. *Gyre* of course means "to spin or twist about"—it is used as a noun in Yeats's "Sailing to Byzantium" (page 753) and the "The Second Coming" (page 718). *Slithy* (sleazy or slovenly), *rath* (an earthen wall), *whiffling* (blowing or puffing), and *callooh* (an arctic duck that winters in Scotland, so named for its call) are legitimate words, too, but Carroll uses them in different senses. *Frabjous* probably owes something to *frab*, a dialect word meaning "to scold, harass, or nag"—as Myra Cohn Livingston points out in her anthology *O Frabjous Day!* (New York: Atheneum, 1977).

Writing in 1877 to a child who had inquired what the strange word meant, Carroll replied:

> I am afraid I can't explain "vorpal blade" for you—nor yet "tulgey wood"; but I did make an explanation once for "uffish thought"—it seems to suggest a state of mind when the voice is gruffish, the manner roughish, and the temper huffish. Then again, as to "burble": if you take the three verbs "bleat," "murmur" and "warble," and select the bits I have underlined, it certainly makes "burble": though I am afraid I can't distinctly remember having made it that way.

(*Uffish* suggests *oafish* too.)

Students can have fun unpacking other portmanteau words: *gimble* (*gamely, gambol*); *frumious* (which Carroll said is *fuming* plus *furious*); *vorpal* (*voracious, purple*), *galumphing* (*galloping in triumph*), and so on. Some of these suggestions come from Martin Gardner, who supplies copious notes on the poem (as well as translations of it into French and German) in *The Annotated Alice* (New York: Bramhall, 1960).

Wallace Stevens, METAMORPHOSIS, page 555

A possible meaning of *metamorphosis* is a sudden change undergone by an insect or a tadpole in the process of maturing. But we think Stevens implies by the falling sky (dead and lying with the worms) and the hanged street lamps that the year is decaying, not changing for the better, but undergoing (as *Webster's New World Dictionary* puts it) "a transformation, especially by magic or sorcery."

Language is, in a sense, also decaying rapidly. The name of the month of October is transformed, too, into a birdcall.

Niz - nil - imbo is a corruption of *November*. It contains the words *nil* (nothingness) and *limbo*.

The first definition: "a sudden twist, turn, or stroke." The season makes such a sudden turn, and so does language. A worm is a "pretty quirk" in that it is physically twisted; also, it happens to appear unexpectedly (in flesh suddenly dead, or fall fruit become overripe.)

E. E. Cummings, ANYONE LIVED IN A PRETTY HOW TOWN, page 556

Trained in the classical languages, Cummings borrows from Latin the freedom to place a word in practically any location within a sentence. The first two lines are easy to unscramble: "How pretty a town Anyone lived in, with so many bells floating up [and] down." The scrambling is artful, and pedestrian words call attention to themselves by being seen in an unusual order.

The hero and heroine of the poem are anyone and noone, whose names recall the pronoun-designated principals in Cummings's play *Him*—hero Him and heroine Me. Are they Everyman and Everywoman? Not at all: they're different, they're strong, loving individuals whom the poet contrasts with those drab women and men of line 5, "both little and small," who dully sow *isn't* (negation) and reap *same* (conformity). Unlike wise noone and anyone, the everyones of line 17 apparently think they're really somebody.

In tracing the history of anyone and noone from childhood through their mature love to their death and burial, Cummings, we think, gives a brief tour through life in much the way that Thornton Wilder does in *Our Town*. But not all readers will agree. R. C. Walsh thinks that, in the last two stanzas, anyone and noone do not literally die but grow into loveless and lifeless adults, whose only hope of rejuvenation is to have children (*Explicator* 22 [May 1964]: item 72). But it seems unlike Cummings to make turncoats of his individualists. Bounded by the passage of the seasons, the rain, and the heavens, the mortal lives of anyone and noone seem concluded in their burial. But in the next-to-last stanza they go on sleeping in love and faith, dreaming of their resurrection.

EXERCISE: *Different Kinds of English*, page 557

Anonymous, CARNATION MILK, page 557

A. R. Ammons, SPRING COMING, page 557

William Wordsworth, MY HEART LEAPS UP WHEN I BEHOLD, page 558

William Wordsworth, MUTABILITY, page 558

Anonymous, SCOTTSBORO, page 559

Students won't need much help to see that "Carnation Milk" is unschooled speech; that Ammons combines technical terms with colloquial speech (*nice*); that Wordsworth's diction in "My heart leaps up" is plain and unbookish (except for *natural piety*), while his language in "Mutability" is highly formal—not only in diction, but in word order ("Truth fails not"); and that "Scottsboro" is a song in the speech of a particular culture (and, by the way, wonderful in its power to express).

About A. R. Ammons: Compare Eberhart's "The Fury of Aerial Bombardment" (page 552) in its use of terms from science. Try applying to Ammons's poem the observation by Samuel Johnson (referred to in question 2 under Eberhart). What might Johnson have thought of "Spring Coming"? (Probably: "Sir, Ammons has written in no language, and no man can utter it.")

On the difficult "Mutability" (in case anyone cares to read it for its sense): "the tower sublime" may refer to the Bastille, suggests Geoffrey Durant in his excellent discussion of the poem in *Wordsworth and the Great System* ([Cambridge: Cambridge UP, 1970] 82–85). For other poems with the theme of mutability, see Shelley's "Ozymandias" (page 756), Shakespeare's "That time of year . . . " (page 871), Auden's "As I Walked Out One Evening" (page 781), Thomas's "Fern Hill" (page 886), and (in this same chapter) Cummings's "anyone lived in a pretty how town."

Repercussions from the Scottsboro case lasted long. In October, 1976, the state of Alabama finally granted full pardon to Clarence Norris, last survivor of the nine "Scottsboro boys," after he had spent sixteen years in prison, five on death row, and thirty-one years as an escaped fugitive. In 1976, following the televised showing of a dramatization, "Judge Horton and the Scottsboro Boys," both alleged rape victims unsuccessfully brought suit against NBC for libel, slander, and invasion of privacy. The last of these suits was dismissed in July, 1977. The last of the Scottsboro boys, Clarence Norris, died on January 23, 1989. Norris spent fifteen years in prison and outlived three death sentences until, in 1976 when he was 64, the Alabama Parole Board unanimously found him innocent. Harper Lee imagines a similar case in Alabama in her novel *To Kill a Mockingbird* (1960), a book which some of your students may know—it is still on some high school reading lists.

16 Saying and Suggesting

John Masefield, Cargoes, page 561

Much of the effect of Masefield's contrast depends on rhythms and word-sounds, not just on connotations. In stanza 2, the poet strews his lines with dactyls, producing ripples in his rhythm: *diamonds, emeralds, amethysts, cinnamon.* In the third stanza, paired monosyllables (*salt-caked, smoke stack, Tyne coal, roadrails, pig-lead, firewood*) make for a hard hitting series of spondees. Internal alliteration helps the contrast, too: all those *m*-sounds in the dactyls; and in the harsher lines "Dirty British coaster with a salt-caked smoke, stack, / Butting," all the sounds of the *r,* the *t,* and the staccato *k.*

"Cargoes" abounds with lively, meaningful music—and yet Masefield is generally dismissed nowadays as a mere balladeer—a jog-trot chronicler of the lives of the poor and unfortunate. In naming him poet laureate, George V (it is said) mistakenly thought him a hero of the working class; and, unluckily for his later fame, Masefield, like Wordsworth, enjoyed a long senility.

William Blake, London, page 562

Blake at first wrote, "I wander through each dirty street, / Near where the dirty Thames does flow." For his equally masterful revisions of the poem's closing lines, see page 730 in Chapter 26, "Alternatives."

Tom Dacre's dream has a basis in reality: in Blake's time, sweeps were often sent up chimneys naked, the better to climb through narrow spaces (and thus saving the expense of protective clothing). Martin K. Nurmi points out this fact in his essay, "Fact and Symbol in 'The Chimney Sweeper' of Blake's Songs of Innocence" (*Bulletin of the New York Public Library* 68 [April 1964]: 249–56). "Naked immersion in soot, therefore, is Tom's normal state now, and naked white cleanliness is its natural opposite."

If Blake were to walk the streets of an American city today, would he find any conditions similar to those he finds in "London"? Is this poem merely an occasional poem, with a protest valid only for its time, or does it have enduring applications?

Wallace Stevens, Disillusionment of Ten O'Clock, page 564

Stevens slings colors with the verve of a Matisse. In this early poem, he paints a suggestive contrast between the pale and colorless homeowners, ghostlike and punctually going to bed at ten and, on the other hand, the dreams they wouldn't dream of dreaming; and the bizarre and exotic scene inside the drunken head of our disreputable hero, the old seafarer. Who in the world would wear a beaded sash or *ceinture?* (A Barbary pirate? An Arabian harem dancer)? Ronald Sukenick has made a terse statement of the poem's theme: "the vividness of the imagination in the dullness of a pallid reality" (*Wallace Stevens: Musing the Obscure* [New York: New York UP, 1967]). Another critic, Edward Kessler, has offered a good paraphrase: "Only the drunkard, the irrational man ('Poetry must be irrational' [*Opus Posthumous* 162]), who is in touch with the unconscious—represented here, and often elsewhere, by the sea—can awake his own passionate nature until his blood is mirrored in the very weather" (*Images of Wallace Stevens* [New Brunswick: Rutgers UP, 1972]).

While they will need to see the contrast between pallor and color, students might be cautioned against lending every color a particular meaning, as if the poem were an allegory.

(text pages) 564–566

Stevens expressed further disappointment with monotonous neighbors in a later poem, "Loneliness in Jersey City," which seems a companion piece to this. In Jersey City, "the steeples are empty and so are the people," who can't tell a dachshund from a deer. Both poems probably owe some of their imagery to Stevens's days as a struggling young lawyer, living in rooming houses in East Orange, New Jersey, and Fordham Heights, in New York City.

Gwendolyn Brooks, THE BEAN EATERS, page 565

This spare, understated poem seems one of the poet's finest portraits of poor blacks in Chicago. While "We Real Cool" (page 646) depicts the young, "The Bean Eaters" depicts the old, whose lives are mainly devoted to memories—some happy (*tinklings*, like the sound of their beads) and some painful (*twinges*).

Details are clearly suggestive. The old people are barely eking out a living, eating beans to save money, from *chipware* (chipped tableware, or tableware that chips easily?) set on a creaky old table, using cheap cutlery that doesn't shine.

The long last line, with its extended list, sounds like one of those interminable Ogden Nash lines that Lewis Turco has dubbed Nashers. Why is its effect touching, rather than humorous? Brooks lists the poor couple's cherished, worthless possessions in detail, and each detail matters. Besides, *twinges / fringes*, a fresh rime, achieves a beautiful closure. We end up sharing the poet's respect and affection for this old pair—no mockery.

Timothy Steele, EPITAPH, page 565

"Silence is golden"—but Sir Tact is obviously a coward, afraid to speak his mind. This epigram is included in Steele's first collection of poems, *Uncertainties and Rest* (Baton Rouge: Louisiana State UP, 1979).

Geoffrey Hill, MERLIN, page 565

There is an incantatory quality to this poem that seems to indicate it is Merlin speaking; but then, nearly all of Hill's richly suggestive and highly formal poems tend to sound this way. The dead who might "come together to be fed" recall the souls encountered by Odysseus in the underworld, to whom he fed blood. Once, the towers of Camelot sheltered Arthur and his associates; now, only the pointed tents of piled-up cornstalks (or perhaps growing cornstalks, flying their silklike pennants) stand over the city of the dead.

Wallace Stevens, THE EMPEROR OF ICE-CREAM, page 566

Choosing this poem to represent him in an anthology, Stevens once remarked, "This wears a deliberately commonplace costume, and yet seems to me to contain something of the essential gaudiness of poetry; that is the reason why I like it." (His statement appears in *Fifty Poets: An American Auto-Anthology* ed. William Rose Benet [New York: Diffield, 1933].)

Some students will at once relish the poet's humor, others may discover it in class discussion. Try to gather the literal facts of the situation before getting into the poem's suggestions. The wake or funeral of a poor old woman is taking place in her own home. The funeral flowers come in old newspapers, not in florists' fancy wrappings; the mourners don't dress up, but wear their usual street clothes; the refreshments aren't catered but are whipped up in the kitchen by a neighbor, a cigar-roller. Like ice-cream, the refreshments are a dairy product. Nowadays they would probably be a sour cream chip-dip; perhaps in 1923 they were blocks of Philadelphia cream cheese squashed into cups for spreading on soda crackers. To a correspondent, Stevens wrote that *fantails* refers not to fans but to fantail pigeons (*Letters* [New York: Knopf, 1966] 341). Such embroidery seems a lowbrow pursuit: the poor old woman's pathetic aspiration toward beauty. *Deal* furniture is cheap.

Everything points to a run-down neighborhood, and to a woman about whose passing nobody very much cares.

Who is the Emperor? The usual guess is Death. Some students will probably see that the Emperor and the muscular cigar-roller (with his creamy curds) suggest each other. (Stevens does not say that they are identical.) Ice cream suggests the chill of the grave— and what besides? Today, some of its connotations will be commonplace: supermarkets, Howard Johnson's. To the generation of Stevens, ice cream must have meant more: something luxurious and scarce, costly, hard-to-keep, requiring quick consumption. Other present-day connotations may come to mind: sweetness, deliciousness, childhood pleasure. Stevens's personal view of the ice cream in the poem was positive: "The true sense of Let be be finale of seem is let being become the conclusion or denouement of appearing to be: in short, ice cream is an absolute good" (*Letters* 341). An absolute good! The statement is worth quoting to students who have doubts about the poet's attitude toward ice cream—as did an executive of the Amalgamated Ice Cream Association, who once wrote to the poet in perplexity (see *Letters* 501–2). If ice cream recalls sweet death, still (like curds) it also contains hints of mother's milk, life, and vitality.

On a visit to Mount Holyoke, XJK was told that, as part of an annual celebration, it is customary for the trustees and the seniors to serve ice cream (in Dixie cups) to the freshman class at the grave of Mary Lyon, founder of the college. In a flash he remembered Stevens's poem, and embraced Jung's theory of archetypes.

Walter De La Mare, THE LISTENERS, page 567

This much-loved old chestnut, once a favorite of anthologists, still seems a wonderful demonstration of the value to poetry of hints. The identity of the listeners is by no means clear: the more literal-minded will probably think of bats, mice, and crickets, while others will think of ghosts. The latter theory gains support from the poem: these listeners are *phantom* (line 13), *not of the world of men* (16), strange and mute (21–22). If the Traveller is "the one man left awake" (32), are the Listeners men who have fallen asleep (in death)?

In line 5, *turret* suggests a fortress or castle; but as line 14 makes clear, the scene is a house. Which to believe? Perhaps the scene is set in a kind of Loire Valley chateau: a mansion with castlelike touches.

Attempts to guess what happened before "The Listeners" opens may well be irrelevant, but if students will try, they may find themselves more deeply involved with the poem. This much seems clear: The Traveller has accepted some challenge to visit the house; he has given his promise to someone (more than one person, and someone other than the Listeners, whom the Traveller charges to convey his message to "them"). Perhaps this act is one of the deeds required to lift a curse from a kingdom, as in a fairy tale; or as in the Arthurian story of Sir Percival (or Parsifal), who must spend a night in the terrifying Chapel Perilous.

Robert Frost, FIRE AND ICE, page 568

In his first line, Frost probably refers to those who accept the Biblical prophecy of a final holocaust; and in his second line, to those who accept scientists' forecasts of the cooling of the earth. We admire that final *suffice*. A magnificent understatement, it further shows the power of a rime to close a poem (as Yeats said) with a click like a closing box.

17 Imagery

Ezra Pound, IN A STATION OF THE METRO, page 569

Pound recalled that at first this poem had come to him "not in speech, but in title splotches of color." His account is reprinted by K. K. Ruthven in *A Guide to Ezra Pound's Personae (1926)* (Berkeley: U of California P, 1969). Students might like to compare this "hokku-like sentence" (as Pound called the poem) with the more suggestive Japanese haiku freely translated in Chapter Five.

For a computer-assisted tribute to this famous poem, see the curious work of James Laughlin and Hugh Kenner, reported in "The Mixpoem Program," *Paris Review* 94 (Winter 1984): 193–98. Following Laughlin's suggestion that the five nouns of "In a Station of the Metro" might interestingly be shuffled, Kenner wrote "a Little program in Basic" that enabled a computer to grind out 120 scrambled versions of the poem, including these:

> The apparitions of these boughs in the face;
> Crowds on a wet, black petal.

> The crowd of these apparitions in the petal;
> Faces on a wet, black bough.

Kenner then wrote a program in Pascal that would shuffle eight words and produce 40,320 different versions. We don't know what it all demonstrates, except that Pound's original version still seems the best possible.

Taniguchi Buson, THE PIERCING CHILL I FEEL, page 569

Harold G. Henderson, who translates this haiku, has written a good terse primer in *An Introduction to Haiku* (Garden City: Anchor, 1958). Most of Henderson's English versions of haiku rime like this one; still, the sense of the originals (as far as an ignorant reader can tell from Henderson's glosses) does not seem greatly distorted.

T. S. Eliot, THE WINTER EVENING SETTLES DOWN, page 571

This is the first of the series of four poems called "Preludes," originally published in the July 1915 issue of Wyndham Lewis's *Blast*. It was written during Eliot's days at Harvard. The "Preludes," writes Grover Smith in *T. S. Eliot's Poetry & Plays* (Chicago: U of Chicago P, 1965), belong to the era of "Prufrock." Of "The winter evening settles down," Smith says:

> The first "Prelude" begins with winter nightfall in an urban back street; from indoor gloom and the confined odor of cooking it moves outside into the smoky twilight where gusts of wind whip up leaves and soiled papers, and a shower spatters the housetops. Such adjectives as "burnt-out," "smoky," "grimy," "withered," "vacant," "broken," and "lonely" carry the tone.

Some students may point out, though, that the lighting of the lamps seems to end the poem on a note of tranquillity.

Theodore Roethke, ROOT CELLAR, page 571

Probably there is little point in spending much time dividing imagery into touches and tastes and smells; perhaps it will be enough to point out that Roethke's knowledgeable poem isn't all picture-imagery. There's that wonderful "congress of stinks," and the "slippery planks" are both tactile and visual. Most of the language in the poem is figurative, most of the vegetation is rendered animal: bulbs like small rodents, shoots like penises, roots like a forgotten can of fishing worms. Roethke doesn't call the roots lovely, but obviously he admires their tough, persistent life.

Elizabeth Bishop, THE FISH, page 572

This poem is made almost entirely of concrete imagery. Except for *wisdom* (line 63) and *victory* (66), there is no very abstract diction in it.

Obviously the speaker admires this stout old fighter. The image "medals with their ribbons" (line 61) suggests that he is an old soldier, and the "five-haired beard of wisdom" (line 63) suggests that he is a venerable patriarch, of whom one might seek advice.

The poor, battered boat has become magnificent for having the fish in it. The feeling in these lines is joy: bilge, rust, and cracked thwarts are suddenly revealed to be beautiful. In a way, the attitude seems close to that in Yeats's "Sailing to Byzantium" (page 753), in which the triumphant soul is one that claps its hands and louder sings for every tatter in its mortal dress. The note or final triumph is sounded in "rainbow, rainbow, rainbow!" (line 75). The connotations of *rainbow* in this poem are not very different from the connotations often given the word by misty-eyed romantic poets such as Rod McKuen, but we believe Bishop because of her absolutely hard-eyed and specific view of the physical world. (She even sees the fish with X-ray imagination in lines 27–33.)

Anne Stevenson says in *Elizabeth Bishop* (New York: Twayne, 1966):

> It is a testimony to Miss Bishop's strength and sensitivity that the end, the revelation or "moment of truth," is described with the same attention to detail as the rest of the poem. The temptation might have been to float off into an airy apotheosis, but Miss Bishop stays right in the boat with the engine and the bailer. Because she does so, she is able to use words like "victory" and "rainbow" without fear of triteness.

Because the fish has provided her with an enormous understanding, the speaker's letting it go at the end seems as act of homage and gratitude.

Compare "The Fish" with the same poet's richly imaged "Filling Station" (page 785).

The poet reads this poem on a recording, *The Spoken Arts Treasury of 100 Modern American Poets*, vol. 10, SA 1049.

Oscar Wilde, SYMPHONY IN YELLOW, page 574

"Symphony" in the title suggests music, but the word also denotes "a consonance or harmony of color (as in a painting)"—*Webster's Collegiate*. The poem arrays yellow things and, at last, adds the green Thames for relief.

Wilde's technique in this poem is to render the unnatural natural and vice-versa: a bus is like a butterfly, fog is like a scarf, the Thames is like a jade rod.

Some students may think that Wilde's reference to the Temple (site of the Inns of Court, housing London's law societies) indicates that the poem takes place in Bali or some such exotic, templed land. (But where is the Thames?)

An Australian magazine first printed "Symphony in Yellow," in 1889. Later, it was reported that Wilde made some disparaging remarks about the land Down Under: "It is the abode of anthropophagi, the abode of lost souls, whither criminals are transported to wear a horrible yellow livery. Even they are called 'canaries.' So I have written for them a Symphony in Yellow—they will feel the homely touch. I rhyme 'elms' with 'Thames.' It is a venial offense in comparison with theirs. A symphony with sympathy—how sweet! Suppose I were to add a stanza: And far in the Antipodes / Where swelling suns have sunk to rest / A convict to his yellow breast / Shall hug my yellow melodies." (Quoted by Richard Ellmann in *Oscar Wilde* [New York: Knopf, 1988] 207.)

(text pages) 575–577

John Haines, WINTER NEWS, page 575

We are struck by "the stiffening dogs" in this poem. It is possible both to picture such stiffening and to feel it in one's muscles and bones. Haines appeals mostly to the sense of sight ("clouds of steaming breath," "the white- / haired children") and sound ("Oil tins bang," "the voice of the snowman"). Clearly the children's hair, far from manifesting premature aging, is merely covered with snow. Is that snowman a surreal monster, or is his voice another name for the wind?

In 1947 John Haines went to Alaska as a homesteader, and began to write poetry there. This is the title poem from his first collection, *Winter News* (Middletown: Wesleyan UP, 1966).

Emily Dickinson, A ROUTE OF EVANESCENCE, page 575

"A Route of Evanescence" will probably inspire a heated guessing contest. Contestants will need to pay attention to Dickinson's exact words.

Enclosing this poem in a letter to Thomas W. Higginson, Dickinson gave it a title: "A Humming-Bird."

The poet's report of the hummingbird's arrival from Tunis is fanciful. Besides, the creature could hardly fly 4,000 miles nonstop in one morning. New England hummingbirds don't need to cross the Atlantic; to find a warmer climate, they migrate south. If it was a ruby-throated hummingbird that the poet saw, though, it might indeed have come a long distance: from a winter in Mexico or Central America.

The poet's ornithology may be slightly cockeyed, but her imagery is accurate. Hummingbird wings appear to rotate, but they aren't seated in ball joints; in actuality, they merely flap fast.

Jean Toomer, REAPERS, page 576

This ominous poem, with its contrasts between sound and silence, possibly contains a metaphor. The black field hands are being destroyed by something indifferent and relentless, much as the trapped rat is slain under the blade. (Or, as in "Scottsboro" [page 559], as a cat stalks a "nohole mouse"?)

A grandson of P. B. S. Pinchback, the black who served for a short time during Reconstruction as acting governor of Louisiana, Toomer had only a brief public career as a writer. His one book, *Cane* (1923), which experimentally combined passages of fiction with poetry, helped to spearhead the Harlem Renaissance. "Reapers" is taken from it.

That Toomer was a man divided between his profound understanding of blacks and his own desire to pass for white emerges in a recent biography, *The Lives of Jean Toomer: A Hunger for Wholeness*, by Cynthia Earl Kerman and Richard Eldridge (Baton Rouge: Louisiana State UP, 1987). *The Collected Poems of Jean Toomer* (Chapel Hill: U of North Carolina P, 1988) is a slim volume of 55 poems, the best of them from *Cane*.

Gerard Manley Hopkins, PIED BEAUTY, page 576

Sumptuously rich in music (rime, alliteration, assonance), this brief poem demands to be read aloud.

Some students might agree with Robert Frost's objection that the poem "disappoints . . . by not keeping, short as it is, wholly to pied things" (1934 letter to his daughter Lesley in *Family Letters of Robert and Elinor Frost* [Albany: State U of New York P, 1972] 162). But, as question 4 tries to get at, Hopkins had more in mind than dappled surfaces. Rough paraphrase of the poem: God is to be praised not only for having created variegation, but for creating and sustaining contrasts and opposites. In lines 5–6, tradesmen's tools and gear, like the plow that pierces and cuts the soil, strike through the surfaces of raw materials to reveal inner beauty and order that had lain concealed.

For a convincing argument that Hopkins in "Pied Beauty," like Dickens in *Hard Times,* complains about a drab, mechanical, industrial-age uniformity in Victorian England, see Norman H. MacKenzie, *A Reader's Guide to Gerard Manley Hopkins* (Cornell UP, 1981) 85–86. Few students will crave to fathom the poet's notions of *instress* and *inscape,* but if you do, see John Pick's unsurpassed *Gerard Manley Hopkins, Priest and Poet,* 2nd ed. (Oxford UP, 1966) 53–56.

The point of question 5 is that if the images of the poem were subtracted, its statement of theme also would disappear.

Hopkins discovered the form of "pied Beauty" and called it the *curtal sonnet* (Curtal, riming with *turtle:* "crop-tailed"). But, remarks MacKenzie, such sonnets are like a small breed of horse: "compressed, not merely cut short." Instead of two quatrains, the form calls for two tercets; then, instead of a sestet four lines and brief line more. (Other curtal sonnets by Hopkins: "Peace" and, even more closely cropped, "Ashboughs.")

About Haiku

Basho's frogjump poem (page 578) may well be the most highly prized gem in Japanese literature: in Japanese there exists a three-volume commentary on it.

For an excellent discussion of the problems of teaching haiku, and of trying to write English ones, see Myra Cohn Livingston's *When You Are Alone / It Keeps You Capone: An Approach to Creative Writing with Children* (New York: Atheneum, 1973) 152–62. Livingston finds it useful to tell students a famous anecdote. Kikaku, a pupil of Basho, once presented his master with this specimen:

Red dragonflies—
Tear off their wings
And you have pepper pods.

As a haiku, said Basho, that's no good. Make it instead:

Red pepper pods—
Add wings
And you have dragonflies.

A moment of triumph, such as all teachers of poetry hope for but seldom realize, has been reported in a letter to XJK from Maurice F. Brown of Oakland University, Rochester, Michigan:

> Last year, teaching W. C. Williams in an "invitational" course for a week, I began with "Red Wheelbarrow" . . . and a student hand went up (class of 100): "That's not a poem! That's junk. What if I say, 'Here I sit looking at a blackboard while the sun is shining outside.' Is that a poem?" It was one of those great teaching moments . . . and I did a quick count and wrote it on the board:

Here I sit looking
　At a blackboard while the sun
is shining outside.

Not only a poem . . . a perfect haiku.

A thorough new guide to this rocky acre of poetry, by William J. Higginson with Penny Harter, is *The Haiku Handbook: How to Write, Share, and Teach Haiku* (New York: McGraw, 1985).

As this example of his work demonstrates, Nicholas A. Virgilio (1928–1989), of Camden, New Jersey, perhaps the most interesting contemporary writer of haiku in English, could give the classic Japanese form a touch of mean American city streets. His *Selected*

(text pages) 577–579

Haiku (2nd ed., 1988) gathers the best of his life's work ($9.95 from Firefly Books, Ltd., 3520 Pharmacy Ave., Unit 1C, Scarborough, Ontario, Canada M1W 2T8).

John Ridland, THE LAZY MAN'S HAIKU, page 578

Why is this haiku the work of a slothful soul? Because, according to the poet (in a letter), "he's too lazy to write the proper number of syllables in any line—or to get rid of that occidental end-rhyme à la Harold Henderson's ever-unconvincing translations in that old Anchor book." Henderson, in *An Introduction to Haiku* (Anchor Books, 1956), forced all the haiku to rime like his version from Buson given at the beginning of this chapter. (Reading it as a poem in English, we find it profoundly convincing, and suspect it took work.)

Richard Brautigan, HAIKU AMBULANCE, page 578

This is a Zen poem poking fun at overly thoughty attempts to write Zen poems. Its satire does not seem in the least applicable to the successful haiku-in-English of Paul Goodman, Gary Snyder, and Raymond Roseliep.

FOR REVIEW AND FURTHER STUDY

John Keats, BRIGHT STAR! WOULD I WERE STEADFAST AS THOU ART, page 579

Unlike Petrarchan poets, Keats isn't making the star into an abstraction (Love); he takes it for a visible celestial body, even though he sees it in terms of other things. His comparisons are so richly laden with suggestions (star as staring eye, waters as priestlike), that sometimes students don't notice his insistent negations. The hermit's all-night vigil is *not* what Keats desires. He wants the comfort of that ripening pillow, and (perhaps aware of his impending death) envies the cold star only its imperishability—oh, for unendurable ecstasy, indefinitely prolonged! Compare this to Keats's "To Autumn" (page 838) in which the poet finds virtue in change.

Many readers find the last five words of the poem bothersome. Students might be asked, Does Keats lose your sympathy by this ending? If so, why? If not, how would you defend it? We can't defend it; it seems bathetic, almost as self-indulgent as Shelley's lines in "Indian Serenade":

> Oh, lift me from the grass!
>   I die! I faint! I fail!
> Let thy love in kisses rain
>   On my lips and eyelids pale.

Thomas Mauch, of Colorado College, intelligently disagrees, and suggests how "or else swoon to death" may be defended. The *or*, he thinks, is what grammarians call an inclusive *or*, not an exclusive.

> I believe that the speaker is saying, not that if he can't be forever in the close company of the beloved he would rather be dead—sort of like what Patrick Henry said about liberty—but rather that, given the closeness to the woman, dying in that condition would be just as good as experiencing it forever, since in either case he would not undergo a separation from her (and still retain his consciousness of it). I think it is the same point he makes in the "Ode to a Nightingale":

> Now more than ever seems it rich to die,
>   To cease upon the midnight with no pain,
>   While thou art pouring forth thy soul abroad
>   In such an ecstasy!

The poem, Mr. Mauch concludes, illustrates the kind of closure that Keats admired when he affirmed that a poem should "die grandly."

Timothy Steele, WAITING FOR THE STORM, page 580

This brief poem evokes at least four sensory experiences: the sight of the "wrinkling" bay, the sensation of cold, the sensation of dampness, the sound of rain.

An interesting irony: an upturned boat gets its dry bottom wet. The phrase "a wrinkling darkness" invites thought.

The title of the poem establishes that something is about to happen. Images detail the steady ("moment by moment") approach of the storm: oncoming darkness, the deepening of the cold, the spread of damp air, and finally the arrival of the first rain.

Like most of Steele's poetry, "Waiting for the Storm" is rich in sound. For anyone who cares for musical language, the poem rewards close study. Besides the skilled employment of rime, there is much alliteration: the *k*-sounds in *wrinkling darkness,* *b*-sounds in the first stanza, *hull . . . head.* The lines abound in assonance: especially the *o*-sounds in *boat, moment by moment,* and *colder.* There is a near-rime in *sand* and *damp.*

Experiment: WRITING WITH IMAGES, page 580

To write a poem full of images, in any form, is probably easier for most students than to write a decent haiku. (On the difficulties of teaching haiku writing, see Myra Cohn Livingston, cited under "About Haiku.") Surprisingly, there is usually at least one student in every class who can't seem to criticize a poem to save his neck, yet who, if invited to be a poet, will bloom or at least bud.

Walt Whitman, THE RUNNER, page 580

Try reading "The Runner" without the adverbs *lightly* and *partially.* Does the poem even exist without those two delicate modifiers?

T. E. Hulme, IMAGE, page 580

Hulme's poems seem always to have been brief. In his own collection *Personae,* Ezra Pound took two pages to include "The Complete Poetical Works of T. E. Hulme" (in which "Image" does not appear). Pound remarked, "In publishing his *Complete Poetical Works* at thirty, Mr. Hulme has set an enviable example to many of his contemporaries who have had less to say."

William Carlos Williams, THE GREAT FIGURE, page 581

Is the figure a symbol? It looks like one—such intent concentration upon a particular. In an otherwise static landscape, only the 5 moves. It's the one moving thing, like in the eye of the blackbird in Stevens's "Thirteen Ways" (page 682). As far as we can see, however, the 5 has no great meaning beyond itself. Williams just rivets our attention on it and builds an atmosphere of ominous tension. The assumption, as in "The Red Wheelbarrow," is that somehow the figure has colossal significance. But "The Great Figure" is a much more vivid poem than "Wheelbarrow" (page 521) and it contains no editorializing ("so much depends"). Like the poems in the section "Literal Meaning," it's useful for discouraging students from spelling out colossal significances.

In his *Autobiography* (New York: Random House, 1951), Williams recalls how the poem came to be written. As he walked West 15th Street in New York on a summer's day, he heard

> a great clatter of bells and the roar of a fire engine passing the end of the street down Ninth Avenue. I turned just in time to see a golden figure 5 on a red background

flash by. The impression was so sudden and forceful that I took a piece of paper out of my pocket and wrote a short poem about it.

Citing this story, Bruce Bawer finds "The Great Figure" a typical poem of Williams's major phase: a quick description of a brief experience. "This is one of many familiar poems in which Williams does not shrink from, but rather observes carefully, celebrates and expresses a childlike wonder at the mechanical, the grubby, the vulgar, the trivial, the homely, the banal, or the grossly physical" ("The poetic legacy of William Carlos Williams," *New Criterion*, Sept. 1988, 17–18).

The poem inspired a painting by Charles Demuth, *I Saw the Figure 5 in Gold* (1928, now in New York's Metropolitan Museum of Art).

Robert Bly, DRIVING TO TOWN LATE TO MAIL A LETTER, page 581

No doubt the situation in this poem is real: Bly, who lives in frequently snowbound Minnesota, emits hundreds of letters. Compare this simple poem to Frost's "Stopping by Woods on a Snowy Evening" (page 812), which also has a speaker who, instead of going home, prefers to ogle snowscapes.

Gary Snyder, MID-AUGUST AT SOURDOUGH MOUNTAIN LOOKOUT, page 581

In brief compass, Synder's poem appeals to the mind's eye (with *smoke haze, pitch glows on fir-cones, rocks, meadows*, and the imagined vista at the end), the sense of moisture (*after five days rain*), of hot (*three days heat*) and of cold (*snow-water from a tin cup*). The *swarms of new flies* are probably both seen and heard.

For more background to this poem and to Snyder's work in general, see Bob Steuding, *Gary Snyder* (Boston: Twayne, 1976). A fictional portrait of Snyder appears in Jack Kerouac's novel *The Dharma Bums* (New York: Viking, 1958).

H. D. [Hilda Doolittle], HEAT, page 582

Heat becomes a tangible substance in this imagist poem, whose power resides mainly in its verbs, all worth scrutiny. Compare this poem and Synder's: how does each convey a sense of warmth?

H. D.'s work has enjoyed a recent surge of critical attention. For a concise, insightful comment on the lyrics and their originality, see Emily Stipes Watts, *The Poetry of American Women from 1632 to 1945* (U of Texas P, 1977) 152–58. Susan Stanford Friedman's *Psyche Reborn: The Emergence of H. D.* (Indiana U P, 1981) is valuable.

More recently, *Agenda*, that excellent, very serious British magazine of poetry, devoted a 200-page special issue to H. D. (vol. 25, nos. 3–4, Autumn/Winter 1987–88). With contributions by the poet herself, Susan Stanford Friedman, Eileen Gregory, Denise Levertov, Alicia Ostriker ("The Open Poetics of H.D.") and others, it is available from 5 Cranbourne Court, Albert Bridge Road, London SW11 4PE (price in U.S. funds $17.50).

Besides, there is now an *H.D. Newsletter*, edited by Professor Eileen Gregory, Dept. of English, University of Dallas, Irving, TX 75062.

Emanuel Di Pasquale, A SENSUAL, FOR EZRA POUND, page 582

Feeding grapes to leopards in the nude seems a pleasure more sensuous than sensual, but let's not quibble. Perhaps the poet thinks of Ezra Pound's early Imagist poems in *Personae*.

Linda Gregg, THE GRUB, page 582

This brief poem might provoke a reaction out of a sleepy class—reading it, one imagines oneself a worm getting his face fried, and practically yelps in pain and disgust. The poem is almost entirely made of vivid images, appealing to sight, hearing, touch, the sensation of burning.

(text pages) 582–583

This is a useful item to spring on anyone who assumes that poetry has to be sweetly reassuring. "The Grub" has a certain nightmarishness, like Frost's "Design," page 1493, though it makes no metaphysical speculation.

Linda Gregg's remarkable first collection, *Too Bright to See* (Port Townsend, WA: Graywolf Press, 1981) was followed by another strong one, *Alma* (New York: Random, 1985).

Mary Oliver, RAIN IN OHIO, page 583

Much of the strength of this poem, from Oliver's 1984 Pulitzer-winning collection, *American Primitive*, derives from its images: the sounds of bird cries, the sights of the halted blacksnake, and the white sky with its whirling storm clouds. But much also derives from metaphors: the thunderheads seen as horses, the snake's "long ladder of muscle," its body seen as a pourable fluid. In this poem, metaphors are stated in images: the two elements seem happily inseparable.

The images aren't constructed from thin air; Mary Oliver knows Ohio. Born in Maple Heights, near Cleveland, she attended Ohio State University and has taught as a visiting professor at Case Western Reserve.

18 Figures of Speech

Why Speak Figuratively?

Alfred, Lord Tennyson, THE EAGLE, page 585

For a hostile criticism of this poem, see Robert Graves, "Technique in Poetry," *On Poetry: Collected Talks and Essays* (New York: Doubleday, 1969) 402–05. Graves finds Tennyson's fragment unable to meet the minimal requirement that a poem should make good prose sense. He complains that if the eagle stands on its hands then its wings must be feet, and he ends up by rewriting the poem the way he thinks it ought to be. Though his remarks are fascinating, Graves reads the poem too literally.

A recent critic has suggested that this poem is a product of Tennyson's hopeless nearsightedness. Celebrating the eagle's 20–20 zoom-lens vision and ability to see a fish from high up, Tennyson yearns for a goal he could not attain: "optical inclusiveness." (See Gerhard Joseph, "Tennyson's Optics: The Eagle's Gaze," *PMLA* 92 [May 1977]: 420–27.)

William Shakespeare, SHALL I COMPARE THEE TO A SUMMER'S DAY?

Howard Moss, SHALL I COMPARE THEE TO A SUMMER'S DAY? page 586

Shakespeare's original—rich in metaphor, personification, and hyperbole—means more, of course, than Moss's tongue-in-cheek desecration. The only figure of speech in Moss's rewrite is the simile in line 1, and even that is denegated ("Who says?"). Moss manages to condense 115 great words to 78, a sonnet to a mere thirteen lines. It took a poet skilled in handling rimes to find such dull ones.

Shakespeare's nautical metaphor in line 8 may need explaining: a beautiful young person is a ship in full sail; accident or age can untrim the vessel. Compare this metaphor to "bare ruined choirs where late the sweet birds sang" "That time of year," page 871).

Jon Stallworthy, SINDHI WOMAN, page 587

The poem contains several examples of figurative language. The woman's grace is compared with that of "the cloth blown back from her face." There is "not a ripple in her tread." But perhaps the most striking figure of speech comes in the last two lines, where *weight* works both literally (referring to the stone jar on the woman's head) and figuratively (in reference to the less tangible weights borne by this woman: filth, poverty, and all the other burdens of slum life). Perhaps the figure of speech can be said to extend to the troubles borne by anyone, anywhere. The weightier they are, the poet suggests, the more character do their bearers display.

Metaphor and Simile

Richard Wilbur, A SIMILE FOR HER SMILE, page 589

Despite the title, it may be necessary to point out that the detailed and extended comparison that occupies Wilbur's poem isn't between the smile and the approach of a riverboat, but between the latter and the speaker's *experience* of his loved one's smile, or his anticipation of it, or his memory of it.

(text pages) 589–591

The graceful ingenuity of this poem, in which the simile is made so explicit, recalls earlier metaphysical poetry. Compare Wilbur's simile with Donne's figure of the two parted lovers in "A Valediction: Forbidding Mourning," (page 801), or to Edmund Waller's central metaphor in "On a Girdle" (page 598).

Alfred, Lord Tennyson, FLOWER IN THE CRANNIED WALL, page 589

Why does Tennyson say "what God and man is" instead of "what God and man *are*"? Apparently, this isn't faulty grammar, but higher pantheism. God and man are one.

Slyvia Plath, METAPHORS, page 590

Students usually are prompt to see that the central fact of the poem is the speaker's pregnancy. The speaker feels herself to be a walking riddle, posing a question that awaits solution: What person is she carrying? The "nine syllables" are like the nine months of gestation. All the metaphors refer to herself or to her pregnancy, except those in lines 4–5, which refer to the unborn baby: growing round and full like an apple or plum, seeming precious as ivory (and with ivory skin?), fine-timbered in sinew and bone like a well-built house.

The tone of the poem is clear, if complicated. Humor and self-mockery are evident in the images of elephant and strolling melon. In the last line, there is a note of wonder at the inexorability of gestation and birth: "The train there's no getting off."

A lively class might be asked to point out any possible connection between what the poem is saying about the arbitrary, fixed cycle of pregnancy and its form—the nine nine-syllable lines.

As Plath records in her Boston journal for 20 March 1959, the pregnancy she had hoped for ended in a miscarriage. Grieving and depressed, she went ahead and finished this poem, then explicitly called "Metaphors for a Pregnant Woman" (*Journals*, New York: Ballantine, 1983) 298–99.

Emily Dickinson, IT DROPPED SO LOW – IN MY REGARD, page 590

The whole poem sets forth the metaphor that someone or something the speaker had valued too highly proved to be like a silver-plated item (a chafing dish? a cream pitcher?) that she had mistaken for solid silver. Its smash revealed that it was made of cheap stuff.

In another version, lines 5–6 read: "Yet blamed the Fate that fractured—*less* / Than I reviled myself." Students may be asked which version they prefer, and why they prefer it. (Personally, we much prefer *reviled* to *denounced* because of its assonance—the sound of the *i*—and its alliteration—the *l* in *reviled* and *self*. Besides, *fractured* seems a more valuable word than *flung*: it gets across the notion of something cracked or shattered, and its *r* sets up an alliterative echo with the words *entertaining*, *Wares*, and *Silver*.)

James C. Kilgore, THE WHITE MAN PRESSED THE LOCKS, page 591

White body (the suburbs) holds black body (the inner city) in a tight grip, as if squeezing the life out of it. Students may be asked to spell out the connotations of certain well-loaded words and phrases. Why is the body in line 3 *darkening*? (Because more and more blacks are gathering in the inner city?) And *blighted* suggests both disease and the familiar phrase "urban blight." In the bloodstream, *white corpuscles* act as aggressive devourers. Possibly the hugging *white arms* suggest the commuter's wife and children, to whom he speeds home—but from the black poet's point of view in this poem, they appear sinister.

Kilgore's poem has been a fixture of this book through five editions, and we valued his friendship. He taught at Cuyahoga Community College for 22 years. His volume of new and selected poems, *African Violet: Poems for a Black Woman*, was published in 1982 by Lotus Press, PO Box 21607, Detroit, MI 48221. We were saddened to learn from a

131

wire service news story on December 18, 1988, that he had died, along with a daughter and a grandson, when a fire during the night consumed his Beachwood, Ohio, home.

Peter Williams, WHEN SHE WAS HERE, LI BO, page 591

Li Bo's lines supply Williams with methods for both parts of his poem. First he gives similes and metaphors to show what the lover's presence was, then adds more similes: terms for her absence. She was beautiful, sensuously appealing, and she afforded her lover satisfaction; her absence is like depressing, standardized food and drink and pop music, like a then-mediocre football team.

In the first part, most students will be more familiar with the term *quadraphonic* than with the name of Gustav Mahler (1860–1911). What "custardy honey from the old art books" might the poet mean? (Rubens's nudes seem more beefy than custardy. Botticelli's? Manet's?)

In the last lines, the poet alludes to a popular "process cheese food" spread, smeared on standard white bread; to singer and TV-show host Bobby Vinton (one of whose hits was "La, La, Chicken Cacciatore, My Memory of Love"), and to an NFL team noted for its checkered career.

The wry, comic tone of Williams's poem—communicated by its hilarious, self-deprecating similes—clearly differs from that of Li Bo's lines. Evidently, the Chinese poet is completely serious.

Ruth Whitman, CASTOFF SKIN, page 592

Apparently "crawled away" means that the old woman died, leaving her body behind as a snake sheds a useless skin. "Paper cheek" seems a fine evocation of snakeskin. The simile in line 2 ("small as a twig") also suggests stiffness and brittleness.

EXERCISE: *What Is Similar?* page 592

We'd suggest that this exercise be run through rapidly. We wouldn't give students much time to ponder, but would briskly call on people, and if anyone hesitated for long, would skip to someone else. Give them time to cogitate about these items, and they are likely to dredge up all sorts of brilliant, reached-for similarities in each pair of things—possibly logical, but having nothing to do with the lines. Immediate flashes of understanding are the goal of this exercise, not ponderous explication. Do this one for fun, and so it might be; do it slowly and seriously, and it could be deadly.

OTHER FIGURES

On the subject of puns, students familiar with *Hamlet* and other classics of the Bard may be asked to recall other puns of Shakespeare (besides the celebrated lines about golden lads and girls). If such a discussion prospers, Dr. Johnson's well-known observation in his preface to Shakespeare's works may provide an assertion to argue with:

> A quibble is to Shakespeare what luminous vapors are to the traveler: he follows it at all adventures; it is sure to lead him out of his way, and sure to engulf him in the mire.... A quibble is the golden apple for which he will always turn aside from his career or stoop from his elevation. A quibble, poor and barren as it is, gave him such delight that he was content to purchase it by the sacrifice of reason, propriety, and truth. A quibble was to him the fatal Cleopatra for which he lost the world, and was content to lose it.

James Stephens, THE WIND, page 593

As a birthday present to Stephens, James Joyce once translated this poem into five other languages (French, German, Italian, Latin, and Norwegian). These versions are reprinted in *Letters of James Joyce,* ed. Stuart Gilbert (New York: Viking, 1957) 318–19.

EXERCISE: *Paradox,* page 595

Chidiock Tichborne, ELEGY, WRITTEN WITH HIS OWN HAND, page 595

"One must admit the possibility that these verses were written by some other poet, rather than by the protagonist himself," note J. William Hebel and Hoyt H. Hudson in *Poetry of the English Renaissance* (New York: Appleton, 1929). Set to music by a later composer, the "Elegy" was sung as a madrigal.

George Herbert, THE PULLEY, page 597

The title may need clarification. Man's need for rest is the pulley by which eventually he is drawn to rest everlasting. The pulley Herbert has in mind is probably not horizontal (like the one with a clothesline), but the vertical kind rigged to hoist a heavy weight. Despite the puns, the tone of the poem is of course devoutly serious, Herbert's concern in it being (in the view of Douglas Bush) "to subdue the wilful or kindle the apathetic self."

Lines 2–10, on the "glass of blessings" and its contents, set forth a different metaphor. As Herbert's editor F. E. Hutchinson (*Works* [Oxford: Oxford UP, 1941]) and others have remarked, "The Pulley" seems a Christian version of the story of Pandora. At her creation Pandora received gifts from all the gods, mostly virtues and graces—though Hermes gave her perfidy. In some tellings of the myth, Pandora's gift box (or vase) held not plagues but further blessings. When she became curious and opened it, they slipped away, all except the one that lay at the bottom—hope.

Herbert's poem, in its fondness for the extended metaphysical conceit, invites comparison with Donne's metaphor of the compasses in "A Valediction: Forbidding Mourning" (page 801). If the instructor cares to discuss metaphysical poetry, "The Pulley" may be taken together with Herbert's "Redemption" (page 707) and "Love" (page 824). ("Easter Wings" [page 694] raises distracting considerations and may be left for a discussion of concrete or graphic poetry.) Other metaphysical poems, by Waller and Roethke, immediately follow "The Pulley" in this chapter. (See note on Roethke's poem.) Other poems of Donne and of Dickinson can be mentioned. For a recent poem that contains extended conceits, see Wilbur's "A Simile for Her Smile," also in this chapter.

Students might be encouraged to see that poets of the seventeenth century had certain habits of thought strikingly different from our own; but that some of these habits—like the fondness for startling comparisons of physical and spiritual things—haven't become extinct. Perhaps the closest modern equivalent to the conceits of Herbert and Donne may be found in fundamentalist hymns. Two earlier twentieth-century illustrations:

> If you want to watch old Satan run
> Just fire off that Gospel gun!

and

> My soul is like a rusty lock.
> Oh, oil it with thy grace!
> And rub it, rub it, rub it, Lord,
> Until I see thy face!

(The first example is attributed to a black Baptist hymn writer; the second, to the Salvation Army, according to Max Eastman in *Enjoyment of Laughter* [New York: Simon, 1936] 79.) An even more recent illustration, probably influenced by fundamentalist hymns, in a country-and-Western song recorded in 1976 by Bobby Bare, "Dropkick Me, Jesus" ("through the goalposts of life").

Edmund Waller, ON A GIRDLE, page 598

Another way to enter this poem might be to ask:

1. In what words does the poet express littleness or constriction? (In the *"slender* waist," the rime *confined/bind,* "pale which *held,*" *"narrow* compass," *"bound."*)
2. In what words does he suggest vastness and immensity? (With any luck, students will realize that the entire poem demonstrates the paradox stated in lines 9–10.)

There may be another pun in line 6: *deer.*

What is the tone of "On a Girdle"? Playful and witty, yet tender. You can have fun with this poem by asserting an overly literal-minded reading: somewhere in the world, isn't there probably some ruler who *wouldn't* abdicate just to put his arms around her? If the speaker rejects all the sun goes round, where will that put him and his loved one? If students have trouble objecting to such quibbles, remind them of the definition of a figure of speech. That should help. An overstatement isn't a lie, it's a means of emphasis.

Compare this to another love poem full of hyperbole: Burns's "Oh, my love is like a red, red rose," at the end of this chapter.

Theodore Roethke, I KNEW A WOMAN, page 598

Both outrageous puns occur in line 15. In question 2, the three lines quoted contain overstatement or hyperbole. The speaker's reference to his whole being as "old bones" is synecdoche. "Let seed be grass, and grass turn into hay" are not metaphors, but literal events the speaker hopes for—unless you take the ripening of the grass to be the passage of time. Metaphors occur also in *the sickle* and *the rake,* and in calling lovemaking *mowing.*

"I Knew a Woman" shows Roethke's great affection for metaphysical poetry in its puns, its brief conceits (sickle and rake), and its lovely image out of geometry—"She moved in circles, and those circles moved." Two metaphysical poets of the seventeenth century come immediately before Roethke in this chapter: Herbert and Waller. Here's a good chance to dwell on metaphysical poetry, or at least to mention it. (See the notes on Herbert's "Pulley.")

For more outrageous puns, compare Robert Graves's "Down, Wanton, Down!" (page 538).

FOR REVIEW AND FURTHER STUDY

Robert Frost, THE SILKEN TENT, page 599

Although the word *as* in the opening line might lead us to expect a simile, "The Silken Tent" is clearly an immense metaphor, comparing woman and tent in a multitude of ways. What are the ropes or cords? Not merely commitments (or promises to keep) to friends and family, but generous sympathies, "ties of love and thought," on the part of a woman who cares about everything in the world.

While paying loving tribute to a remarkable woman, the poem is also a shameless bit of showing off by a poet cocksure of his technical mastery. Managing syntax with such grace that the poem hardly seems contrived, Frost has sustained a single sentence into an entire sonnet. "The whole poem is a performance," says Richard Poirier, "a display for the beloved while also being an exemplification of what it is like for a poem, as well as a tent or a person, to exist within the constrictions of space ('a field') and time ('at midday') wherein the greatest possible freedom is consistent with the intricacies of form and inseparable from them" (*Robert Frost: The Work of Knowing* [New York: Oxford UP, 1977] xiv–xv). Poirier points out, too, that the diction of the poem seems Biblical, perhaps echoing "The Song of Songs" (in which the bride is comely "as the tents of Kedar") and Psalm 92 (in which the godly "grow like a cedar in Lebanon"). Not only does the "central cedar pole" signify the woman's spiritual rectitude, it points toward heaven.

In teaching this poem, one can quote Frost's remark to Louis Untermeyer, "I prefer the synecdoche in poetry, that figure of speech in which we use a part for a whole." In 1931 Frost recalled that he had called himself a Synecdochist back when other poets were calling themselves Imagists: "Always, always a larger significance. A little thing touches a larger thing" (qtd. in Elizabeth Shepley Sergeant, *Robert Frost: The Trial by Existence* [New York: Holt, 1960] 325).

Denise Levertov, LEAVING FOREVER, page 600

The man seems glad to go: "stones rolling away" suggests the shedding of some great weight, or possibly even a resurrection (an echo of the rolling away of the stone from the Easter tomb?). But in the woman's view the mountain seems like someone rejected and forlorn. The woman's view, expressed in a metaphor and given force by coming last, seems stronger than the man's simile.

Another question: is the poet right to repeat *away, way, away, away*? The sound reverberates with a terrible flat monotony, it is true—but apparently that is the effect necessary.

Jane Kenyon, THE SUITOR, page 600

This economical poem moves from simile to simile: (1) "like the chest of someone sleeping" (steadily rising and falling); (2) "like a school of fish" (flashing their pale bellies); and (3) "like a timid suitor" (hesitant, drawing back, reluctant to arrive).

Kenyon lives in Danbury, New Hampshire. "The Suitor" is from her first collection, *From Room to Room* (Cambridge: Alice James, 1978).

Richard Wilbur, SLEEPLESS AT CROWN POINT, page 601

In Wilbur's implied metaphor, promontory and wind are to each other as bull is to matador.

Robert Frost, THE SECRET SITS, page 601

Besides its personification of the sitting Secret, Frost's poem contains an implied metaphor. To dance round in a ring is to make futile efforts to penetrate a secret—merely going around in circles.

Margaret Atwood, YOU FIT INTO ME, page 601

The first two lines state a simile. In the second couplet, *hook* and *eye* turn out (to our surprise) to be puns.

Grace Schulman, HEMISPHERES, page 601

Like Margaret Atwood in "You fit into me," Grace Schulman in this poem is talking about the fit of two bodies engaged in sexual intercourse. Atwood's of course is a hate poem, while Schulman's is a love poem. Schulman compares the fit of the lovers' bodies to the perfect fit of the pieces of a mended terra cotta vase and then, more extravagantly, to the

(text pages) 601–603

fit of the continents before they were split apart. Both these poems can profitably be compared with Nancy Willard's "Marriage Amulet" (page 892).

John Tagliabue, MAINE VASTLY COVERED WITH MUCH SNOW, page 602

This poem starts with a comparison that keeps expanding. First, we get a simile ("as busy as monks"), then a metaphor (the seeds are capsules containing the Dead Sea scrolls, or something), then another simile (theology like the flourish), and finally an implied metaphor (continuing the idea that the squirrels are scholarly monks.)

The poet, retired from teaching at Bates College in Maine, has traveled extensively in the Orient.

John Ashberry, THE CATHEDRAL IS, page 602

Slated is a pun.

W. S. Merwin, SONG OF MAN CHIPPING AN ARROWHEAD, page 602

The poem contains an apostrophe to the chips of flint or stone. That the chips are "little children" is also a metaphor; so is "the one you are hiding"—the emerging arrowhead.

Robert Burns, OH, MY LOVE IS LIKE A RED, RED ROSE, page 603

Figures of speech abound in this famous lyric: similes (lines 1–2, 3–4), a metaphor (*sands o' life*, 12), overstatement (8 and 9, 10), and possibly another overstatement in the last line.

See other professions of love couched in hyperbole, among them Waller's "On a Girdle" (page 598), Roethke's "I Knew a Woman" (page 598), Marvell's "To His Coy Mistress" (page 846), and Auden's "As I Walked Out One Evening" (page 781). Are the speakers in these poems mere throwers of blarney, whom no woman ought to trust?

For a discussion of this poem that finds more in it than figures of speech, see Richard Wilbur, "Explaining the Obvious," in *Responses* (New York: Harcourt, 1976). Burns's poem, says Wilbur, "forsakes the lady to glory in Love itself, and does not really return. We are dealing, in other words, with romantic love, in which the beloved is a means to high emotion, and physical separation can serve as a stimulant to ideal passion." The emotion of the poem is "self-enchanted," the presence or absence of the lady isn't important, and the very idea of parting is mainly an opportunity for the poet to turn his feelings loose. Absurd as this posture may be, however, we ought to forgive a great songwriter almost anything.

19 Song

SINGING AND SAYING

Most students who wrote comments about this book in its last edition said that this chapter was the one that most appealed to them. "It shows that poetry isn't all found in books," was a typical comment; and many were glad to see song lyrics that they knew.

Even if there is not time for a whole unit on song, the instructor who wishes to build upon this interest can use at least some of this chapter to introduce the more demanding matters of sound, rhythm, and form (treated in the chapters that follow). Some instructors take the tack that lyric poetry begins with song, and so begin their courses with this chapter, supplemented by folk ballads elsewhere in the text.

Besides Ben Johnson's classic nondrinking song, many other famous poems will go to melodies. The tradition of poems set to music by fine composers is old and honorable. For lists of such poems with musical settings (and recordings), see *College English* for February 1985 and December 1985.

If a class seems in grave need of livening, it might be fun to sing a poem or two to an unsuitable tune: Robert Frost's "Stopping by Woods on a Snowy Evening" to the tune of "Hernando's Hideaway" (a combination originally discovered by waiters at the Bread Loaf Writers Conference, who sang it to Frost); or Wordsworth's "She Dwelt Among the Untrodden Ways" to the tune of "Yankee Doodle" (which tune supplies a beautifully flip ending: "But she is in her grave, and oh, / The difference to me!")

Ben Jonson, TO CELIA, *page 605*

Students may not know that in line 2 *I will pledge* means "I will drink a toast." Also, *I would not change for thine* (line 8) in modern English becomes "I would not take it in exchange for yours."

To demonstrate that "To Celia" is a living song, why not ask the class to sing it? Unfortunately, you can no longer assume that the tune is one that everyone knows, so you may need to start them off.

Anonymous, THE CRUEL MOTHER, *page 606*

Some versions of this ballad start the narrative at an earlier point in time, with a woman discovering that she is pregnant by the wrong man when she is about to marry another. See Alan Lomax's Notes to *The Child Ballads,* vol. 1, Caedmon, TC 1145, which record contains an Irish version.

If the instructor cares to discuss the bottomless but student-spellbinding topic of archetypes, this ballad will serve to illustrate an archetype also visible in the stepmother figure of many fairy stories.

Edwin Arlington Robinson, RICHARD CORY, *page 610*

Paul Simon, RICHARD CORY, *page 610*

This pair sometimes provokes lively class discussion, especially if someone in the class maintains that Simon converts Robinson into fresh, modern terms. Further discussion may be necessary to show that Robinson's poem has a starkly different theme.

(text pages) 610–615

Robinson's truth, of course, is that we envy others their wealth and prestige and polished manners, but if we could see into their hearts we might not envy them at all. Simon's glib song does not begin to deal with this. The singer wishes that he too could have orgies on a yacht, but even after he learns that Cory died a suicide, his refrain goes right on, "I wish that I could be Richard Cory." (Live rich, die young, and make a handsome corpse!)

Some questions to prompt discussion might include:

1. In making his song, Simon admittedly took liberties with Robinson's poem. Which of these changes seem necessary to make the story singable? What suggestions in the original has Simon picked up and amplified?

2. How has Simon altered the character of Richard Cory? Is his Cory a "gentleman" in Robinson's sense of the word? What is the tone of Simon's line, "He had the common touch"? Compare this with Robinson: "he was always human when he talked." Does Robinson's Cory have anything more than "Power, grace and style"?

3. In the song, what further meaning does the refrain take on with its third hearing, in the end, after the news of Cory's suicide?

4. What truth about life does Robinson's poem help us see? Is it merely "Money can't make you happy" or "If you're poor you're really better off than rich people"? Does Simon's narrator affirm this truth, deny it, or ignore it?

Frank J. D'Angelo has noticed that the name Richard Cory is rich in connotations. It suggests Richard Coeur de Lion, and other words in Robinson's poem also point to royalty: *crown, imperially, arrayed, glittered, richer than a king.*

BALLADS

Anonymous, BONNY BARBARA ALLAN, page 612

Despite the numerous versions of this, the most widespread of all traditional ballads in English, most keep the main elements of the story with remarkable consistency. American versions tend to be longer, with much attention to the lovers' eventual side-by-side burial, and sometimes have Barbara's mother die of remorse, too! Commentators since the coming of Freud have sometimes seen Barbara as sexually frigid, and Robert Graves once suggested that Barbara, a witch, is killing Sir John by sorcery. An Irish version makes Barbara laugh hideously on beholding her lover's corpse.

To show how traditional ballads change and vary in being sung, a useful recording is *The Child Ballads,* vol. 1, Caedmon, TC 1145, containing performances collected in the field by Alan Lomax and Peter Kennedy. Six nonprofessional singers are heard in sharply different versions of "Barbara Allan," in dialects of England, Scotland, Ireland, and Wales.

Dudley Randall, BALLAD OF BIRMINGHAM, page 615

Randall's poem is an authentic broadside ballad: it not only deals with a news event, it was once printed and distributed on a single page. "I had noticed how people would carry tattered clippings of their favorite poems in their billfolds," the poet has explained, "and I thought it would be a good idea to publish them in an attractive form as broadsides" (Interview in *Black World,* Dec. 1971). "Ballad of Birmingham" so became the first publication of Randall's Broadside Press, of Detroit, which has since expanded to publish books and issue recordings by many leading black poets, including Gwendolyn Brooks, Don L. Lee, and Nikki Giovanni.

The poem seems remarkably fresh and moving, though it shows the traits of many English and Scottish popular ballads (such as the questions and answers, as in "Edward," and the conventional-sounding epithets in stanza 5). Randall presents without comment the horror of the bombing—in the mother's response and in the terrible evidence—but we are clearly left to draw the lesson that if the daughter had been allowed to join the open protest, she would have been spared.

Four black girls were killed in 1963, when a dynamite blast exploded in Birmingham's Sixteenth Street Baptist Church. In September, 1977, a Birmingham grand jury finally indicted a former Ku Klux Klansman, aged 73, on four counts of first-degree murder.

For Review and Further Study

John Lennon and Paul McCartney, Eleanor Rigby, page 616

"Eleanor Rigby," we think, is a poem. Although swayed by the superstition that priests are necessarily lonely because celibate, Lennon's portrait of Father McKenzie and of Eleanor have details that reflect life. Both music and words contain an obvious beat, and if students pick out those syllables in long lines 4 and 7, 14 and 17, and 24 and 27, they will be getting into the subject of meter. (Each of the lines contains a stressed syllable followed by four anapests.)

Exercise: *Songs or Poems or Both?* page 617

Anonymous, Fa, mi, fa, re, la, mi, page 618

Anonymous, The silver swan, who living had no note, page 618

Bruce Springsteen, Born to Run, page 618

Of these four song lyrics, only "Fa, mi" seems completely unreadable: its words do little more than fill up a tune.

Comparing the two madrigals, the instructor can raise the question, Why is "The silver swan" a memorable poem while "Fa, mi" isn't? With luck and a little prompting, students may see that poetry isn't mere musical noise: poems say something. They also tend to contain figurative language, such as the metaphor "unlocked her silent throat" and the apostrophe to death; and interesting sounds—the subtle *silver / living* and other alliterations. What, by the way, is a "swan song"? Some students may need an explanation.

Springsteen's early lyric, which helped make him famous, seems to us better when heard than when read on the page. Still, it has elements of poetry: strong feeling; metaphors in "Baby this town rips the bones from your back" (also hyperbole there) and in lines 12–13 (those sexual innuendoes), 16 (the lovers compared to high-wire walkers), and 36 ("we'll walk in the sun"); some vivid imagery, most realistic in stanza 3: girls combining their hair in rearview mirrors, kids huddled on the beach in a mist; rime, rhythm, internal rime in lines 7, 24, and 33.

20 Sound

SOUND AS MEANING

Alexander Pope, TRUE EASE IN WRITING COMES FROM ART, NOT CHANCE, *page 621*

Nowadays, looking at the pages of an eighteenth-century book of poetry, we might think the liberal capitalization and use of italics merely decorative. But perhaps Pope wished to leave his readers little choice in how to sound his lines. Most of his typographical indications seem to us to make sense—like a modern stage or television script with things underlined or capitalized, lest the actors ignore a nuance.

Line 12 is deliberately long: an alexandrine or twelve-syllable line that must be spoken quickly in order to get it said within the time interval established by the other shorter, pentameter lines.

William Butler Yeats, WHO GOES WITH FERGUS? *page 623*

Originally a song in Yeats's play *The Countess Cathleen,* this famous lyric overflows with euphony. Take just the opening question (lines 1–3): the assonance of the various *o*-sounds; the initial alliteration of *w, d,* and *sh;* the internal alliteration of the *r* in *Fergus, pierce,* and *shore*—musical devices that seem especially meaningful for an invitation to a dance. The harsh phrase *brazen cars* seems introduced to jar the brooding lovers out of their reveries. Unless you come right out and ask what brazen cars are, a few students probably will not realize that they are brass chariots. In ancient Ulster, such chariots were sometimes used for hunting deer—though how you would drive one of them through the deep woods beats us.

If you discuss meter, what better illustration of the power of spondees than "And the WHITE BREAST of the DIM SEA"?

The last line of the poem, while pleasingly mysterious, is also exact. The personification "dishevelled wandering stars" makes us think of beautiful, insane, or distracted women with their hair down: Ophelia in Olivier's film *Hamlet.* That they are wandering recalls the derivation of the word *planet:* Greek for "wanderer." In what literal sense might stars look disheveled? Perhaps in that their light, coming through the atmosphere (and being seen through ocean spray) appears to spread out like wild long hair. For comparable figures of speech, see Melville's "The Portent" (page 543), in which John Brown's beard is a meteor; and Blake's "Tyger," (page 788), in which the personified stars weep and throw spears.

EXERCISE: *Listening to Meaning, page 624*

John Updike, WINTER OCEAN, *page 624*

Frances Cornford, THE WATCH, *page 624*

William Wordsworth, A SLUMBER DID MY SPIRIT SEAL, *page 624*

Emanuel di Pasquale, RAIN, *page 625*

Aphra Behn, WHEN MAIDENS ARE YOUNG, *page 625*

Onomatopoeia is heard in Updike's *scud-thumper* and *pusher* (echoing the boom and rush of the waves), and is heard even more obviously in Cornford's watch-ticks and Behn's

(text pages) 625–628

hum-drum. In di Pasquale's lines, the *s*-sounds fit well with our conception of rain, and *hushes* is an especially beautiful bit of onomatopoeia. By the way, di Pasquale's poem is especially remarkable in view of the fact that the poet, born in Sicily, did not learn English until he was sixteen.

In Wordsworth's Lucy poem, sound effects are particularly noticeable in the first line (the soporific *s*'s) and in the last two lines (the droning *r*'s and *n*'s). If students go beyond the sound effects and read the poem more closely, they might find problems in the first stanza. Is the poet's *slumber* a literal sleep or a figurative one? That is: is Wordsworth recalling some pleasant dream of Lucy (whether the living Lucy he used to know, or the dead Lucy in Eternity), or is he saying that when she was alive he was like a dreamer in his view of her? If so, he was deluded in thinking that she would always remain a child; he had none of the usual human fears of death or of growing old. However we read the poem, there is evidently an ironic contrast between the poet's seeing Lucy (in stanza 1) as invulnerable to earthly years, and his later view that she *is* affected, being helplessly rolled around the sun once a year with the other inanimate objects. And simple though it looks, the poem contains a paradox. The speaker's earlier dream or vision of Lucy has proved to be no illusion but an accurate foreshadowing. Now she *is* a "thing," like rocks and stones and trees; and she cannot feel and cannot suffer any more from time's ravages.

Aphra Behn was the first English woman to earn a living by her pen. *Oroonoko* (1688), a tale of slavery in Surinam, is sometimes called the first true English novel. Her colorful life, mostly spent in London's literary bohemia, included a hitch in Holland as a spy for the Crown. In her destitute late years she was pilloried in lampoons ("a lewd harlot"), perhaps because she remained faithful to the Stuarts. She was buried in Westminster Abbey under her poetic pen name—Astrea Behn. Nowadays, she seems to be enjoying a respectful dusting-off. See the extensive treatment given her in *Kissing the Rod: An Anthology of 17th Century Women's Verse*, ed. Germaine Greer and others (London: Virago Press, 1988) 240–60.

Alliteration and Assonance

A. E. Housman, Eight O'Clock, page 627

The final *struck* is a serious pun, to which patterns of alliteration begun in the opening line (*st . . . st, r*), and continued through the poem, have led up. The ticking effect of the clock is, of course, most evident in *the clock collected*.

Compare Housman's strapped and noosed lad with the one in Hugh Kingsmill's parody of Housman (page 737).

Robert Herrick, Upon Julia's Voice, page 627

Julia, apparently, is singing, for the lutes provide accompaniment. At the beginning and end, this brief poem is particularly rich in music: the sibilance of *so smooth, so sweet, so silv'ry*; and the alliteration (both initial and internal) of the *m*- and *l*-sounds in the last line.

The second line seems a colossal hyperbole, meaning, "Julia, your singing is sweet enough to make the damned in Hell forget to wail." In Herrick's poems, such flattery is never out of place. This is the quality of Herrick's work—a lovely deliberate absurdity—XJK has tried to echo (feebly) in a two-line parody:

> When Vestalina's thin white hand cuts cheese,
> The very mice go down upon their knees.

Janet Lewis, Girl Help, page 628

A sensitive comment on this poem is that of the poet's husband, Yvor Winters:

> There is almost nothing to it, really, except the rich characterization of a young girl with her life before her and the description of a scene which implies an entire

(text pages) 628–631

way of life. The meter is curious; one is tempted to call it irregular three-beat accentual, but it seems to be irregular iambic trimeter. On the second basis, the first line starts with a monosyllabic foot, and the fourth line has two feet, both iambic. But *wide* and *broom* are almost evenly accented, and both are long; If one is moved by this to choose the accentual theory, *her* is too lightly accented to count, although it would count as the accented syllable of an iambic foot. There are a few other such problems later in the poem. But there is no problem with the rhythm; the fourth line virtually gives us the movement of the broom, and the seventh and eighth give us the movement of the girl; the movement of the poem is that of indolent summer in a time and place now gone; the diction, like the rhythm, is infallible.

Winters (1900–1968)—see his own poem on page 895—wrote this analysis in *Forms of Discovery* (Denver: Swallow, 1967) 332.

The poet reads "Girl Help" on her recording *Janet Lewis Reading at Stanford,* News & Publications Service of Stanford University, 1975; available from Serendipity Books, 1970 Shattuck Ave., Berkeley CA 94709. "Domestic poetry" is her province, remarks Kenneth Fields in the notes on that record album's sleeve. He cites (in order to disagree with) a review of Lewis's work by Theodore Roethke: "The nursery, the quiet study, the garden, the graveyard do not provide enough material for talent of such a high order."

Janet Lewis, known for *The Wife of Martin Guerre, The Trial of Soren Quist,* and other novels, has brought out *Poems Old and New: 1918–1978* (Athens: Swallow-Ohio UP, 1981).

EXERCISE: *Hearing How Sound Helps,* page 628

Wyatt's version surpasses Surrey's for us, especially in Wyatt's opening line, with its remarkable use of assonance—a good instance of the way vowel-sounds can slow our reading of a line and make us linger over it. The initial alliteration and internal alliteration in these first two lines help besides. In its rhythm, Surrey's version sounds a bit singsong by comparison. The inverted word order in the last line seems awkward.

Alfred, Lord Tennyson, THE SPLENDOR FALLS ON CASTLE WALLS, page 629

If read aloud rapidly, this famous lyric from Tennyson's *The Princess* will become gibberish; and the phrase *Blow, bugle, blow,* a tongue twister. But if it is read with any attention to its meaning, its long vowels and frequent pauses will compel the reader to slow down. Students may want to regard the poem as mellifluous nonsense, but may be assured that the poem means something. In fact, it is based on a personal experience of the poet's. Visiting the lakes of Killarney in 1848, Tennyson heard the bugle of a boatman sound across the still water, and counted eight distinct echoes. "The splendor falls" is the poet's attempt to convey his experience in accurate words.

RIME

William Cole, ON MY BOAT ON LAKE CAYUGA, page 630

This is one of a series of comic quatrains, "River Rhymes," first printed in *Light Year '85* (Case Western Reserve U: Bits Press, 1984).

Kenneth Burke, THE HABIT OF IMPERFECT RHYMING, page 631

What is this poem saying? Something like: It is too bad that certain fine, positive things often find themselves cheek by jowl with certain dreadful things (passion with nuclear fission, beauty with shoddy), but that's the way of the contemporary world. Still, such rude juxtapositions are necessary; we couldn't do without them. We wouldn't want to insist on nothing but neat, exact juxtapositions of wonderful/positive/living things with dreadful things (like loneliness and death). That would be worse.

We have long admired Burke's satiric poetry. Most of it has been gathered in his *Collected Poems 1915–1967* (Berkeley: U of California P, 1968).

Brad Leithauser, TRAUMA, page 633

Brilliant riming here, we think. What does this terse poem mean? Students ought to know what a *trauma* is (from the Greek: "wound"), and a *suture*. Is this trauma physical or psychic? We aren't told. The suture (surgical thread) suggests a physical wound or incision; but in a metaphoric sense a psychic wound, too, can be closed and made to heal. That the amassed past becomes the future is a classic idea, perhaps "ne'er so well expressed."

Leithauser is a writer to watch. His books of poems *Hundreds of Fireflies* (New York: Knopf, 1982) and *Cats of the Temple* (Knopf, 1986), and his novels *Equal Distance* (Knopf, 1984) and *Hence* (Knopf, 1989), have attracted much critical attention.

William Butler Yeats, LEDA AND THE SWAN, page 634

The deliberately awful off-rime *up / drop* ends the sonnet with an appropriately jarring plop as the God-swan discards the used Leda and sinks into his postejaculatory stupor.

Other questions that can be raised:

1. What *knowledge* and *power* does Yeats refer to in line 14?

2. Do the words *staggering* (line 2) and *loosening* (line 6) keep to the basic meter of the poem or depart from it? How does rhythm express meaning in these lines? (It staggers on *staggering* and loosens on *loosening*.)

3. Compare this poem to Donne's sonnet "Batter my heart" (page 540). Is the tone of Yeats's sonnet—the poet's attitude toward this ravishing—similar or dissimilar?

For an early draft of the poem, see Yeats's *Memoirs*, ed. Denis Donoghue (New York: Macmillan, 1973) 272–74.

Gerard Manley Hopkins, GOD'S GRANDEUR, page 635

Students who think Hopkins goes too far in his insistence on rimes and other similar sounds will have good company, including Robert Bridges, William Butler Yeats, and Yvor Winters. Still, it is hard not to admire the euphony of the famous closing lines—that ingenious alternation of *br* and *w*, with a pause for breath at that magical *ah!*—and the cacophony of lines 6–8, with their jangling internal rimes and the alliteration that adds more weight to *smeared, smudge,* and *smell.* For Hopkins, of course, sound is one with meaning; and the cacophonous lines just mentioned are also, as John Pick has pointed out, "a summary of the particular sins of the nineteenth century." For a brilliant demonstration that sound effects in Hopkins's poetry have theological meaning, see J. Hillis Miller, *The Disappearance of God* (Cambridge: Harvard UP, 1963) 276–317. Miller finds the poet's theory revealed in his sermons and journals: "Any two things however unlike are in something like"; therefore, "all beauty may by a metaphor be called rhyme."

In the text, it seemed best not to bury the poem under glosses, but to let the instructor decide how thoroughly to explicate it. Here are a few more glosses in case they seem necessary:

7. *man's smudge:* the blight of smoke and ugliness cast over the countryside by factories and mines. As a student for the priesthood in North Wales and as a parish priest in London and Liverpool, Hopkins had known the blight intimately. Another suggestion in the phrase: nature is fallen and needs to be redeemed, like man, who wears the smudge of original sin. 12. *morning . . . springs:* The risen Christ is like the sun at dawn. Eastward

is the direction of Jerusalem, also of Rome. (Hopkins cherished the hope that the Church of England and the Pope would one day be reconciled.) 13–14: *bent / World:* Perhaps because of its curvature the earth looks bent at the horizon; or perhaps the phrase is a transferred epithet, attributing to the earth the dove's bent-over solicitude. (And as the world seems to break off at the horizon, line 13 breaks at the word *bent.*) 14. *broods:* like a dove, traditional representation of the Holy Ghost.

For still more suggestions, see Pick, *Gerard Manley Hopkins, Priest and Poet,* 2nd ed. (Oxford: Oxford UP, 1966) 62–4; Paul L. Mariani, *Commentary on the Complete Poems of Gerard Manley Hopkins* (Ithaca: Cornell UP, 1970); and (not least) the poet's "Pied Beauty."

A sonnet by Wordsworth (page 717) also begins "The world is," and Hopkins no doubt knew of it. In their parallel (though different) complaints against trade and commerce, the two deserve to be compared. Both poets find humanity artificially removed from nature: this seems the point of Hopkins's observation in lines 7–8 that once soil was covered (with grass and trees) and feet were bare; now soil is bare and feet are covered. Clearly we have lost the barefoot bliss of Eden; but in answer to Wordsworth, one almost expects Hopkins to cry, "Great God! I'd rather be a Christian." (Wordsworth by *world* means "worldliness.")

James Whitehead, THE COUNTRY MUSIC STAR BEGINS HIS POLITICS, page 636

Is it the audience's love (line 6) or the country music star's "troubled pride" (line 8) that is "The size of death and a revival tent?" Or does the lack of punctuation around this phrase signal the poet's deliberate ambiguity about what it modifies? Clearly the speaker in the poem is troubled by his twin perceptions: that "everyone is going broke / Investing in good times and sentiment"— and that he earns his wages by supplying the "songs that help grown children play." He sees society's insatiable demand to be entertained as deadly. "The thing I do has prospered and gone wrong."

The inconclusive and jarring off-rimes in the poem make palpable the sense of discord and general decay the poet apparently wishes to convey. The Biblical echo in line 13, the images of waste and death seem to lend the country music star the aura of a prophet crying in the wilderness.

James Whitehead teaches in the MFA program at the University of Arkansas. Two collections of his poems are now available in a single volume, *Local Men and Domains* (Urbana: U of Illinois P, 1987), a book containing more than its share of unorthodox sonnets. He is no less well known for his remarkable novel, *Joiner* (New York: Knopf, 1971), about a young man's growing up on and off the football field.

Robert Frost, DESERT PLACES, page 636

Some possible answers:

1. Terrible pockets of loneliness.

2. The word *snow,* occurring three times. Other *o*-sounds occur in *oh, going, showing, no, so,* and *home.* The *l* of *lonely* is echoed by alliteration in *looked, last,* and *lairs.*

3. It makes us feel a psychic chill! Yet the feminine rime lightens the grim effect of what is said and gives it a kind of ironic smirk.

READING AND HEARING POEMS ALOUD

Here is a fresh comment by William Stafford on why certain poets read their poems with apparent carelessness. Poets spend their energies in writing poems and are not effective public speakers. Unlike the Russian poet Andrei Voznesensky, a great performer, Stafford remarks,

> Most of the poets I know would feel a little guilty about doing an effective job of reading their poems. They throw them away. And I speak as one who does that. It feels fakey enough to be up there reading something as though you were reading it for the first time. And to say it well is just too fakey. So you throw it away.
>
> (Interview in *The Literary Monitor* 3, no. 3/4, [1980]).

This comment raises provocative questions for discussion. What is the nature of a poetry reading? Should it be regarded as a performance, or as a friendly get-together?

For a symposium on poetry readings, with comments by Allen Ginsberg, James Dickey, Denise Levertov, and twenty-nine other poets, see *Poets on Stage* (New York: Some/Release, 1978).

A rich and convenient source of recorded poetry, no doubt the best in the country, is the Poets' Audio Center, a nonprofit service to the literary community and academe operated by the Watershed Foundation. The center can supply on cassette more than 700 readings by great and minor poets living and dead, both commercial and noncommercial productions. They issue a free list of 200 of their most popular recordings, also a complete catalog. (This last costs $3, but is free to acquisitions librarians who write on a letterhead.) Among the cassette offerings are their own Watershed Tapes, an extremely well produced and interesting series of readings by contemporaries. Address: Poets' Audio Center, P.O. Box 50145, Washington, DC 20004.

A 1989 catalogue of about 250 radio broadcasts on cassette, including programs featuring poets John Ashbery, Gwendolyn Brooks, John Ciardi, Rita Dove, Allen Ginsberg, Anthony Hecht, Philip Levine, Paul Zimmer, and many others reading and talking about their work, is available from New Letters on the Air, University of Missouri Kansas City, 5100 Rockhill Road, Kansas City, MO 64110, phone (816) 276-1159.

EXERCISE: *Reading for Sound and Meaning*, page 638

Michael Stillman, IN MEMORIAM JOHN COLTRANE, page 639

William Shakespeare, FULL FATHOM FIVE THY FATHER LIES, page 639

A. E. Housman, WITH RUE MY HEART IS LADEN, page 639

T. S. Eliot, VIRGINIA, page 640

In Michael Stillman's tribute to the great jazz saxophonist, *coal train* is not only a rich pun on Coltrane's name, it also becomes the poem's central image. The poet has supplied this comment:

> One thing about that poem which has always pleased me—beyond its elegiac strain—is the way the technique of the lines and phrases corresponds to a musical effect on Coltrane's playing. He was known for his ability to begin with a certain configuration of notes, then play pattern after pattern of variations. The repetition of "Listen to the coal . . . listen to the . . . listen to . . . listen" was one way to capture a feature of his playing. The image of the coal train disappearing into the night comes, particularly, from a place on the James River, west of Richmond, where I happened to be when I heard of Coltrane's death. Like all jazz musicians, I felt the loss very deeply.

Shakespeare's song contains an obvious illustration of onomatopoeia (the bell's sound), obvious alliteration in the *f*-full first line, and (less obviously) internal alliteration (note the *r* and *n* sounds), and assonance galore. Like a drowned man's bones, ordinary language becomes something "rich and strange" in this song.

(text pages) 639–640

Housman's lyric will reward the same sort of scrutiny, not that a class will sit still for much scrutiny of this sort!

Eliot's "Virginia" is an experiment in quantitative verse, according to George Williamson (*A Reader's Guide to T. S. Eliot* [New York: Noonday, 1957]). You might read aloud "Virginia" and Campion's quantitative "Rose-cheeked Laura" (page 653), and ask the class to detect any similarity. Ted Hughes has written of "Virginia" with admiration. How is it, he wonders, that Eliot can create so vivid a landscape without specific images? "What the poem does describe is a feeling of slowness, with a prevailing stillness, of suspended time, of heat and dryness, and fatigue, with an undertone of oppressive danger, like a hot afternoon that will turn to thunder and lightning" (*Poetry Is* [New York: Doubleday, 1967]).

21 Rhythm

STRESSES AND PAUSES

In the first section of this chapter, rhythm is discussed with as few technicalities as possible. For the instructor wishing to go on to the technicalities, the second part of the chapter, "Meter," will give the principles of scansion and the names of the metrical feet.

Except for one teacher at the University of Michigan, James Downer, who would illustrate the rhythms of Old English poetry by banging on his desk for a drum, we have never known anyone able to spend whole classes on meter without etherizing patients. Meter, it would seem, is best dealt with in discussing particular poems.

EXERCISE: *Appropriate and Inappropriate Rhythms*, page 643

The rollicking anapests of Poe and Cook seem inappropriate, ill-suited to the poets' macabre subject matter.

Tennyson's monosyllabic words, further slowed by pauses, convey not only the force of the tide, but its repetitiveness.

The rhythms of these lines by Keeler and Shakespeare seem suitably rollicking.

Gwendolyn Brooks, WE REAL COOL, *page 646*

The poet might have ended every line with a rime, as poets who rime usually do:

> We real cool.
> We left school.

The effect, then, would have been like a series of hammer blows, because of so many short end-stopped lines and so many rimes in such quick succession. But evidently Brooks is after a different rhythm. What is it? How to read the poem aloud? Let members of the class take turns trying and compare their various oral interpretations. If you stress each final *We*, then every syllable in the poem takes a stress; and if, besides, you make even a split-second pause at every line break, then you give those final *We's* still more emphasis. What if you don't stress the *We's*, but read them lightly? Then the result is a skipping rhythm, rather like that of some cool cat slapping his thighs.

After the class has mulled this problem, read them Brooks's own note on the poem (from her autobiography *Report from Part One* [Detroit: Broadside, 1972] 185):

> The ending WEs in "We Real Cool" are tiny, wispy, weakly argumentative "Kilroy-is-here" announcements. The boys have no accented sense of themselves, yet they are aware of a semi-defined personal importance. Say the "We" softly.

Kilroy, as students may need to know, was a fictitious—even mythical—character commemorated in graffiti chalked or penciled by U.S. soldiers wherever they traveled in World War II. KILROY WAS HERE was even scrawled in the sands of Anzio. A small testimonial that the graffitist is a person.

As a student remarked about the tone and theme of this poem, "She doesn't think they're real cool, she thinks they're real fool—to die so young like that."

Brooks has recorded her own reading of the poem for *The Spoken Arts Treasury of 100 Modern American Poets*, vol. 13, SA 1052.

(text pages) 646–649

Robert Frost, NEVER AGAIN WOULD BIRDS' SONG BE THE SAME, page 646

He is Adam. In line 9, you say "as MAY be," because the iambic rhythm wants you to. The effect of Frost's closing string of sixteen monosyllables is quietly powerful, in keeping with his final understatement. Much of our pleasure in those lines comes from hearing ordinary, speakable phrases so beautifully accommodated.

Sometimes teachers and critics try to account for the unique flavor of Frost's language by claiming that it is all a matter of his vocabulary. But what rural words are there in this poem, a poem as Yankee-sounding as they come, despite some highly literate diction ("Admittedly an eloquence")? No, what makes the lines memorable is that they embody what Frost called the "sound of sense." He was convinced that certain phrases, customarily spoken with feeling, have a pattern of intonation so distinctive that we can recognize them and catch their meaning even if we hear only the drone of them from behind a closed door. (See Frost's explanations and illustrations in *Selected Letters,* ed. Lawrance Thompson [New York: Holt, 1964] 79–81 and 111–13.)

Reading this poem to an audience in Ann Arbor in March 1962 when he was 88, Frost delivered its closing line with the air of a slowball pitcher hurling a flawless third strike into the catcher's glove to retire the side. The audience caught it and burst out clapping.

Frost's line of monosyllables, by the way, doesn't seem dull either.

Ben Jonson, SLOW, SLOW, FRESH FOUNT, KEEP TIME WITH MY SALT TEARS, page 647

O sounds slow the opening line, whose every word is a monosyllable. Further slowing the line, eight of the ten monosyllables take heavy beats. "Drop, drop, drop, drop" obviously racks up still more stresses, as do the spondees that begin lines 4, 5, and 6. The entire effect is that we are practically obliged to read or sing the poem slowly and deliberately — as befits a lamentation.

Alexander Pope, ATTICUS, page 647

Generosity, kindness, courage, wholeheartedness, and humility seem mainly what Atticus/Addison lacks. As this famous dissection shows, Pope's custom is to break each line after a word that completes not only the line but also a syntactical unit. And so the argument advances in couplets like locked boxes, each box neatly packed with phrases, dependent clauses, things in series, antitheses. Only line 3 is a run-on line, if you go by the book's simple advice that a run-on line doesn't end in a punctuation mark. A sensitive reader, reading the passage aloud, would probably pause slightly after *fires* anyway, for these couplets enforce a rhythm hard to ignore.

For a definition of the heroic couplet and a brief discussion of it, students may be directed to pages 662–663.

EXERCISE: *Two Kinds of Rhythm,* page 648

Sir Thomas Wyatt, WITH SERVING STILL, page 648

Dorothy Parker, RÉSUMÉ, page 649

These two poems differ in rhythm: Wyatt compels a heavy pause only at the end of every quatrain, while Parker end-stops every line. Students may be shown that pauses and meanings go together. Both poems are cast in two sentences, but Wyatt develops one uninterrupted statement throughout the entire poem (in sonnet fashion: first the summary of the speaker's problem in the opening three stanzas, then the conclusion beginning with "Wherefore all ye"). "Résumé," as its punctuation indicates, makes a new self-contained statement in every line.

A question on meaning: Must light verse necessarily be trivial in its theme? State Parker's theme in "Résumé." Surely, it isn't trivial. At least in theme, the poem seems comparable to Hamlet's soliloquy, "To be or not to be. . . . "

After *Not So Deep as a Well,* her collected poems of 1936, Parker brought out no more poetry collections. "My verses," she insisted to an interviewer. "I cannot say poems. Like everybody was then, I was following in the exquisite footsteps of Miss Millay, unhappily in my own horrible sneakers" *(Writers at Work: The Paris Review Interviews,* 1st ser. [New York: Viking, 1959]). Parker's wit, acerbic and sometimes macabre, is as clear from "Résumé" as it is from her celebrated remark on being informed that Calvin Coolidge had just died: "How do they know?"

METER

XJK used to think of a meter as a platonic ideal norm from which actual lines diverge, but J. V. Cunningham's essay "How Shall the Poem be Written?" changed his mind. Metrical patterns (in the abstract) do not exist; there are only lines that poets have written, in which meters may be recognized. "Meter," declares Cunningham, "is perceived in the actual stress-contour, or the line is perceived as unmetrical, or the perceiver doesn't perceive meter at all" *(The Collected Essays of J. V. Cunningham* [Chicago: Swallow, 1976] 262).

Max Beerbohm, ON THE IMPRINT OF THE FIRST ENGLISH EDITION OF THE WORKS OF MAX BEERBOHM, page 649

John Updike has paid tribute to this brilliant bit of fluff:

> The effortless a-b-a-b rhyming, the balance of "plain" and "nicely," the need for nicety in pronouncing "iambically" to scan—this is quintessential light verse, a twitting of the starkest prose into perfect form, a marriage of earth with light, and quite magical. Indeed, were I a high priest of literature, I would have this quatrain made into an amulet and wear it about my neck, for luck.
>
> ("Rhyming Max," a review of Beerbohm's collected verse reprinted in *Assorted Prose* [New York: Knopf, 1965].)

Thomas Campion, ROSE-CHEEKED LAURA, COME, page 653

Campion included this famous lyric in his polemic *Observations on the Art of English Poesie* (1602), in which he argued that English poets ought to adopt the quantitative meters of Greek and Latin. "This cannot be done in English," says John Hollander, "with its prominent word stress, save by assigning Latin vowel lengths to the written English, and simply patterning what amounts to a typographical code which cannot be heard as verse. . . . 'Rose-cheekt Laura' is therefore merely an unrhymed English trochaic poem, perfectly plain to the ear" (introduction to *Selected Songs of Thomas Campion,* selected by W. H. Auden [Boston: Godine, 1973]).

Walter Savage Landor, ON SEEING A HAIR OF LUCRETIA BORGIA, page 654

In first printing Landor's poem in 1825 in the *New Monthly,* Leigh Hunt explained, "A solitary hair of the famous Lucretia Borgia . . . was given me by a wild acquaintance who stole it from a lock of her hair preserved in the Ambrosian Library at Milan." (Lord Byron could very well have been the wild acquaintance.) According to Hunt, when he and Landor had met in Florence, they had struck up a conversation over the Borgia hair "as other acquaintances commence over a bottle."

(text pages) 655–657

EXERCISE: *Meaningful Variation*, page 655

Aside from minor variations from metrical norm (such as the substitution of a trochee for an iamb), the most meaningful departures in these passages seem to occur in these words or phrases:

1. Dryden: *deviates*. (Now there's a meaningful deviation!)
2. Pope: the spondees *snake*, *drags*, and *slow length*.
3. King: *like a soft drum*.
4. Longfellow: *autumnal*, and the last line, which we would scan, "The CAT a ract of DEATH FAR THUN der ing from the HEIGHTS." Wonderful arrangement of unstressed syllables in that line! Its rhythm is like that of an avalanche bumbling around for a while before rumbling down.
5. Stevens: *spontaneous, casual, ambiguous*.

EXERCISE: *Recognizing Rhythms*, page 656

Edna St. Vincent Millay, COUNTING-OUT RHYME, page 656

A. E. Housman, WHEN I WAS ONE-AND-TWENTY, page 656

William Carlos Williams, THE DESCENT OF WINTER (SECTION 10/30), page 657

Walt Whitman, BEAT! BEAT! DRUMS! page 657

Langston Hughes, DREAM BOOGIE, page 657

Probably it is more important that students be able to recognize a metrical poem than that they name its meter. The Millay and Housman poems are thoroughly metrical; the Williams and Whitman are not, but include metrical lines when the poets are describing or imitating the sound of something with a regular rhythm: the clank of freight car wheels, the whistle's *wha, wha,* the beating of drums. In Whitman's poem, besides the refrain (lines 1, 8, and 15) there are the primarily iambic lines that end each stanza. Hughes's "Dream Boogie" starts out with a metrical beat, then (deliberately) departs from it in the italicized interruptions.

22 Closed Form, Open Form

Beginning students of poetry often have a hard time appreciating either a sonnet or a poem in open verse because they have yet to distinguish one variety of poetry from the other. On first meeting an unfamiliar poem, the *experienced* reader probably recognizes it as metrical or nonmetrical from its opening lines—perhaps even can tell at first glance from its looks on the page (compact sonnet or spaced-out open verse). Such a reader then settles down to read with appropriate expectations, aware of the rules of the poem, looking forward to seeing how well the poet can play by them. But the inexperienced reader reads mainly for plain prose sense, unaware of the rhythms of a Whitmanic long line or the rewards of a sonnet artfully fulfilling its fourteenth line. Asked to write about poetry, the novice reader may even blame the sonnet for being "too rigid," or blame William Carlos Williams for "lacking music" (that is, lacking a rime scheme), or for "running wild." Such readers may have a right to their preferences, but they say nothing about a poem, nor about the poet's accomplishments.

That is why this chapter seems essential. To put across to students the differences between the two formal varieties, it isn't necessary to deal with every last fixed form, either. One can do much by comparing two poems (closed and open) on the theme of sorrow— Elizabeth Barrett Browning's fine sonnet "Grief" and Stephen Crane's astonishing "The Heart." Before taking up closed form, you might care to teach some song lyrics—those in Chapter 7, or a couple of traditional folk ballads. That way the student isn't likely to regard fixed forms as arbitrary constructions invented by English teachers. A stanza, you can point out, is the form that words naturally take when sung to a tune. That is how stanzas began. Sing a second round of a song, and you find yourself repeating the pattern of it.

CLOSED FORM: BLANK VERSE, STANZA, SONNET

John Keats, THIS LIVING HAND, NOW WARM AND CAPABLE, page 662

After Keats's death, these grim lines were discovered in the margin of one of his manuscripts. Robert Gittings has pointed out that the burden of the poem is much like that of two letters Keats wrote late in life to Fanny Brawne, charging her conscience with his approaching death and blaming her for enjoying good health. "This," says Gittings, "marks the lowest depths of his disease-ridden repudiation of both love and poetry" (*John Keats* [Boston: Atlantic-Little, 1968] 403). To discuss: can a repudiation of poetry nevertheless be a good poem?

John Donne, SONG ("GO AND CATCH A FALLING STAR"), page 664

Maybe it is worth pointing out that, in bringing together short stanzas to make one longer one, Donne hasn't simply joined quatrain, couplet, and tercet like a man making up a freight train by coupling boxcars. In sense and syntax, each long stanza is all one; its units would be incomplete if they were separated.

Ronald Gross, YIELD, page 665

Fitting together drab and prosaic materials, Gross leaves them practically unaltered. What he lends them are patterns that seem meaningful. By combining traffic-sign messages in "Yield," he implies that the signs insistently pressure us with their yips and barks. "Yield" states its theme implicitly: we are continually being ordered to conform, to give in, to

(text pages) 665–669

go along with laws laid down for us. We must heed the signs in order to drive a car, but perhaps it is chilling to find their commands so starkly abstracted. Students might wish to discuss whether it is reading too much into the poem to suspect that this theme applies to other areas of our lives, not only to driving.

A discussion of Gross's work may be one of those rare sessions that end with the students' realization that to remove speech from its workaday contexts and to place it into lines is, after all, what most poets do. Many poems, not only found poems, reveal meanings by arranging familiar things into fresh orders.

Nancy Adams Malone of Mattatuck Community College in Waterbury, Connecticut contributes an insight about "Yield": "Everybody seems to think it's a comment on social conformity, but it seems to me easier and more fun to read it as a seduction."

Experiment: FINDING A POEM, page 667

Timothy F. Walsh, of Otero Junior College in Colorado, discovered another found poem in this book's previous edition:

"The sonnet,"
in the view of Robert Bly,
"is where old professors
go to die."

"It was fun," he writes, "to discuss found poems following my students' reading about them on page 664, then point to one I found just two pages later."

Michael Drayton, SINCE THERE'S NO HELP, COME LET US KISS AND PART, page 668

Nay, yea, wouldst, and *mightst* are the only words that couldn't equally well come out of the mouth of a lover in the twentieth century.

There seems to be an allegorical drama taking place, as Laurence Perrine has pointed out in "A Drayton Sonnet," *CEA Critic* 25 (June 1963): 8. Love is also called Passion; and apparently his death is being urged along by the woman's infernal Innocence.

Elizabeth Barrett Browning, GRIEF, page 669

The octave may be forgettable, but the sestet is perhaps one of the eight or ten high moments in English sonnetry. "Grief" antedates the more famous *Sonnets from the Portuguese* (mostly composed in 1845 and 1846). In 1842, when working on "Grief" and some sonnet exercises, Mrs. Browning wrote to her friend Mary Mitford, "The Sonnet structure is a very fine one, however imperious, and I never *would* believe that our language is unqualified for the very strictest Italian form." But as Alethea Hayter has noticed, "Grief" isn't at all in the strictest Italian form; the shift in its argument comes in the middle of line 8. In her early sonnets, says Hayter, the poet seems to be "arranging roses now in a tall vase, now in a flat bowl, but always in something either too tall or just too shallow" (*Mrs. Browning* [New York: Barnes, 1963]). Still, in gratitude for the sestet of "Grief," we can accept many awkward roses.

R. S. Gwynn, SCENES FROM THE PLAYROOM, page 669

This ironic, outrageous poem might be valuable to turn to on a day when a class needs waking up.

How much do we know about these people? That this family is wealthy is suggested by the fact that it's the cook's night off. That these brats are treated indulgently is hinted by the children's having apparently killed a series of unlucky goldfish before this latest one. If Mother were alert enough to notice the monstrous acts of her Gestapo brood, much more than onions would make her cry.

(text pages) 669–674

Like the reader, the poet deplores both the children and their parents. His attitude is made plain by his drawing the children as Nazis, their parents as blind or foolish.

Of the younger formalist poets now coming to prominence, Gwynn may be the keenest satirist. This sonnet is from his collection *The Drive-In* (Columbia: U of Missouri P, 1986). He has also written an extended satire on current poets and poetry in the great tradition of Pope's *Dunciad: The Narcissiad* (New Braunfels, TX: Cedar Rock P, 1981). He frequently reviews poetry for *Texas Review, Sewanee Review,* and other literary journals, and is currently an associate professor of English at Lamar University in Beaumont, Texas.

Alexander Pope, Epigram Engraved on the Collar of a Dog, page 670

Students may be asked: What's the point? Pope makes a devastating comment on society. With few exceptions (such as His Royal Highness), every man is a dog: owned by somebody, accepting handouts, licking his master's hand, learning to heel.

Martial, Sir John Harrington, William Blake, E. E. Cummings, Langston Hughes, J. V. Cunningham, John Frederick Nims, Stevie Smith, Robert Crawford, Paul Ramsey, R. L. Barth, Bruce Bennett, A selection of epigrams, pages 671–673

Highly various, these twelve examples illustrate the persistence of the epigram since Martial. Blake and Barth offer definitions of the epigram, written as epigrams. Not all epigrams come in rimed couplets, as Langston Hughes and Bruce Bennett demonstrate. Whether the form of an epigram is closed or open, its essense consists of brevity and a final dash of wit.

Besides writing "Of Treason," called the best epigram in English, Harrington has another claim to immortality: he invented the water closet.

Cunningham, American master of the verse epigram in our time, has had few recent rivals. Instructors who wish further examples of this fixed form will find many to quote in his *Collected Poems and Epigrams* (Chicago: Swallow, 1971).

Nims, closest rival to Cunningham, has collected his epigrams, including "Contemplation," in *Of Flesh and Bone* (New Brunswick: Rutgers UP, 1967). When first printed, in *The New Yorker,* this poem was called "A Thought for Tristram"—suggesting that *you* means Isolde, betrothed of King Mark, with whom Tristan/Tristram shares a love potion.

If the haikulike brevity of epigrams tempts you to ask your class to write a few, resist the temptation. Even from a bright class the results are likely to depress you. A successful epigrammatist needs, besides the ability to condense, the ability to deliver that final rapier thrust of nastiness. A talented creative writing class, after tackling poems in a few of the less demanding forms (ballads, villanelles, sestinas) might try epigrams, either rimed or rimeless.

Dylan Thomas, Do not go gentle into that good night, page 674

No mere trivial exercise (as a villanelle tends to be), Thomas's poem voices his distress at the decline and approaching death of his father. At the time, the elder Thomas was a semi-invalid, going blind, suffering from the effects of tongue cancer. As a teacher of English at Swansea Grammar School, the poet's father had ruled his class with authority; but those who knew him only in his last years knew a different, humbled man. (See Constantine FitzGibbon, *The Life of Dylan Thomas* [Boston: Atlantic-Little, 1965] 294–95.)

Like many other Thomas poems, this one contains serious puns: *good night, grave.* "Another assumption in this poem," says Amy Mulvahill (in a student paper written at Tufts), "may be Thomas's own self-destructive drive that led him to drink himself to death. It's possible that he preferred to taunt death with his boisterous life—to go down unrepentant and brawling."

(text pages) 674–681

Repetitious as a villanelle is, the form suits this poem, making its refrains sound like prayers said over and over. If you have any student poets, they might be challenged to write villanelles of their own. The hard part is to make the repeated lines occur naturally, to make them happen in places where there *is* something to be said. But the repetitious form is helpful. Write the two refrain lines and already your labors are eight nineteenths over.

For another instance of Thomas's fondness for demanding, arbitrary forms, see the poem "Prologue," at the beginning of Daniel Jones's edition of *The Poems of Dylan Thomas* (New York: New Directions, 1971). A poem of 102 lines, its first and last lines rime with each other, as do lines 2 and 101, 3 and 100, 4 and 99, and so on, until two riming lines collide in the poem's exact center. Except for that inmost pair of lines, no reader is likely to notice the elaborate rime scheme since rimes so far apart can't be heard; but apparently it supplied the poet with obstacles to overcome and with a gamelike pleasure.

OPEN FORM

E. E. Cummings, BUFFALO BILL'S, page 679

Cleanth Brooks and Robert Penn Warren have taken this poem to be an admiring tribute to William Cody (*Understanding Poetry*, 3rd ed. [New York: Holt, 1960]). But Louis J. Budd, in an interesting dissent, thinks Cummings is satirizing the theatricality of the old sideshow straight shooter and finds Mister Death "a cosmic corporal gathering up defunct tingods and stuffed effigies" (*The Explicator* 11 [June 1953]: item 55).

Emily Dickinson, VICTORY COMES LATE, page 680

Of all Dickinson's poems, this is the most formally open, and as a result it sounds almost contemporary. Clearly the poet is in charge of her open form, for each line-break comes after a word that (being essential to meaning) easily stands for the special emphasis. We could imagine a few other possible locations for line-breaks. (After *Love* in the last line?—or would that make the poem seem to end in a riming couplet?) But no doubt the poet knew her own mind.

Dickinson's outsized hyphens (or delicate minidashes) usually seem to indicate pauses, little hesitations, as if to give a word or phrase (whether following the dash or preceding it) special emphasis.

If there is time to work further with this splendid poem, both language and meaning deserve a deeper look. Note the pun in line 3: *rapt* for "wrapped." In the next-to-last-line, *keep* seems a subjunctive verb: the statement may be a prayer. What sort of "Victory" does the poet mean? Thomas H. Johnson thought she referred to a Civil War battle, but we think this is one of her several poems concerned with reputation and fame. Perhaps the sparrows are minor poets, like Dickinson, unprinted and unacclaimed. Perhaps the Eagle is a famous bard glutted with renown: Ralph Waldo Emerson, maybe. The conclusion seems a bitter acceptance, as if to say: "God is wise not to feed me too generously." Ironically, the poem today seems literally true: Dickinson's victory has indeed come, although she cannot taste it.

William Carlos Williams, THE DANCE, page 681

Scanned, the poem is seen to abound in pairs of unstressed syllables. The result is a bouncing rhythm—anapestic or dactylic, depending on where one wishes to slice the lines into feet. This rhythm seems appropriate to a description of frolicking dancers and helps establish the tone of the poem, which is light, however serious.

Williams severs his units of sense again and again in midphrase: placing his linebreaks after *and, the, about, thick, those, such.* In this poem run-on lines predominate, and this is not only a technical device but a way of underlining the poem's meaning. Williams conveys a sense of continuous movement in a syntax that keeps overflowing line units.

By repeating its opening line, the poem, like Breughel's dancers, comes round in a circle to where it began. Another metaphor is possible: like a painting enclosed in a frame, the poem encloses its central scene in a frame of words.

Williams first saw Breughel's painting in Vienna in 1924, but wrote this poem in 1942, some eighteen years later. A French critic, Jacqueline Saunier-Ollier, has speculated on the curious fact that the poem, in describing a vividly colorful tableau, omits all color images. Her work on Williams's Breughel poems is summed up in *William Carlos Williams: Man and Poet*, ed. Carroll F. Terrell (Orono: National Poetry Foundation, 1983) 528–29.

Stephen Crane, THE HEART, page 681

Walt Whitman, CAVALRY CROSSING A FORD, page 186

Two nineteenth-century American poems, the pair seem comparable mainly in brevity and use of narration. The assonance and internal alliteration in Whitman's phrase *silvery river* are echoed in the poem's very opening line: the assonance of the *i*-sound in *line, wind, islands;* the internal alliteration of the *r* in *array, where, green.* But any line of this short poem will repay such inspection. Crane's "The Heart" is obviously less heavy on verbal music, although *Held his heart in his hands* is heavily alliterative; and the second stanza favors the letter *b.* There is rime, too: *it/bitter, bitter/heart.*

Whitman seems to lambaste his poem with sound-effects in his enthusiasm for his grand military spectacle. Crane cares for music, too, and yet his is a subtler, harsher one. Although longer in words, Whitman's "Cavalry" contains fewer pauses than "The Heart" (fifteen compared to Crane's seventeen, if every comma and line-end counts as a pause). The result is, in Crane's poem, a much more hesitant, start-and-stop movement—appropriate, perhaps, to a study of self-immolation. Whitman apparently wants an expansive, continuous progress in his syntax, as in his cavalry.

Wallace Stevens, THIRTEEN WAYS OF LOOKING AT A BLACKBIRD, page 682

Suggestive as blackbirds may be, the theme of the poem is, "Pay attention to physical reality." Stevens chides the thin, ascetic men of Haddam who would ignore good blackbirds and actual women for golden phantasms. He also chides that asinine aristocrat who rides about Connecticut (of all places) in a glass coach as if thinking himself Prince Charming. The poem ends in a section whose tone is matter-of-fact flatness, rather as though Stevens were saying, "Well, here's the way the world is; if you don't like it, go read newspapers." Taken as a series of notes for an argument for literalism, this much-discussed poem seems to have unity and to lead to a definite conclusion. For another (and more complicated) view of it, see Helen Hennessy Vendler, *On Extended Wings* (Cambridge: Harvard UP, 1969).

Way-of-looking number 5 recalls Keats's "Grecian Urn": "Heard melodies are sweet...."

Way number 10 eludes final paraphrase. Are the "bawds of euphony" supposed to be, perhaps, crass ex-poets who have sold out their Muses, who utter music to please the box office instead of truth? But blackbirds flying in a green light are so strikingly beautiful that even those dull bawds would be moved to exclaim at the sight of them.

Gary Gildner, FIRST PRACTICE, page 685

Some possible answers to the questions:

1. Hill is a sadist. He is determined to render defenseless boys as bitter as he is.

2. The speaker doesn't want to identify himself with such a world-view; apparently, he wishes to remove it from himself by thrusting the hideous Hill away off in the third person.

3. The broken line indicates a pause—while the boy athletes suck in their breaths and don't dare answer.

4. Closure.

5. The rewrite would make Gildner's appalling poem seem neat, swingy, and jingly.

6. Poems don't have to traffic in moonlight and roses. Nothing in human experience need be alien to them—not even Hill.

In a recent comment on "First Practice," Gildner recalls that the poem came straight from his boyhood in the 1950s. When his parochial school (Holy Redeemer, grades one through eight) started a football team, an ex-Marine who had played ball in service volunteered to coach it. Practice was held in The Bomb Shelter, a thick-walled basement with a low ceiling. "Some of us later felt slimy and ashamed at hitting out like that—at being *moved* by Cliff's speech. And I know that many of us were more afraid to lose a game than to break a bone. . . . For years I tried to write a short story about this experience, but the attempts always sounded wrong, false. One day I decided to simply 'list' the story's important elements. Except for one unnecessary word, the poem 'First Practice' appeared" (comment in *Poetspeak: In their work, about their work,* ed. Paul B. Janeczko [Scarsdale: Bradbury, 1983] 39–41).

Michael Heffernan, LIVING ROOM, page 686

Although its open form makes it look casual in design, although its flat lines often seem close to prose, "Living Room" is carefully made: there's a firm hand at the helm. Heffernan's lines break on nouns, strong verbs, the ends of phrases. This story makes a point, and the poem tells a complete story.

This poem is clearly set in a small town, where everybody knows everybody else: John and Phil's hardware store, Uncle Bud the jeweler.

An oddly powerful poem, and an infectiously funny one. We can't help feeling that there's something sinister about cooking a wild animal in the backroom and washing it down with bourbon. This is a secret feast to which only the good customers, the initiates, are invited—and invited to risk food poisoning. It would be easy to read too much into the poem, but it is hard not to suspect the poet of playing myth against myth: secret feast versus Feast of the Nativity. The wild animal stewed, its paw severed and placed on the scalepan—there's some pagan ritual here, a hint of black magic and alchemy. (Also a pun on *nails*.) There's something anti-Christmas in Phil's jokingly placing that sinister paw on bread (making a disgusting mock hors d'oeuvre). A feast of rotten raccoon is an odd way to celebrate the feast of the Nativity!

But in the end the poem develops a contrast between darkness and light, between the hardware store where toilet parts hobnob with the poor beast's parts and (once the speaker has "stepped into the light") the jewelry shop where delighted Uncle Bud decrees Kathy's happiness. It moves from a room full of death (the backroom) and the "Living Room" of the title. All the stomach-turning detail about spoiled raccoon and the poor beast's cut-off paw gives way—surprise!—to a closing image of love and loveliness. Kathy, the speaker's wife, delighted with her new watch, dances to a song that happens to be on the air. The demons of the severed paw have lost. Maybe "nobody's gods win," and yet the demons don't win, either—Kathy gets her watch, and joy triumphs.

The theme is nutshelled in stanza five, but these lines will be hard to paraphrase. A stab at rewording the theme: Things on earth are *not* as they are in Heaven. In the ordinary human world "down here," darkness and light mingle. Good and evil are never absolutes. Christmas and rotten raccoon coexist; one trip downtown (with a stop at the Underworld of the raccoon-cooker) makes the dance in the Living Room possible.

Michael Heffernan, who teaches in the creative writing program at the University of Arkansas, has published three highly accomplished collections: *The Cry of Oliver Hardy*

(U of Georgia P, 1979), *To the Wreakers of Havoc* (U of Georgia P, 1984), and *The Man at Home* (U of Arkansas P, 1988), from which "Living Room" is taken.

For Review and Further Study

Leigh Hunt, Rondeau, *page 687*

Jenny (so the story goes) was Jane (Mrs. Thomas) Carlyle, who gave Hunt a buss when he brought word that one of Carlyle's books had been accepted by a publisher.

A true French rondeau has fifteen lines, and follows rules more ingenious than those Hunt set for himself. For specifications, see Lewis Turco, *The Book of Forms* (Hanover: UP of New England, 1986) 215–216.

Keith Waldrop, Proposition II, *page 687*

Like an epigram, this poem is concise, and the whole poem leads up to a conclusive word: *wind*. Unlike an epigram, its subject isn't human folly, and its apparent purpose may be to express an insight, not to deliver a witty blow.

Elizabeth Bishop, Sestina, *page 688*

We would answer the questions like this.

1. That some terrible loss—a death in the family?—causes the grandmother to weep seems a guess that fits the poem. The old woman tries to hide her grief from the child (lines 6, 10, 31–32); she thinks it was somehow foretold (9).

2. We have no authority to read this poem as autobiography, but the figure of the grandmother—the most important person in Bishop's early life—and the stormy setting (such as we might find in a village on the Nova Scotia coast) invite us to do so. The source of grief may have been the death of the poet's father (hence, an irony that the child draws a man with tear-shaped buttons) or it may have been the illness of her mother, hospitalized several times for a mental disorder. When Bishop was eight months old her father died, and according to Robert Giroux, "The first real home Elizabeth knew was in the coastal town of Great Village, Nova Scotia, where her widowed mother returned in order to be with her parents" (introduction to Bishop's *Collected Prose* [New York: Farrar, 1984]). When the poet was five, her mother had a final breakdown, leaving the girl in the care of her grandmother. Apparently Bishop looked back on her days in Nova Scotia with affectionate yearning. When she was six, her father's wealthy parents moved her to Worcester, Massachusetts, for a less happy stay.

3. Small round pieces of paper. Almanacs (such as *The Old Farmer's*) come with punched holes to make them easy to string and hang on a hook or a nail.

4. The playful ingenuity of the sestina, like that of the villanelle, tempts a poet to wax clever; yet Bishop is writing a deeply felt, moving poem in it. The tone is lightly serious, compassionate—yet with touches of gentle humor: the Little Marvel Stove, the child's drawings. Irony, too, informs the poem: a contrast between the grandmother's sorrow and the child's innocent ignorance.

5. Nims's comment seems an apt description of "Sestina." In the six repeated words, we are given the setting *(house)* the characters *(grandmother, child)*, and key symbols *(stove, almanac, tears)*. "Sestina" weaves all six into a subtle relationship. This poem is full of things that suggest magic: the prophetic almanac, the teacup (with which fortune-tellers divine), the "marvellous stove." It also is full of secret-keepers: the grandmother, the almanac with its powers of prophecy, the concluding reference to the "inscrutable" house." The repetitions are worth tracing: *tears*, in particular, accumulates an effect. In stanza 2 the tears

(text pages) 688–692

arrive like an equinoctial storm; in 3, the kettle also weeps; in 4, tea is tears; in 5, the man in the child's drawing wears tears; in 6, the almanac weeps paper tears; and finally, in the envoy, tears are flowers. "Time to plant tears" may be literal quotation from the almanac, *tears* being (if memory serves) the name of a small white flower favored by rock gardeners.

Bishop's *Complete Poems* contains another intriguing sestina: "A Miracle for Breakfast." At the time it was written Bishop remarked (in a 1937 letter to Marianne Moore):

> It seems to me that there are two ways possible for a sestina—one is to use unusual words as terminations, in which case they would have to be used differently as often as possible—as you say, 'change, of scale.' That would make a very highly seasoned kind of poem. And the other way is to use as colorless words as possible—like Sidney, so that it becomes less of a trick and more of a natural theme and variations. I guess I have tried to do both at once. (Quoted by Nims in his essay, cited in question 5.)

In the later "Sestina," the terminal words seem deliberately usual ones.

EXERCISE: *Urgent Repetition*, page 689

This experiment just might leave you surprised at the quality of some of its results. Whoever writes a sestina has a powerful ally—the form—on his or her side.

In a tour de force, a student I knew in a poetry workshop at Tufts once wrote a fairly successful sestina taking *one, two, three, four, five* and *six* for its repeated words. The result seemed only mildly boring and mechanical!

Geoffrey Chaucer, YOUR YEN TWO WOL SLEE ME SODENLY, page 690

It can be great fun for students to learn (well, more or less) how to pronounce Chaucer's English, provided one has the time and strength to help them make the attempt. One does much better by Chaucer's lines if one puts on an Irish brogue. (A couple of Guinness stouts before class usually helps.)

Some scholars doubt that Chaucer himself wrote this poem; but if he did not, someone who thoroughly knew Chaucer's work probably did.

"Since I escaped from love, I've grown so fat . . . " is, of course, a crude modernization of another poem from the "Merciles Beaute" series. Carlos Baker offers another modern American version of it in his book of poems, *A Year and a Day* (Nashville: Vanderbilt UP, 1963).

EXERCISE: *Seeing the Logic of Open Form Verse*, page 690

E. E. Cummings, IN JUST-, page 691

Linda Pastan, JUMP CABLING, page 691

Donald Finkel, GESTURE, page 692

Charles Olson, LA CHUTE, page 692

Cummings's poem is one of his "Chansons Innocentes," little songs for children. In it, however, we meet a poet who is familiar with the classics and who naturally associates spring with goat-footed Pan. In Greek mythology, the god's pipes heralded the return of Persephone, and caused birds and beasts to start up at his call. In Cummings's view, he seems a kind of Pied Piper who brings children running.

158

Line-breaks and capital letters in the poem seem designed to emphasize particulars. *Just-spring*, capitalized, is the name of a holiday: the moment when spring begins. Dividing its name with a line-break gives it more importance, perhaps; and *mud- / luscious* similarly takes emphasis. Why are the children's names telescoped (*eddieandbill, bettyandisbel*)? So that these names will be spoken rapidly, pell-mell, the way their owners run and the way children speak about their friends. And when the lame balloonman completes his transformation into Pan, the word *goat-footed* is framed with white space on a line by itself. Except by putting it in capitals, the poet could hardly have thrown more weight on it.

Pastan, in telling us of two cars, divides each line and builds separate columns of words. The last line, all one, fits the proposed action (that the two merge) to the words. Jump cabling we take to be a metaphor for coitus. (For a similar comparison, see E. E. Cummings's well-known "she being Brand / -new," in his *Complete Poems*.) The playful "Jump Cabling" is starkly different in tone from Pastan's "Ethics" (page 855); it was first published in an anthology of light verse, Robert Wallace's *Light Year '85* (Cleveland: Bits, 1984).

Finkel's "Gesture" arranges lines so that the reader's following eyes make movements like those described. The poem centers on a witty metaphor: a poet presenting a poem to a reader is like someone cracking a whip. Indentations in lines 6–9 roughly depict the path of the whip in uncoiling. Then the eye-arresting block of type—

snaps
softly

—brings the reader's eyes to a momentary halt, suggesting the crack of the whip. There's a small pause before the whip straightens and the poem goes on to end. Obviously, to be appreciated, this poem needs to be seen in print.

Charles Olson's "La Chute," like "in Just-," can be effective even if merely heard aloud. So many times repeated, the words *drum* and *lute* acquire impressive emphasis; and by repeating phrases and hesitating (*who / will bring it up, my lute / who will bring it up where it fell*), Olson makes his poem sound unrehearsed and conversational. White space seems used somewhat cryptically and arbitrarily, but it does throw great weight on *my lute* (line 7) and *They* (line 10), essential words that can take the emphasis. What is the poem saying? We take it to be (like "Gesture") a poem about poetry, with lute and drum representing the poet's talents. Well aware that he was a forefather of a poetic school (the Black Mountain poets), Olson may have foreseen his death and wondered who would carry on his work for him.

23 Poems for the Eye

For more examples of graphic poetry, see the anthologies edited by Klonsky and Kostelanetz (cited in footnotes to this chapter). Other useful anthologies include Emmett Williams's *Anthology of Concrete Poetry* (New York: Something Else, 1967), Eugene Wildman's *Chicago Review Anthology of Concretism* (Chicago: Chicago Review, 1967), Mary Ellen Solt's *Concrete Poetry: A World View* (Bloomington: Indiana UP, 1969), and Emmett Williams's selection of "Language Happenings" in *Open Poetry: Four Anthologies of Expanded Poems*, ed. Ronald Gross and George Quasha (New York: Simon, 1973).

George Herbert, EASTER WINGS, page 694

John Hollander, SWAN AND SHADOW, page 695

The tradition of the shaped poem, or *Carmen figuratum*, seems to have begun in Renaissance Italy, and the form flourished throughout Western Europe in the seventeenth century. English practitioners of the form, besides Herbert, included Robert Herrick (in "The Pillar of Fame") and George Puttenham.

Of "Easter Wings," Joan Bennett has remarked, "The shape of the wings on the page may have nothing but ingenuity to recommend it, but the diminuendo and crescendo that bring it about are expressive both of the rise and fall of the lark's song and flight (Herbert's image) and also the fall of man and his resurrection in Christ (the subject that the image represents)" (qtd. by F. E. Hutchinson in his edition of Herbert's *Works* [Oxford: Oxford UP, 1941]). Visual shape and verbal meaning coincide strikingly when the second stanza dwindles to *Most thin*.

Like Herbert, Hollander clearly assumes that a word-shape has to have a meaningful relation to what is said in it. His reflected swan is one of twenty-five shaped poems collected in *Types of Shape* (New York: Atheneum, 1969). Other graphic poems in the book include a car key, a goblet, a beach umbrella, an Eskimo Pie, and the outline of New York State. Paul Fussell, Jr., discussing "Easter Wings" and Hollander's shaped poems, expresses reservations about this kind of poetry. Most shaped poems, he finds, are directed more to eyes than ears—"or better, we feel that the two dimensions are not married: one is simply in command of the other." But the greatest limitation in the genre is that there are few objects that shaped poems can effectively represent: "their shapes can reflect the silhouettes of wings, bottles, hourglasses, and altars, but where do we go from there?" (*Poetic Meter and Poetic Form* [New York: Random, 1965] 185–87). Students might be told of Fussell's view and asked to comment. A further disadvantage of most shaped poetry is that it cannot be heard aloud without loss.

William Blake, A POISON TREE, page 697

Reproduced in approximately the same size as the original, Blake's engraving first appeared in his *Songs of Experience* (1794). According to the poet, he had long sought the best method for presenting his poems. At last it was revealed to him by his dead brother Robert in a dream. He was to engrave both poem and design on a copper plate and, when the pages were printed, outlines would be filled in with watercolors. Each copy would be a bit different from every other. Many editions of Blake's poems and drawings show this method. An inexpensive color facsimile of the first edition of *Songs of Innocence* has been issued by Dover (New York, 1971).

"A Poison Tree" seems one of Blake's more fortunate marriages of visual art with poetry. His illustration for his celebrated "Tyger" (page 788), also from the *Songs of Experience,* shows a beast much less formidable than the one we imagine from the words alone. A topic for discussion of writing: Should poetry be illustrated? or should a poem be left to the eye of the mind?

Edwin Morgan, SIESTA OF A HUNGARIAN SNAKE, page 699

While too much can be made of this work, evidently it has sound values as well as eye appeal. Some brave volunteer might read it aloud. The initial *sz* sounds like the word *Siesta.* Perhaps the artist reverses the letters S and Z to make his line sound more like a snore (inhale: SZ—exhale: ZS). SZ looks and sounds like the end of an Eastern European name; perhaps that is why the snake is Hungarian.

Why the alternation of large and small letters? To look like stripes in the middle of the snake at his fattest part.

If a poem by definition is made of words, it may be questioned whether this concrete poem may be called a poem.

Dorthi Charles, CONCRETE CAT, page 700

This trifle first appeared in the second edition of *An Introduction to Poetry* and has been retained out of loyalty to the past. While hunting for an illustration of the sillier kind of concrete poem that simply and unfeelingly arranges words like so many Lincoln Logs, XJK found the very thing in one of William Cole's anthologies of humorous poetry: "Concrete Poem" by the British wit Anthony Mundy. Mundy's work repeats *miniskirt* several times in the form a miniskirt, and tacks on a couple of *leglegleglegs*. No doubt he was parodying concrete poetry, too. But the cheapskate in XJK rebelled at the thought of paying for permission to reprint such a simple doodad, so decided to cut and paste together a homemade specimen. While constructing the cat, he started having some fun with it, making the tongue a *U,* and so on. As far as we know, however, the pun in the cat's middle stripe (*tripes*) is the only place where language aspires toward poetry and becomes figurative.

24 Symbol

T. S. Eliot, THE BOSTON EVENING TRANSCRIPT, page 702

To help a class see the humor of Eliot's poem, try reading it aloud and pronouncing the name of the newspaper slowly and deliberately, in the dullest tones you can muster. This small gem can serve effectively to introduce an early, longer Eliot poem of spiritual desolation, "The Love Song of J. Alfred Prufrock" (page 805).

Emily Dickinson, THE LIGHTNING IS A YELLOW FORK, page 703

Perhaps the poet would have added more punctuation to this poem had she worked longer on it; a rough penciled draft is its only surviving manuscript. Students may ask, Isn't the fork a symbol? No, it is the other half of a metaphor: what the lightning is like. The lightning (like most literary symbols) is a physical thing or event, reportedly seen. The Apparatus of the Dark (neither fork nor lightning) is whatever dimly glimpsed furniture this cosmic house may hold. The fork seems too simple an instrument to deserve the name of Apparatus. The lightning is doing the revealing, not itself being revealed.

Thomas Hardy, NEUTRAL TONES, page 705

Students usually like to sort out the poem's white, gray washed-out, and ashy things. Can anyone think of a more awful description of a smile than that in lines 9–10? The God in line 2 seems angry and awe-inspiring. He has chided or reproved the sun, and caused it to turn pale in fear (like a schoolboy before a stern headmaster).

Line 8 is a stickler. In Hardy's first draft it read, "On which was more wrecked by our love." Both versions of the line seem awkward, and the present version is obscure, but probably the sense of this and the previous line goes: we exchanged a few words about the question, Which one of us had lost (suffered) the more by our love affair? (That is, after *which* we should mentally insert "of the two of us.")

For speculation about the facts behind "Neutral Tones," see Robert Gittings's fine biography *Young Thomas Hardy* (Boston: Little, 1975) 86–93. Much has been guessed about the possible love affair between young Hardy and his cousin Tryphena Sparks; but if the woman in "Neutral Tones" was indeed real, no one has identified her for sure.

Similar in imagery to "Neutral Tones" is this horrific line from Hardy's novel *The Woodlanders*, chapter 4, when a poverty-stricken woman, Marty South, sees her last hopes expire: "The bleared white visage of a sunless winter day emerged like a deadborn child" (cited by F. B. Pinion in *A Commentary on the Poems of Thomas Hardy* [New York: Barnes, 1977]).

George Herbert, REDEMPTION, page 707

The old burdensome lease that the speaker longs to cancel is original sin, which Christ, by his sacrifice (lines 12–14) allows humankind to throw off. Who is the speaker? Humanity—or perhaps (like Bunyan's Pilgrim) an individual soul in search of salvation.

What figure of speech is "a ragged noise"? (It's like the "Blue—uncertain stumbling buzz" of the Emily Dickinson poem discussed below.)

Emily Dickinson, I HEARD A FLY BUZZ — WHEN I DIED, page 708

Plump with suggestions, this celebrated fly well demonstrates a symbol's indefiniteness. The fly appears in the room—on time, like the Angel of Death—and yet it is decidedly ordinary. A final visitor from the natural world, it brings to mind an assortment of suggestions, some offensive (filth, stenches, rotting meat, offal, and so forth). But a natural fly is a minor annoyance; and so is death, if one is certain of Eternity. Unsure and hesitant in its flight, the fly buzzes as though faltering. It is another failing thing, like the light that comes through the windows and through the eyes (which are, as a trite phrase calls them, "the windows of the soul"). For other transferred epithets (like "Blue—uncertain stumbling Buzz"), see pages 595–596 of the text, in the chapter on figures of speech.

Most students will easily identify "Eyes around" as those of surrounding friends or relatives, and "that last Onset" as death throes. Is "the King" Death or Jesus? It seems more likely that the friends and relatives will behold death. What is the speaker's assignable portion? Physical things: keepsakes bequeathed to friends and relatives; body, to the earth.

Discussion will probably focus on the final line. It may help students to remember that the speaker is, at the present moment of the poem, in Eternity. The scene she describes is therefore a vision within a vision. Perhaps all the last line means is (as John Ciardi has argued), "And then there was no more of me, and nothing to see with." But the last line suddenly thrusts the speaker to Heaven. For one terrible moment she finds herself, with immortal eyes, looking back through her mortal eyes at a blackness where there used to be light.

Robert Frost, THE ROAD NOT TAKEN, page 708

Stanley Burnshaw writes, in *Robert Frost Himself* (New York: Braziller, 1986), that Frost often said "The Road Not Taken" was about himself combined with Edward Thomas, a Welsh poet and good friend. Knowing this, Burnshaw confessed, didn't contribute much to his understanding of the poem. Still, the story is tantalizing. In *Robert Frost: The Years of Triumph* (New York: Holt, 1970) biographer Lawrance Thompson tells about the "excruciations through which this dour Welshman [Thomas] went each time he was required to make a choice." This amused Frost, who once said to Thomas, "No matter which road you take, you'll always sigh, and wish you'd taken another." "The Road Not Taken" (originally called "Two Roads") was apparently written to poke quiet fun at this failing. When Frost sent the poem in a letter to Thomas, the Welshman apparently missed the joke. He assumed, as have many readers since, that the speaker in the poem was Frost himself. Disappointed, Frost (according to Thompson) "could never bear to tell the truth about the failure of this lyric to perform as he intended it."

Despite the ambiguity that surrounds the poet's intent, the poem succeeds. The two roads are aptly symbolic of the choices we have to make almost every day of our lives. Still, perhaps the poem's essential playfulness is evident in the dramatic "sigh" with which the speaker expects some day to talk about his choice, and in the portentousness of the last line, which seems a bit exaggerated considering that the two roads were "really about the same."

Christina Rossetti, UPHILL, page 709

This allegorical poem develops a conventional simile: life is like a journey (shades of *Pilgrim's Progress!*). The road is the path of life; the day, a lifespan; the inn at the end of the road, the grave; other wayfarers, the dead; the door, the mouth of the grave (or perhaps the gate of Heaven); the beds, cold clay (or perhaps Heavenly rest). The title suggests another familiar notion: that life is a struggle all the way.

One possible way to paraphrase line 14: "You'll find the end result of your lifelong strivings: namely, death, and the comfort of extinction." A more happily Christian paraphrase is possible, for Rossetti professed herself a believer: "Your labor shall bring you to your goal, the sight of the Lord." Without admitting the possibility of such a faith, the poem will seem grimmer and more cynical than it is.

Do these two characters seem individuals? Not in the least. This is a straight question-and-answer poem, a dialogue between two stick figures.

(text pages) 709–712

"Oh No" (page 521) seems another poem about where you arrive when you die. Creeley, we suspect, kids a conventional notion of Heaven: he makes it smug, artificial place where the saved sit around smirking at one another.

Gjertrud Schnackenberg, SIGNS, page 710

Signs, in the poet's sense of the word, seem plain indicators of approaching doom, hardship, or loss. In the opening stanza, lines in a hand await a palm-reader's interpretation. Students should have little trouble seeing what the other signs foretell.

We wouldn't call these items symbols: their meanings seem too specific and obvious. Dickinson's fly carries a larger cargo of suggestions. (This comparison is not meant to put down "Signs," a poem with a different purpose.)

To date, Schnackenberg has published two fine collections, *Portraits and Elegies* (Boston: Godine, 1982) and *The Lamplit Answer* (New York: Farrar, 1985). "Signs," though it is an early poem, comes from the more recent.

Sir Philip Sidney, YOU THAT WITH ALLEGORY'S CURIOUS FRAME, page 710

This sonnet seems a classic expression of a poet's resentment toward readers and critics who read needlessly profound meanings into his lines—a valuable warning for students to heed when studying symbolism.

The second line may refer to a traditional folk belief. Like an evil fairy, the hunter after allegory spirits away a natural child and leaves a changeling—an odd, ill-favored supernatural brat—in its place.

EXERCISE: *Symbol Hunting,* page 710

William Carlos Williams, POEM, page 711

Theodore Roethke, NIGHT CROW, page 711

Wallace Stevens, ANECDOTE OF THE JAR, page 712

"Night Crow" and "Anecdote of the Jar" contain central symbols; "Poem" is to be taken literally.

Students familiar with Stevens sometimes reason, "The jar is a thing of the imagination, that's why it's superior to the wilderness—it makes order out of formless nature, the way Stevens thinks art is supposed to do." But Stevens is constantly warning us of the dangers of mind divorced from the physical world, and we think he means this gray, bare, dominion-taking jar to be ominous. Who could think a wilderness *slovenly* before it came along? Some critics take the phrase *of a port in air* to mean a portal, "an evanescent entry . . . to order in a scene of disorder" (Ronald Sukenick, *Wallace Stevens: Musing the Obscure* [New York: New York UP, 1967]). We read it differently: *portly,* imposing, pompous. Although it is true that Stevens frequently raises the same philosophic or aesthetic questions, from poem to poem he keeps supplying very different answers. See the brilliant essay on Stevens by J. Hillis Miller in *Poets of Reality* (Cambridge: Harvard UP, 1965).

Jerald Bullis has written an intriguing poem in response to "Anecdote of the Jar." Thanks to Peter A. Fritzell of Lawrence University for discovering it.

BUCK IT

    Take a shot-up bucket in a swale of woods—
    For years "things" have been adjusting to it:
    The deer have had to warp their whylom way

(text pages) 710–712

Through the fern to honor the order in their blood
That says not to kick it; the visiting woodcock

Probably take it for some kind of newfangled stump,
And doubtless welcome any addition that offers
Additonal cover—especially if its imposition
Provides a shelving stay for worm-rich mulch;
A rivulet of breeze low-eddying the swale

Breaks around it much the way a stress's
Flow gets an increment of curvature
From encounter with an old Singer
Sewing machine; the ferns thereabout have turned
A bit more plagiotropic; if it's upright

And the lid's off it's an urn for leaves, bark-bits,
Bird droppings; but in the scope of the whole
Forty-acre woodpatch is it likely
To take dominion everywhere? no more
Than a barbed-wire tangle of words or a good jar.

25 Myth

Besides the poems in this chapter, other poems in the text will readily lend themselves to the study of myth and its pervasiveness in poetry.

Personal myths may be found in the poems of Blake; in Hardy's "Convergence of the Twain" (page 816); and in certain poems of Yeats outside this chapter, such as "Leda and the Swan" (page 634) and "Sailing to Byzantium" (page 753).

Poems containing central references to familiar classical myths are Cummings's "in Just—" (page 691), with its reincarnation of the Great God Pan; Mark Alexander Boyd's "Cupid and Venus" (page 789); and Allen Ginsberg's "A Supermarket in California" (page 813). Christian mythos is of course inseparable from the devotional poems of Donne and Herbert; from Hopkins's poems and G. K. Chesterton's "The Donkey" (page 795); from Eliot's "Journey of the Magi" (page 804) and Yeats's "The Magi" (page 902); from Paul Zimmer's "The Day Zimmer Lost Religion" (page 520); from Milton's sonnets and from many more.

In this chapter, Thomas Hardy (in "The Oxen") and William Wordsworth (in "The World Is Too Much with Us") sadly contemplate myths in decline—a theme found also in William Stafford's "At the Klamath Berry Festival" (page 880).

D. H. Lawrence, BAVARIAN GENTIANS, page 715

Written in 1929 when Lawrence was ill and nearing death, this splendid poem has been read as a kind of testament. As Keith Sagar has paraphrased it, "the poet's soul has been invited to the nuptials and accepts with joy." Dissolution offers not mere oblivion but the promise of renewed life, the cyclical rebirth of both the gentians and Persephone. (*The Art of D. H. Lawrence* [Cambridge: Cambridge UP, 1966], 244–45.) Another famous poem of Lawrence's last months, "The Ship of Death," may be read as a companion to this.

Why is "Bavarian Gentians" a better title for the poem than Lawrence's first thought, "Glory of Darkness"?

Thomas Hardy, THE OXEN, page 716

The legend that farm animals kneel on Christmas Eve is widespread in Western Europe. Hardy takes it to suggest the entire Christian mythos, which "in these years" (since Darwin) few embrace as did the "flock" of children and old people remembered in the opening stanza. The *gloom* in line 15 may resemble the gloom of the unbeliever, and its doleful sound is enforced by its riming with *coomb*—like *barton*, a word from older rural speech.

The tone of "The Oxen" is not hostility toward faith, but wistfulness. Not exactly the village atheist Chesterton said he was, Hardy in late life kept going to church and hoping for a reconciliation between the Church of England and science-minded rationalists.

William Wordsworth, THE WORLD IS TOO MUCH WITH US, page 717

As its sense and its iambic meter indicate, the opening line calls for a full stress on the *with*.

Wordsworth isn't arguing, of course, for a return to pagan nature worship. Rather like Gerard Manley Hopkins blasting trade in "God's Grandeur" (page 635), he is dismayed that Christians, given to business and banking, have lost sight of sea and vernal woods. They should pay less heed to the world, more to the earth. What "powers" have they laid waste? The ability to open themselves to nature's benevolent inspirations. Modestly, the poet includes himself in the *us* who deserve reproof. The impatient outburst ("Great God!")

is startlingly unbookish and locates the break in sense between octave and sestet in an unconventional place.

Compare Wordsworth's "Composed upon Westminster Bridge" (page 896) for a somewhat similar theme. For another comment on the decline of certain traditional myths, see William Stafford's "At the Klamath Berry Festival" (page 880).

William Butler Yeats, THE SECOND COMING, page 718

The brief discussion in the book leaves several points untouched. Students may be asked to explain Yeats's opening image of the falcon and the falconer; to discuss the meaning of the *Blood-dimmed tide* and the *ceremony of innocence;* to explain how the rocking cradle at Bethlehem can be said to "vex" twenty centuries to nightmare; and to recall what they know about the sphinx.

In *A Vision*, Yeats sets forth his notion of the two eras of history (old and new) as two intertwined conelike gyres, revolving inside each other in opposing directions. He puts it succinctly in a note for a limited edition of his poem *Michael Robartes and the Dancer* (1921):

> The end of an age, which always receives the revelation of the character of the next age, is represented by the coming of one gyre to its place of greatest expansion and of the other to that of its greatest contraction. At the present moment the life gyre is sweeping outward, unlike that before the birth of Christ which was narrowing, and has almost reached its greatest expansion. The revelation which approaches will however take its character from the contrary movement of the interior gyre.

Students can be asked to apply this explanation to "The Second Coming." (In fact, this might be a writing assignment.)

For other evidence of Yeats's personal mythology, direct students to "Leda and the Swan" (page 634) and "Sailing to Byzantiium" (page 753). For alternative versions of "The Second Coming," see Yeats's worksheets for the poem as transcribed by John Stallworthy in *Between the Lines: Yeats's Poetry in the Making* (Oxford: Oxford UP, 1963).

Dick Allen, NIGHT DRIVING, page 719

Perhaps we are all, in a sense, night driving—making our way dimly through life buoyed up and kept going by myths old and new, "all the lies / Not really lies." Is this what Allen's poem is about? It can of course be taken literally. The speaker, tired, drives at night from Bridgeport and remembers how his father used to say that the lights of the other cars on the road were cats' eyes. Somehow it gives him pleasure, at least for a while, to contemplate the "images which make / The world come closer, cats' eyes up ahead."

Dick Allen, who teaches at the University of Bridgeport, included this poem in *Flight and Pursuit* (Louisiana State UP, 1987).

Harvey Shapiro, NATIONAL COLD STORAGE COMPANY, page 719

The name of this actual company provides the poet with a hint-filled central symbol. But like any good symbol, the National Cold Storage Company doesn't literally stand for death or anything else. It takes into itself not only the dead of many wars (and President John F. Kennedy) but the poet's personal little deaths or frustrations: "Midnight tossings, plans for escape, the shakes" (8). In the course of the poem, a dull factory-front becomes a colossal devourer. All beautiful things, all human life, are swallowed and placed in cold storage at the end.

Brooklyn Bridge does indeed resemble a harp, as Shapiro observes in line 13. One of the poets who crept over it was Hart Crane, who lived in an apartment overlooking the bridge while writing his long poem *The Bridge* and who once clambered its

(text pages) 720–726

superstructure on a magazine writing assignment. In lines 13–15, Shapiro is probably thinking of Crane's early suicide by drowning.

Many of Shapiro's best poems are urban landscapes; he has lived in New York for about thirty years, mostly working as an editor for the *New York Times*.

The poet chose the title of this poem for his volume of new and selected poems, *National Cold Storage Company* (Wesleyan UP, 1988).

John Milton, LYCIDAS, page 720

"Lycidas" for many students is formidable, and before teaching the poem, at least a half hour of class time will probably be needed for preparation. At least a smattering of information on the time and place of Milton's elegy is helpful, if students are to see that their own lives (and friendships) and Milton's life at Cambridge are not completely remote from each other. A book useful for background is Louis Potter's *A Preface to Milton* (New York: Scribner's, 1971), especially 122–25. Many students are intrigued by mythology and find that to read up on a flock of classical myths reveals much to them, besides making them able to follow Milton's allusions. You may wish to allow two class hours to the poem itself, enough time to deal with only a few passages. Milton's powerful condemnation of the false shepherds (lines 119–31) is usually a high point of the poem for students, and students generally end with at least some respect for Milton as a mythmaker.

Michael Fixler has seen the poem's multifold allusions to myth as one with its "unheard" or "unexpressive" music. "In myth Orpheus sings even as his severed head, resting on his silent lyre, races down the swift Hebrus. But in 'Lycidas' that song is 'unexpressive' in yet a further sense, being implicit, a part of the allusion" (" 'Unexpressive Song': Form and Enigma Variations in *Lycidas*, a New Reading," *Milton Studies* 15 [1981]: 213–55).

For the instructor who wishes further aid in reading the poem, the amount of available criticism is, of course, vast. Modern studies we have found valuable include Rosemond Tuve's "Theme, Pattern, and Imagery in *Lycidas*," in *Images and Themes in Five Poems of Milton* (Cambridge: Harvard UP, 1957); Jon S. Lawry's " 'Eager Thought': Dialectic in *Lycidas*," in *Milton: Modern Essays in Criticism;* ed. Arthur E. Barker (Oxford: Oxford UP, 1965); and Michael Fixler's discussion in *Milton and the Kingdoms of God* (Evanston: Northwestern UP, 1964) 56–60.

26 Alternatives

THE POET'S REVISIONS

William Butler Yeats, THE OLD PENSIONER, *page 728*

William Butler Yeats, THE LAMENTATION OF THE OLD PENSIONER, *page 728*

The first pensioner is a lackluster old coot; the later one, strong-willed and defiant. In the "Lamentation" the speaker sees his own ruin clearly and coldly: in line 14, he himself becomes the broken tree. In context, *transfigured* is a splendid word; it helps to transfigure the original feeble refrain, and in the new refrain Time becomes a flesh-and-blood enemy. The 1890 version is full of end-stopped lines; in the final poem, syntax tends to be seamless: stanzas are whole sentences. About all that Yeats kept from the 1890 version is the chair by the fire and one tree, the dominant rime-sound and the stanza pattern. Perhaps by aging (and by learning more of love and politics), Yeats came to know at first hand how an old man feels.

Another early poem Yeats completely recast is "The Sorrow of Love." See the *Variorum Edition of the Poems of W. B. Yeats* (New York: Macmillan, 1957).

Walt Whitman, A NOISELESS PATIENT SPIDER, *page 730*

Walt Whitman, THE SOUL, REACHING, THROWING OUT FOR LOVE, *page 731*

In revising "The Soul, reaching," Whitman scrapped all but the first two lines; then extended their metaphor into a whole poem. In doing so, he slightly changed his original conception and in its finished form, "A Noiseless Patient Spider" isn't a poem about human beings reaching out in love to other human beings, but about the soul trying to form contact with higher reality.

"A Noiseless Patient Spider" may be used effectively in discussing symbolism in poetry, as well as figures of speech. Whitman's poem is open in form, and yet certain lines fall into traditional measures (almost into rime, too), as would be indicated by rearranging them:

> Till the bridge you will need be form'd
> Till the ductile anchor hold,
> Till the gossamer thread you fling
> Catch somewhere, O my soul.

TRANSLATIONS

Federico García Lorca, LA GUITARRA (GUITAR), *page 732*

The translator's liberty with Lorca's lines 23–24 seems well taken: "on the branch" would be weaker as a line by itself than "and the first dead bird on the branch."

Lorca's poem comes from his *Poema del cante jondo* (1921), an early sequence of brief lyrics based on traditional folk music. The *cante jondo* ("deep song") was an Andalusian form related to *flamenco;* and in 1922 Lorca and composer Manuel de Falla organized a festival of the cante jondo, offering prizes for new songs in the old tradition.

Carl W. Cobb has suggested that the "Heart heavily wounded / by five sharp swords" is the guitar itself, struck by the player's five fingers (*Federico García Lorca* [New York: Twayne, 1967]).

EXERCISE: *Comparing Translations*, page 733

Horace, ODES I (38), page 733

William Cowper, SIMPLICITY, page 734

Hartley Coleridge, FIE ON EASTERN LUXURY! page 734

Eugene Field, THE PREFERENCE DECLARED, page 734

Cowper's neoclassical translation seems fair both to the tone of Horace's poem and to its sense. Cowper rearranges ideas to make his rimes come out right (ringing in "Thus outstretched beneath my vine" rather early) and to emphasize the boy rather than the drinker. He does not adorn, however, and he preserves the simplicity of style of the original.

The work of Hartley Coleridge (the son of Samuel Taylor Coleridge) had moments of felicity, but this translation is not one of them. He expands the original eight lines to twelve, falls into awkward syntax in the entire second stanza, and chooses a diction that exhibits the "studious pomp" Horace would avoid *(toilsome pain, mis-seems)*.

Vigorous and direct, Field's irreverent version is a small masterpiece of speaking colloquially inside tight metrical verse. Obviously, however, Field shatters the tone of the original. That the author of such sentimental mawkishness as the famous "Little Boy Blue" often could write this well is also evident in Field's other translations from Horace, published with his brother Roswell Martin Field in *Echoes from the Sabine Farm* (New York: Scribner's, 1896).

Interesting translations of this famous ode are plentiful: the *Poems of Gerard Manley Hopkins* contains a bookish one ("Ah child, no Persian—perfect art! / Crowns composite and braided bast . . ."). Cowper made another version of the poem, apparently, like Campion's "Rose-cheeked Laura," an attempt to write English Sapphics (the Greek measures that Horace was imitating):

> Boy! I detest all Persian fopperies,
> Fillet-bound garlands are to me disgusting,
> Task not thyself with any search, I charge thee,
>   Where latest roses linger;
> Bring me along (for thou wilt find that readily)
> Plain myrtle. Myrtle neither will disparage
> Thee occupied to serve me, or me drinking
>   Beneath my vine's cool shelter.
>
>   (from Cowper's *Poems* of 1815, ed. John Johnson)

Charles Baudelaire, RECUEILLEMENT (MEDITATION), page 734

Lord Alfred Douglas, PEACE, BE AT PEACE, O THOU MY HEAVINESS, page 735

Robert Bly, INWARD CONVERSATION, page 735

Robert Lowell, MEDITATION, page 735

Richard Howard, MEDITATION, page 735

These four translations from Baudelaire exhibit divergent fashions. Douglas keeps the music while high-handedly changing the sequence of ideas; his version has a most Swinburnean fin-de-siecle tinge ("To pluck the fruits of sick remorse and fear"). Personally, we like Douglas's version the best of the four—considered as English poetry. The opening line does not say what Baudelaire says, but it is splendid in music and sense. The ending, too, seems an inspired distortion, worthy of Edward FitzGerald.

Bly's translation is a combination of invention and fidelity. The opening line seems wordy, gauche, and inaccurate. Bly's *rotten herds* are far cruder than Baudelaire's *multitude vile*, and we confess ourselves unable to understand what a *lyncher without touch* is. Bly's version improves as it goes along, both as a translation and as an English poem. The *sense of loss* in line 11 is a fine way to render *Regret*, and the last three lines, lovely in English, seem to come closer to the tone of Baudelaire's original than either Douglas or Lowell does. On general principle, Bly refuses to translate rimed metrical verse into rimed metrical verse. At the end he is able to keep the sense of Baudelaire's poem though the music is drastically altered.

Lowell's version is from his collection of *Imitations*, where it is offered not as a faithful translation but as a free adaptation. It is very close in sense, however, to the original. Here and there, Lowell adds a characteristic stroke of his own: the addition of the metaphor in *coffined* (line 3), the blunt rendering of *robes surrannés* as *old clothes* (line 10). Though a close cognate of *marche*, the last word, *march*, with its military connotations, seems an unfortunate choice. But Lowell's translation occupies the middle ground between fidelity and free improvisation and manages to convey some sense of the music of the original.

Richard Howard's new rendering of *Les Fleurs du Mal* (Boston: Godine, 1982) has proved a surprise best-seller for its publisher and has won generous praise. A professional translator of Gide, Camus, Genet, de Beauvoir, Perse, Barthes—about 150 French books altogether—Howard has given us what may be the most finely accurate translation of the nasty *Fleurs* to date, not counting prose paraphrases. Beautifully, Howard communicates the sense of the poems, that is, their silliest part, while conveying little sense of the most impressive part: Baudelaire's music. By and large, Howard ignores the rime and meter of the original, on occasion letting a rime occur but usually going out of his way to avoid any. The product is a Baudelaire who sounds like most current contributors to *American Poetry Review*, a helpful and sensitive crib for readers who know some French.

These general reservations aside, Howard's "Meditation" seems among his more satisfying versions, done with something of the rhythm of the original. The opening lines are brisk and colloquial: the ending is lovely—how much better to make the night arrive than, as Lowell does, to make it march! And "dear departed dowdy years" is an inspired touch—one of the few places where Howard adds something, and makes up for some of what, in any translation, must inevitably be lost.

PARODY

Ezra Pound, in his *ABC of Readings*, urges students of poetry to write parodies of any poems they find ridiculous, then submit their parodies to other students to be judged. "The gauging pupil should be asked to recognize what author is parodied. And whether the joke is on the parodied or the parodist. Whether the parody exposes a real defect, or merely makes use of an author's mechanism to expose a more trivial content."

T. E. Brown, MY GARDEN, page 737

J. A. Lindon, MY GARDEN, page 737

With a few thrusts of his rake, Lindon punctures Brown's high-falutin language (*wot* and *grot* are the most flagrantly "poetic"), his overwrought reliance on exclamation points, and the shaky logic he uses to prove the existence of God. Of the two, the parody is unquestionably the better poem.

Hugh Kingsmill, WHAT, STILL ALIVE AT TWENTY-TWO? page 737

Kingsmill's insistence on dying young suggests "To an Athlete Dying Young" (page 832), but the parodist grossly exaggerates Housman's hint of nihilism. Like bacon, Kingsmill's lad will be "cured"—of the disease of life. (And how often Housman himself says *lad,* by the way.) His metaphysical conceit of ink and blotting pad coarsens Housman's usual view of night and day. (Some comparable Housman lines, from "Reveille": "Wake: the silver dusk returning / Up the beach of darkness brims, / And the ship of sunrise burning / Strands upon the eastern rims.")

Kenneth Koch, MENDING SUMP, page 738

Perhaps it would be well to read aloud Frost's "Mending Wall" before having students read Koch's hatchet job. To appreciate Koch's allusions to Frost, students also might hear a little of "The Death of a Hired Man." It is often included in high-school textbooks but may not be familiar to all students.

George Starbuck, MARGARET ARE YOU DRUG, page 739

"Margaret Are You Drug" is a deliberately crass, lowbrow, American version of Hopkins's "Spring and Fall" (page 829).

27 Evaluating a Poem

Telling Good from Bad

Ezra Pound long argued for the value of bad poetry in pedagogy. In his *ABC of Reading*, Pound declared that literary education needs to concentrate on revealing what is sham, so that the student may be led to discover what is valid. It is a healthy gesture to let the student see that we don't believe everything contained in a textbook to be admirable. Begin with a poem or two so outrageously awful that the least sophisticated student hardly can take it seriously—some sentimental claptrap such as Cook's "The Old Arm-Chair." From these, you can proceed to subtler examples. It is a mistake to be too snide or too self-righteous toward bad poems, and it is well to quickly turn to some excellent poetry if the classroom starts smelling like a mortuary. There is a certain sadness inherent in much bad poetry; one can readily choke on it. As Allen Tate has said, the best attack upon the bad is the loving understanding of the good. The aim in teaching bad poetry has to be the admiration of good poetry, not the diffusion of mockery.

One further suggestion on bad poetry: a program of really execrable verse orated with straight faces by a few students and members of the faculty can be, with any luck, a fine occasion. For bad poems to work on besides those offered in this chapter, see the dustier stacks in a library or the following anthologies: *Heart Throbs* and *More Heart Throbs*, ed. Joe Mitchell Chapple (New York: Grosset, 1905 and 1911 respectively; many later editions); *The Stuffed Owl: An Anthology of Bad Verse*, ed D. B. Wyndham Lewis and Charles Lee (London: Dent, 1930; reprinted in the United States by Capricorn paperbacks); *Nematodes in My Garden of Verse*, ed. Richard Walser (Winston-Salem: Blair, 1959); *Worst English Poets*, ed. Christopher Adams (London: Wingate, 1958); *Pegasus Descending: A Book of the Best Bad Verse*, ed. James Camp, X. J. Kennedy, and Keith Waldrop (New York: Macmillan, 1971); and *The Joy of Bad Verse*, ed. Nicholas T. Parsons (London: Collins, 1988).

Anonymous, O Moon, when I gaze on thy beautiful face, page 742

Glorious behind seems inexact, and so does *boundaries* for "boundlessness."

Grace Treasone, Life, page 742

Treasone's poem develops a central metaphor, but its language is wildly imprecise. Is the tooth "that cuts into your heart" one's own or somebody else's? (It is probable that the poet means not tooth but "toothache.") Anatomically, the image seems on a par with the "heart's leg" of the tradesman poet quoted by Coleridge. Through the murk of her expression, however, the poet makes clear her theme: the familiar and sentimental notion that life is really all right if you see it through (or have a competent dentist).

Treasone's item first adorned a Dover, New Jersey, newspaper column of local poets called "This Way to Parnassus."

Stephen Tropp, My Wife Is My Shirt, page 743

To give this item the benefit of doubt, its metaphor is elaborated consistently but is hard to see anatomically. To compare the shirt-sleeves to the wife's armpits and the shirt's neck to her mouth, rather than to other parts of her, seems arbitrary. In a personification an inanimate object is seen as human. In "My Wife Is My Shirt" a person is seen as an inanimate object.

(text pages) 743–748

If this paraphrase of his idea is a fair one, Tropp does not get it across at all. The buttoning of the blood—a thoughty and ingenious figure—seems merely horrible. Tenderness is lost.

This poem appears in *Beat Coast East: An Anthology of Rebellion* (New York: Excelsior, 1960) and may be a relic of a kind of work once fashionable, in which the poet tries to turn off his feelings and to "play it cool." Whatever the poet's intentions, the result is bathetic. All we know about Tropp comes from a note in a poetry magazine that said he lived in New York and had a little boy named Tree.

William McGonagall, THE ALBION BATTLESHIP CALAMITY, page 743

Many consider McGonagall the greatest bad poet in the language; certainly "The Albion Battleship Calamity" stands high among his chefs d'oeuvre. This poem should bring out what Ezra Pound has called (admiringly) "the natural destructiveness of the young." Students should have little trouble spotting the poem's most glaring faults. McGonagall's diction stays vague and general because he is always more interested in his own opinions than in reporting sense-testimony. His opinions tend to be thoughtless: "But accidents will happen without any doubt . . . " This item takes up too much space in the book, but could not be cut without hurting its maundering repetitiveness, which adds so greatly to its comic effect. The poem reads well aloud and can send a crowd into gales of laughter. Then the monotony of its rimes (-ay, -ay) becomes obvious.

An anecdote told of McGonagall illustrates his longwindedness. At a Dundee performance of *Macbeth* in which he played the title role, he refused to die when stabbed, but held the stage, rattling on, improvising—until a disgusted Macduff picked him up and carried him off, to the delight of the audience.

The text leaves unretouched the grammar and spelling (*who's* for *whose*, etc.) of the original.

McGonagall: A Library Omnibus (London: Duckworth, 1980) contains the Dundee bard's complete poems.

Emily Dickinson, A DYING TIGER – MOANED FOR DRINK, page 745

This is not, by any stretch of critical imagination, a good poem. Besides the poet's innocent lack of perception that *His Mighty Balls* can suggest not eyeballs but testicles, the concluding statement (that the fact that the tiger was dead is to blame) seems an unDickinsonian failure of invention. Perhaps the poet intended a religious allegory (Christ the Tiger). Her capitalization of *He* in the last line doesn't seem sufficient proof of such intent, for her habits of capitalization cannot be trusted for consistency.

The failures of splendid poets are fascinating. As in this case, they often seem to result from some tremendous leap that sails over and beyond its object, causing the poet to crash to earth on the other side.

EXERCISE: *Seeing What Went Wrong*, page 746

J. Gordon Coogler, a printer by trade, was said to have displayed a sign in the window of his print-shop in Columbia, S.C.: "POEMS WRITTEN WHILE YOU WAIT."

For Byron's rousing lines, we have to thank Walter H. Bishop of Atlanta, for whom this discovery won first prize in a contest to find the worst lame verse by a well-known poet, conducted by John Shelton Reed in *Chronicles*. (The results were reported in the magazine's issue of November 1986.)

Mattie J. Peterson has attracted fierce partisans, some of whom see her battling Julia A. Moore for the crown of Queen of American Bad Verse. Richard Walser has brought out of a modern facsimile of her *Little Pansy, A Novel, and Miscellaneous Poetry* (originally 1890; Charlotte: McNally & Loftin, 1967).

Francis Saltus Saltus is the rediscovery of Nicholas T. Parsons in *The Joy of Bad Verse* (London: Collins, 1988). We lifted these two excerpts from Mr. Parson's splendid anthology. It was a temptation to lift a third:

> Oh! such a past can not be mute,
>   Such bliss can not be crushed in sorrow,
> Although thou art a prostitute
>   And I am to be hanged tomorrow.

Saltus regarded himself as a rakehell. As C. T. Kindilien has observed, "Although he idealized cigarette-smoking women, looked for pornography in the Bible, and honored Baudelaire, Gerard de Nerval, and Le Marquis de Sade, he never escaped the tone of the boy who expected any moment to be caught smoking behind the barn" (*American Poetry in the Eighteen-Nineties*, Brown UP, 1956, 188–9).

Rod McKuen, THOUGHTS ON CAPITAL PUNISHMENT, *page* 748

William Stafford, TRAVELING THROUGH THE DARK, *page* 749

McKuen is still popular with some students, and any dogmatic attempt to blast him may be held against you. There may be value in such a confrontation, of course; or you can leave evaluation of these two works up to the class. Just work through McKuen's effusion and Stafford's fine poem, detail by detail, in a noncommittal way, and chances are good that Stafford will win the contest.

It may not be apparent that Stafford's poem is ordered by a rime scheme from beginning to end: *abcb* stanzas and a final couplet. Stafford avoids obvious rimes in favor of the off-rimes *road / dead* and *engine / listen* and the cutoff rimes *killing / belly, waiting / hesitated,* and *swerving / river*—this last a device found in some folk ballads. McKuen's poem announces an obvious rime scheme but fails to complete it. Unlike Stafford, he throws rime out the window in the end, with the effect that his poem stops with a painful inconclusiveness.

Stafford contributes a long comment on his poem to *Reading Modern Poetry: A Critical Introduction,* ed. Paul Engle and Warren Carrier (Glenview: Scott, 1968).

EXERCISE: *Fine or Shoddy Tenderness, page* 750

Julia A. Moore, LITTLE LIBBY, *page* 750

Bill Knott, POEM, *page* 751

Dabney Stuart, CRIB DEATH, *page* 751

Leo Connellan, SCOTT HUFF, *page* 751

Ted Kooser, A CHILD'S GRAVE MARKER, *page* 752

One horrendously bad poem, followed by four good ones. All five poets tackle a difficult subject: the death of a child or young person, but only Julia Moore ("the Sweet Singer of Michigan") throws off restraint and wallows in a bath of emotion. Perhaps she is sincere, but apparently she has no idea that writing a powerful poem requires anything more than a wish to share a good cry. In truth, lines like "Her Savior called her, she had to go," and "While eating dinner, this dear little child / Was choked on a piece of beef" demonstrate an alarming insensitivity to audience (and to the sound of words). The poet's uncertain handling of rime and rhythm add to the ludicrousness.

Mrs. Moore was the author of *The Sentimental Songbook* (1876), which Mark Twain said he carried with him always. She may have inspired the portrait of poet Emmeline

(text pages) 752–753

Grangerford and her elegies in *Huckleberry Finn*. Her rimes, though, could be surprising. Given the first three-and-a-half lines of this quatrain, how would you expect it to end? You might read it to the class, and ask them to guess how it finishes:

> Many a man joined a club
> That never drank a dram,
> Those noble men were kind and brave,
> They do not care − −.

(Wrong. The last two words are "for slang.")

In Bill Knott's poem, on the other hand, freedom from artifice seems to help what is said: a direct, seemingly casual and yet playful statement of grief.

Dabney Stuart focuses quietly but powerfully on the dread which the living feel when they confront a child's death. Perhaps few students will notice that "Crib Death" is an acrostic (we didn't, until the poet pointed out this fact). In a letter to XJK, Stuart wrote, "The poem is particularly close to me since it . . . honors the daughter my wife lost 19 years or so ago. Not my child, as we are newly married (1983), but part of my grief nonetheless."

Leo Connellan's "Scott Huff" may seem to skirt closer to sentiment than the other four recent poems. Like Mrs. Moore's, this poem is a direct expression of grief. We do not know anything about Scott Huff nor why we should grieve for him, except that he is only sixteen. "Yellow buttercups" might seem at first a merely conventional posy; yet, come to think about them, buttercups are a fresh choice. And in his images ("the weight of his foot" that causes the car to spring awake, the salmon in the rivers and the fresh cold springs), the poet shows more concern for the specific, physical world than truly sentimental poets ever do.

Ted Kooser imagines his way into a grief expressed seventy years ago. The speaker in his tender poem, moved by a clumsy but touching grave marker, acknowledges the love and longing that inspired it. The passage of time supplies the emotional distance necessary.

KNOWING EXCELLENCE

William Butler Yeats, SAILING TO BYZANTIUM, *page 753*

Has XJK implied that this poem is a masterpiece so far beyond reproach that no one in his right mind can find fault with it? That is, of course, not the truth. If the instructor wishes to provoke students to argument, he might read them the withering attack on Yeats's poem by Yvor Winters (*Forms of Discovery* [Chicago: Swallow, 1967] 215–16). This attack really needs to be read in its entirety. Winters is wrong, we believe, but no one can begin to answer his hard-headed objections to the poem without being challenged and illuminated.

Other discussions of the poem, different from XJK's and also short, include Richard Ellmann's in *Yeats: The Man and the Masks* (New York: Macmillan, 1949) and John Unterecker's in *A Reader's Guide to William Butler Yeats* (New York: Noonday, 1959). Those who wish to go deeper still and to read a searching examination (informed by study of Yeats's manuscripts) can be directed to Curtis Bradford, "Yeats's Byzantium Poems," *PMLA* 75 (Mar. 1960): 100–25. For those interested in alternatives, Jon Stallworthy reprints nearly all the legible manuscript versions in *Between the Lines: Yeats's Poetry in the Making* (Oxford: Clarendon, 1963) 87–112.

A deconstructionist reading of "Sailing to Byzantium," subjecting the poem to relentless questioning, showing where it fails to make sense and how it doesn't work, is offered by Lawrence I. Lipking in "The Practice of Theory" (in *Profession 83: Selected articles from the Bulletins of the Association of Departments of English and the Association of Departments of Foreign Languages*, MLA, 1983). But in his role as a poststructuralist, Lipking confesses himself "a sheep commissioned to say something sympathetic about wolves." He finds deconstructionist tactics offending his students, especially bright idealistic ones who expect their teachers to show them why certain works are great, and who wish poems to "make sense" and to relate to their own lives.

Jean Bauso has used a writing assignment to introduce this challenging poem. "I want you to pretend that you're an old person—someone in his or her eighties," she tells a class. "You've got arthritis, so buttoning or zipping your clothes is slow. Now you will write for ten minutes nonstop in the voice of this old person that you've made yourself into. You want to follow your person's stream of consciousness as he or she sits there thinking about the human condition, about the fact that we human beings have to die." After the students free-write for ten minutes, they read a few of the results, and she picks up on any comments about wishing for immortality. Then she asks for responses to the name Byzantium, perhaps holds up some pictures of the Santa Sophia mosaics. She then reads "Sailing to Byzantium" aloud, gives out reading sheets with points for reading it alone, and dismisses the class. For Bauso's detailed account of this lesson plan and her reading sheets, see "The Use of Free-Writing and Guided Writing to Make Students Amenable to Poems," *Exercise Exchange* (Spring 1988).

EXERCISE: *Two Poems to Compare*, page 755

Arthur Guiterman, ON THE VANITY OF EARTHLY GREATNESS, page 755

Percy Bysshe Shelley, OZYMANDIAS, page 756

The title of Guiterman's bagatelle playfully echoes that of a longer, more ambitious poem: Samuel Johnson's "The Vanity of Human Wishes." If Guiterman's achievement seems smaller than Shelley's in "Ozymandias," still, it is flawless. "Ozymandias," although one of the monuments of English poetry, has a few cracks in it. Many readers find line 8 incomplete in sense: the heart that fed what, or fed on what? From its rime scheme, we might think the poem a would-be Italian sonnet that refused to work out.

Nevertheless, Shelley's vision stretches farther than Guiterman's. Ozymandias and his works are placed at an incredibly distant remove from us. The structure of the poem helps establish this remoteness: Ozymandias's words were dictated to the sculptor, then carved in stone, then read by a traveler, then told to the first-person speaker, then relayed to us. Ironies abound, more subtle than Guiterman's. A single work of art has outlasted Ozymandias's whole empire. Does that mean that works of art endure (as in "Not marble nor the gilded monuments")? No, this work of art itself has seen better days, and soon (we infer) the sands will finish covering it. Obviously, the king's proud boast has been deflated, and yet, in another sense, Ozymandias is right. The Mighty (or any travelers) may well despair for themselves and their own works, as they gaze on the wreckage of his one surviving project and realize that, cruel as Ozymandias may have been, time is even more remorseless.

What are the facts behind Shelley's poem? According to the Greek historian Diodorus Siculus, Ozymandias was apparently a grand, poeticized name claimed for himself by the Egyptian pharaoh Rameses II. Diodorus Siculus saw the king's ninety-foot-tall statue of himself, carved by the sculptor Memnon, in the first century B.C. when it was still standing at the Ramesseum in Thebes, a mortuary temple. Shelley and his friend Horatio Smith had read a description of the shattered statue in Richard Pococke's *Description of the East* (1742). Smith and Shelley wrote sonnets expressing their imagined views of the wreckage, both of which Leigh Hunt printed in his periodical *The Examiner* in 1818. This is Smith's effort, and students might care to compare it with Shelley's in quality:

> On a Stupendous Leg of Granite, Discovered
> Standing by Itself in the Deserts of Egypt

(text pages) 756–758

>In Egypt's sandy silence, all alone,
>   Stands a gigantic leg, which far off throws
>   The only shadow that the desert knows.
>'I am great Ozymandias,' saith the stone,
>   'The king of kings: this mighty city shows
>   The wonders of my hand.' The city's gone!—
>   Nought but the leg remaining to disclose
>The site of that forgotten Babylon.
>
>We wonder, and some hunter may express
>Wonder like ours, when through the wilderness,
>   Where London stood, holding the wolf in chase,
>He meets some fragment huge, and stops to guess
>   What powerful but unrecorded race
>   Once dwelt in that annihilated place.

For more background to the poem, see H. M. Richmond, "Ozymandias and the Travelers," *Keats-Shelley Journal* 11 (1962): 65–71.

William Shakespeare, My mistress' eyes are nothing like the sun, page 756

Have students state positively each simile that Shakespeare states negatively, and they will make a fair catalog of trite Petrarchan imagery. Poking fun at such excessive flattery is a source of humor even today, as in an old wheeze: "Your teeth are like the stars—they come out at night."

Thomas Campion, There is a garden in her face, page 757

In tone, Campion's lyric is admiring and tender, and yet there are ironies in it. The street vendor's cry, as Walter R. Davis has pointed out, "undercuts, with its earthy commercialism, the high Petrarchan style of the rest of the song." In Campion's society, a girl of marriageable age was, in a sense, on sale to the highest bidder.

For Campion's music to his song—incorporating the set melody of the street cry "Cherry ripe, ripe, ripe!"—see Davis's edition of Campion's *Works* (New York: Doubleday, 1967). Campion has been called (by W. H. Auden) "the only man in English cultural history who was both a poet and a composer."

Some students, taking the figures of speech literally, may find the last stanza absurd or meaningless. They can be led to see that Campion's angels with bended bows enact his theme that the young girl's beauties are defended. Who defends them? She herself, by her nay-saying frowns and by her immaturity. Throughout the poem run hints of Eden. The garden, that *heav'nly paradise,* holds *sacred cherries*—forbidden fruit—and so the guardian angels seem traditional. John Hollander thinks the Petrarchan cliché of bowlike eyebrows "redeemed" by its new associations. "The courtly compliment now turns out to be found in beautiful sexual attainment, in the plucking of cherries that are not forbidden apples, and just for that reason, such attainment isn't always easy" (introduction to *Selected Songs of Thomas Campion* [Boston: Godine, 1973]).

Walt Whitman, O Captain! My Captain! page 758

This formerly overrated poem is uncharacteristic of Whitman in its neatly shaped riming stanzas and in its monotonously swinging observation of iambic meter, so inappropriate to a somber elegy. The one indication that an excellent poet wrote it is the sudden shift of rhythm in the short lines that end each stanza—particularly in line 5, with the unexpected turning-on of heavy stresses: "O heart! heart! heart!"

Carl Sandburg, Fog, page 759

Like Pound's "In a Station of the Metro" (page 569), Sandburg's celebrated minipoem is all one metaphor, all imagery. Not closely detailed, it seems vague when set next to Eliot's agile fog-cat in "Prufrock" (page 806). Eliot can depict even fog without vagueness; evidently cats are right up his alley. Students also might enjoy a look at his *Old Possum's Book of Practical Cats*.

Thomas Gray, Elegy Written in a Country Churchyard, page 760

Like other critics, students may disagree widely in their statements of Gray's theme. Roger Lonsdale (cited below) sees the main preoccupation of the poem to be "the desire to be remembered after death, a concern which draws together both rich and poor, making the splendid monuments and the 'frail memorials' equally pathetic." Concern with being remembered after death informs the Epitaph as well, making it seem intrinsic to the poem (despite Landor's objections). But Gray also seems to suggest that there is positive virtue in remaining little known.

About question 10: most students will readily see that Gray, Shelley, and Guiterman all state (however variously) a common theme—the paths of glory lead but to the grave. In ranking the three poems in order of excellence, however, they risk getting into a futile debate unless they can agree that (1) Guiterman's excellent comic poem need not be damned for not trying to be an elegy; and (2) Gray's poem *is* more deep-going, moving, musical, and ultimately more interesting in what it says than Guiterman's.

Of three stanzas found in the earliest surviving version of the poem (the Eton manuscript) and deleted from the published version of 1753, one has been much admired. It followed line 116, coming right before the Epitaph:

> There scatter'd oft, the earliest of the year,
> By hands unseen, are showers of violets found:
> The Red-breast loves to build, and warble there,
> And little footsteps lightly print the ground.

Topic for discussion: Should Gray have kept the stanza?

Topic for a paper of moderate (600- to 1,000-word) length: "Two Views of Anonymity: Gray's 'Elegy' and Auden's 'Unknown Citizen.'"

Twentieth Century Interpretations of Gray's Elegy, ed. H. W. Starr (Englewood Cliffs: Prentice, 1968), is a convenient gathering of modern criticism. A good deal of earlier criticism is summarized by Roger Lonsdale in his edition of *The Poems of Gray, Collins, and Goldsmith* (New York: Norton, 1969), which provides extensive notes on texts and sources. For a modern poem at least partly inspired by the "Elegy," see Richard Wilbur's "In a Churchyard" in *Walking to Sleep* (New York: Harcourt, 1969). Among the countless parodies, there is an anonymous "Allergy in a Country Churchyard" that begins, "The kerchoo tolls, Nell's 'kerchief swats away."

Exercise: *Evaluating the Unfamiliar*, page 764

Lorine Niedecker, Popcorn-can Cover, I take it slow, Sorrow moves in wide waves, page 765

James Hayford, Behind the Wall, Under the Holy Eye, Dry Noon, page 766

We aren't alone in admiring the artistry of these two underappreciated poets. William Carlos Williams called Niedecker's work "good poetry. It is difficult and warm. It has life to it." Robert Frost said Hayford has "valid spirituality. And in the shortest possible space [he] manages to come out somewhere."

Niedecker, though she published five collections with small presses here and in the U.K., is only rarely met with in anthologies, and the new wave of feminist criticism seems not yet to have caught up with her. *Niedecker,* a play by Kristine Thatcher based on the poet's quiet life, was staged in March, 1989, at Apple Corps, an Off-Broadway theater.

It's a glum reflection on the state of publishing that James Hayford has written excellent poetry for half a century without (until lately) finding a commercial house willing to take a chance on him. He self-published his last two books. In the *New York Times* for June 26, 1988, Jules Older interviews him and tells the story of Hayford's discouraging rejections.

This is the first time this book has tried hard to cope with the problem of evaluating unfamiliar poetry. If you have time for this matter, we think a part of a class spent with these two poets will be rewarding. The work of both is quiet and unflamboyant, but students should find things to admire. Any snob who, on reading Niedecker's biographical sketch and learning that she once worked as a hospital cleaning person, might think she couldn't be a poet—and that would be a great chance to enlighten him.

Questions on Niedecker:

What do we know about the people who live in the house with the "Popcorn-can cover"? (They're poor.)

In "I take it slow," is *alone* (line 2) a misprint for *along* or does it make sense? (It could well be an error—Niedecker's texts are in some disarray—but it makes sense.)

"Sorrow moves in wide waves": What details make the dying woman real to you? (The thimble, her turning blue, her final plea. This is an objective poem, seeing another person's decline from the outside, and so contrasts sharply with Dickinson's subjective "I heard a Fly buzz—when I died.")

Questions on Hayford:

In "Behind the Wall," what suggestions do you find in the sealed space and in the carpenter's hammer? (There's something mythic about that hammer—it's "old as the ark"—and yet it's a smaller, more commonplace find than those expected in the second stanza. No secret staircase, no Secret Garden. The hammer is associated with a ghost: it belonged to a man long dead, who finally fell down himself, like the hammer his hand let fall.)

In "Under the Holy Eye," what belief is the poem about? (That God sees everything—"His eye is on the sparrow.")

In "Dry Noon," what is the tone, the poet's attitude toward his material? (Wry, amused, sympathetic. The old man's breeches aren't a comic touch, but a matter of fact. Somehow, this fine, dry, still hour and even the immobile pants reflect the farmer's serene, inactive old age.)

Neither poet's collected works will be found in most bookstores, but Niedecker's *From This Condensery: the Complete Writings* may be ordered for $30 from Inland Book Co., 22 Hemingway Ave., East Haven, CT 06512, distributor for the Jargon Society; and Hayford's *Star in the Shed Window: Collected Poems 1933–1988* may be ordered from the New England Press, P.O. Box 575, Shelburne, VT 05482 ($14.95 paper, $24.95 cloth, plus $1.50 for postage).

28 What Is Poetry?

Robert Francis, CATCH, page 769

Putting a spin on his poem-ball, hurling high ones and sometimes grounders, the poet keeps the reader on his toes, yet sometimes provides an immediate reward. Most students, though they might paraphrase this poem differently, will find this much of Francis's comparison easy to pick up—although they might object that the reader doesn't get to hurl one back at the poet. And how does a poet "outwit the prosy?" Playing with swagger and style, the poet doesn't want to communicate in a dull, easily predictable way. A magician of the mound, he or she will do anything to surprise, to make poetry a high-order game in which both poet and reader take pleasure.

29   Poems for Further Reading

Anonymous, EDWARD, page 774

"Edward," with its surprise ending in the last line, is so neatly built that it is sometimes accused of not being a popular ballad at all, but the creation of a sophisticated poet, perhaps working from a popular story. Students might be asked to read the information about ballads in Chapter Seven, "Song," either before or after reading this ballad.

QUESTIONS FOR DISCUSSION:

1. In "Edward," why is the first line so effective as an opening? What expectations does it set up in the hearer's mind? How does the last line also display the skill of a master storyteller?

2. Reading a ballad such as "Edward" on the printed page, we notice that its refrains take up much room. Do you find this repetitiousness a hindrance to your enjoying the poem? Is there anything to be said in favor of the repetitiousness?

3. What is the value to the poem of the question-and-answer method of storytelling? ("Edward" proceeds like a courtroom cross-examination in which the mother, by pointed questioning, breaks down her son's story. Dramatically, the method is powerful; and it holds off until the very end the grimmest revelation.)

4. What else could the author of "Edward" conceivably have told us about this unhappy family? Do you find it troublesome that Edward and his mother behave as they do without our quite knowing why? Might the story have suffered if told by a storyteller who more deeply explored the characters' motivations?

Anonymous, THE THREE RAVENS, page 775

Anonymous, THE TWA CORBIES, page 776

QUESTIONS FOR DISCUSSION:

1. In "The Three Ravens," what is suggestive in the ravens and their conversation? How are the ravens opposed in the poem by the hawks and the hounds? (The ravens are selfish eaters of carrion, but the hawks and hounds are loyally standing guard over their dead master's body. Their faithfulness also suggests that of the fallow doe.)

2. Are you persuaded by Friedman's suggestion (quoted in the note under "The Three Ravens") that the doe is a woman who is under some enchantment? What other familiar fairy tales or stories of lovers transformed into animals do you recall?

3. Do you agree that "The Twa Corbies" is "a cynical variation of 'The Three Ravens,'" as it has been called? Compare the two poems in their comments on love and faithfulness.

4. For all the fantasy of "The Three Ravens," what details in the ballad seem realistic reflections of the natural world?

Anonymous, SUMER IS ICUMEN IN, page 777

If you are not an expert in Middle English pronunciation, ask a colleague who is to give you a briefing and take a stab at reading "Summer is icumen in" aloud to the class. Even the experts are only guessing (educatedly) how thirteenth-century English sounded, and your reading need do no more than suggest that the language has not always stood still.

(text pages) 777–778

This slight, exquisite, earthy song may be compared with Shakespeare's "When daisies pied," in which the song of the cuckoo suggests cuckoldry (as it doesn't in this innocent, humorless lyric).

For a parody, see Ezra Pound's "Ancient Music," beginning "Winter is icummen in," in his collection *Personae* (New York: New Directions, 1949), and in *Selected Poems* (from the same publisher, 1957).

Anonymous, I SING OF A MAIDEN, page 777

How could Christ be conceived and Mary remain a virgin? This medieval lyric appears to reply to the question: as easily and naturally as the dew falls.

The cesura in each line shows that English poetry, as late as the fifteenth century, had not quite broken away from the Old English two-part line.

"I sing of a maiden" can be used to remind students of the value to poetry of figures of speech. In line 1, *makeless* (meaning both "matchless" and "without a mate") is a serious pun. Each of the poem's three middle couplets contains a simile.

Anonymous, WESTERN WIND, page 778

Originally a song for tenor voices, "Western Wind" probably was the work of some courtier in the reign of Henry VIII. Untouched by modernization, it reads (in its one surviving manuscript):

> Westron wynde when wyll thow blow
> the smalle rayne douune can rayne
> Chryst yf my love wer in my armys
> and I yn my bed agayne

This famous poem, according to contemporary poet Deborah Digges, "demands that we create a context for this speaker who is far from home, waiting, it would appear, for the weather to break so that he can return. Is he at sea? Is he lost on the landscape? Is he dying? The poem refuses to answer, lives only by its breath, its longing" ("Lyrics and Ballads of the 15th Century," *Poetry Pilot*, [monthly newsletter of the Academy of American Poets], January 1989, page 4).

QUESTIONS FOR DISCUSSION:

1. In reading the poem, how does it help to know that the moist, warm west wind of England brings rain and is a sign of spring?

2. What do the first pair of lines and the last pair have to do with each other?

3. Do you agree with a critic who suggested that the speaker is invoking Christ, asking for help in obtaining sex? "By a blasphemous implication Christ is in effect assigned the role of a fertility spirit" (F. W. Bateson, *English Poetry: A Critical Introduction* [London: Longmans, 1966]).

4. Consider another critic's view: the unhappy speaker is stressing his (or her) longing to go to bed with his (or her) loved one, and so the word *Christ* is an exclamation. (We prefer this view; see Arthur O. Lewis, Jr., writing in *The Explicator* 15 Feb. 1957: item 28.)

Matthew Arnold, DOVER BEACH, page 778

Arnold and his family did such an efficient job of expunging the facts of his early romances that the genesis of "Dover Beach" is hard to know. Arnold may (or may not) have been in love with a French girl whom he called Marguerite, whose egotistic gaiety made her difficult. See Lionel Trilling's discussion of the poem and of Arnold's Marguerite poems in his biography *Matthew Arnold* (New York: Columbia UP, 1949). Marguerite, Trilling suspects, viewed the world as much more various, beautiful, and new than young Arnold did.

(text pages) 778–781

A sympathetic reading of "Dover Beach" might include some attention to the music of its assonance and alliteration especially the *s*-sounds in the description of the tide (lines 12–14). Line 21 introduces the central metaphor, the Sea of Faith. Students will probably be helped by a few minutes of discussion of the historical background of the poem. Why, when the poem appeared in 1867, was religious faith under attack? Darwin, Herbert Spencer, and Victorian industrialism may be worth mention. Ignorant armies (line 37) are still with us. Arnold probably had in mind those involved in the Crimean War of 1853–56, perhaps also those in the American Civil War. For sources of the poem, see C. B. Tinker and H. F. Lowry, *The Poetry of Matthew Arnold* (New York: Oxford UP, 1940) 173–78.

A dour view of the poem is taken by Donald Hall in "Ah, Love, Let Us Be True" (*American Scholar* Summer 1959). Hall finds "love invoked as a compensation for the losses that history has forced us to sustain," and adds, "I hope there are better reasons for fidelity than disillusion. . . . Like so many Victorian poems, its negation is beautiful and its affirmation repulsive." This comment can be used to provoke discussion. A useful counterfoil to "Dover Beach" is Anthony Hecht's satiric poem "The Dover Bitch," in his collection *The Hard Hours* (New York: Atheneum, 1960) and in many anthologies. For other critical comment, see William E. Cadbury, "Coming to Terms with 'Dover Beach,'" *Criticism* 8 (Spring 1966): 126–38; James Dickey, *Babel to Byzantium* (New York: Farrar, 1968) 235–38 (a good concise general essay); and A. Dwight Culler, *Imaginative Reason: The Poetry of Matthew Arnold* (New Haven: Yale UP, 1966).

Compare Hardy's attitude in "The Oxen" (page 716) with Arnold's wistful view of the Sea of Faith.

John Ashbery, AT NORTH FARM, page 779

It is never easy to decide what an Ashbery poem "means." This one is rich with suggestions about which students may be invited to speculate. Who is this threatening catlike "someone" for whom we set out milk at night and about whom we think "sometimes, / Sometimes and always, with mixed feelings?" Is it the Grim Reaper? And yet Death always knows where to find the person he's looking for. And what are we to make of lines 7–11? How can the granaries be "bursting with meal, / The sacks of meal piled to the rafters" if "Hardly anything grows here"? The poet hints at a terrible sterility underlying the visible abundance at North Farm. Perhaps the farm can be regarded as, among other things, a paradigm of the world, rich in material things but spiritually empty. But that is to reduce the poem to flat words. Because such paraphrases tend to slip from Ashbery's poems like seals from icebergs, this poet's work is a current favorite of critics. It challenges them to make subtler and stickier paraphrases.

Margaret Atwood, ALL BREAD, page 780

Rich in imagery and metaphor, Atwood's poem may seem at first like wild and whirling words, but a closer look will show that it says something. Atwood sees the earth as the ancestor shared by both eater and eaten, by the bread and the person who bakes and eats it. The dirt that is made of plant and animal remains "flows through the stems into the grain, / into the arm." In "All Bread," images of death ("Live burial under a moist cloth . . . the row / of white famine bellies / swollen and taut in the oven") coexist with images of life, perhaps to emphasize the intimate connection between the two. Eating becomes a holy act: "to know what you devour / is to consecrate it, / almost."

W. H. Auden, AS I WALKED OUT ONE EVENING, page 781

This literary ballad, with its stark contrast between the innocent song of the lover and the more knowing song of the clocks, affords opportunities to pay close attention to the poet's choice of words. Auden selects words rich in connotations: the *brimming* of the river (which suggests also the lover's feelings), the *crooked* neighbor (with its hint of dishonesty and corruption as well as the denotation of being warped or bent by Time, like the "diver's brilliant bow"). Figures of speech abound: the opening metaphor of the crowds like wheat (ripe and ready to be scythed by Time the Reaper), the lover's extended use of hyperbole in lines 9–20, the personifications of Time and Justice, the serious pun on *appalling* in line 34 (both awe-inspiring and like a pall or shroud, as in Blake's "London," page 562), the final reconciliation in metaphor between the original "brimming river" and the flow of passing Time. Auden's theme appears to be that as young lovers grow old, their innocent vision is smudged and begrimed by contact with realities—and yet "Life remains a blessing" after all.

The lover's huge promises in stanzas 3 and 4 ("I'll love you Till China and Africa meet . . . ") have reminded Richard Wilbur of the hyperbolic boasts of the speaker in Burns's "Oh, my love is like a red, red rose" (page 603). Burns speaks for the romantic lover, wrapped in his own emotions, but Auden's view of romantic love is skeptical. "The poem then proceeds to rebut [the lover's] lines, saying that the human heart is too selfish and perverse to make such promises" (*Responses* [New York: Harcourt, 1976] 144).

This poem may appear to have too little action in it to resemble folk ballads in more than a few touches. Auden himself, according to Monroe K. Spears, did not call this a ballad but referred to it as "a pastiche of folk-song."

"As I Walked Out One Evening" is one of the "Five Lyrics" included in *W. H. Auden Reading* (Caedmon recording TC 1019). For comparison with the poet's own modest delivery, *Dylan Thomas Reading*, vol. 4 (Caedmon, TC 1061) offers a more dramatic rendition.

W. H. Auden, Musée des Beaux Arts, page 783

In Breughel's *Landscape with the Fall of Icarus* reproduced on page 783, students may need to have their attention directed to the legs disappearing in a splash, one quarter inch below the bow of the ship. One story (probably apocryphal) is that Breughel's patron had ordered a painting on a subject from mythology, but the artist had only this landscape painting completed. To fill the order quickly, Breughel touched in the little splash, gave the picture a mythological name, and sent it on its way. Question: How does that story (if true) make Breughel seem a shallower man than Auden thinks he is?

Besides the *Landscape,* Auden apparently has in mind two other paintings of Pieter Breughel the Elder: *The Census,* also called *The Numbering at Bethlehem* (Auden's lines 5–8), and *The Massacre of the Innocents* (lines 9–13). If the instructor has access to reproductions, these works might be worth bringing in; however, the *Landscape* seems central to the poem. This painting seems indebted to Ovid's *Metamorphoses,* but in Ovid the plowman, shepherd, and fisherman looked on the fall of Icarus with amazement. The title of Auden's poem, incidentally, is close to the name of the Brussels museum housing the *Landscape*: the Musées Royaux des Beaux Arts.

Edward Mendelson has remarked on the poem in *Early Auden* (New York: Viking, 1981):

> The poetic imagination that seeks out grandeur and sublimity could scarcely be bothered with those insignificant figures lost in the background or in the crowd. But Auden sees in them an example of Christianity's great and enduring transformation of classical rhetoric: its inversion of the principle that the most important subjects require the highest style. If the sufferings of a carpenter turned preacher mattered more to the world than the doom of princes, then the high style, for all its splendor, was a limited instrument. . . . These casually irregular lines make none of the demands for action and attention that marked Auden's earlier harangues on the urgency of the times, yet beneath the apparent surface disorder a deeper pattern of connectedness gradually makes itself felt. The unassertive rhymes, easily overlooked on a first reading, hold the poem together.

(text pages) 783–785

Yet another device of language helps bring unity to Auden's meditation, in P. K. Saha's view. Four clauses begin with *how* and one phrase begins with *anyhow* (line 11). These *hows* vary in meaning; still, the repeated *how* is the crucial word in the linguistic pattern of the poem ("Style, Stylistic Transformations, and Incorporators," *Style* 12 [1978]: 18–22).

Jimmy Santiago Baca, SPLICED WIRE, page 784

Who is the speaker in this poem? A real wire that has been spliced? A disappointed lover with delusions of grandeur? His muse? Students will enjoy coming up with various more or less cogent interpretations. We think the poem revolves around a central metaphor. The speaker sees himself as the electric power that used to activate his former love's whole life: he supplied the heat, toasted the toast, turned on the radio. But he couldn't (or wouldn't) set her up in roomy style. His leaving her knocked out the power, "pulling the plug." Other men will plug into her, but they'll supply weak, fluctuating current. Her lights and heat won't work so well from now on. This serves her right. (The more we reread this poem, the smugger it seems!)

Baca is a Native American born in Santa Fe, whose work has been published in Mexico and Germany, and in this country, by New Directions, Louisiana State University Press, and small presses. "One of the most naturally gifted poets I've ever known," blurbs Denise Levertov on the collection this poem comes from, *What's Happening* (Willimantic, CT: Curbstone, 1982). He's worth watching.

R. L. Barth, THE INSERT, page 785

We would sum up Barth's theme: war blunts the sensibilities of the participant. Like one who begins taking heroin, the new soldier is at first eager for excitement; but with every use, the thrill is of shorter duration.

We didn't know exactly what an *insert* was, and asked the poet to define it. He replied: "Dropping troops, in this case recon troops, into an area by helicopter — one never knew if it would be a 'hot' LZ or not and was always ready for the worst."

Barth's poem invites comparison with other poems by another poet with field experience: Wilfred Owen's "Dulce et Decorum Est" (page 528) and "Anthem for Doomed Youth" (page 854). Barth's protest is stated not in the abstract, but in what lies before the observer's senses. Owen, by contrast, for all his vivid detail, seems continually to be editorializing.

Elizabeth Bishop, FILLING STATION, page 785

QUESTIONS FOR DISCUSSION:

1. What is the poet's attitude toward the feeble attempts at beautification detailed in lines 23–33? Sympathy, contempt, or what? How is the attitude indicated? (The attempts are doomed, not only by the gas station's being saturated with oil, but by the limitations of the family, whose only reading appears to be comic books and whose tastes run to hairy plants and daisy-covered doilies. In line 20, *comfy* is their word, not the poet's own. But the tone of the poem seems to be good-humored amusement. The sons are "quick and saucy" — likable traits. The gas station can't be beautiful, but at least its owners have tried. In a futile gesture toward neatness, they have even arranged the oil cans in symmetry.)

2. What meanings do you find in the last line? (Somebody has shown love for all motorists by arranging the oil cans so beautifully that they spell out a soothing croon, such as what one might say over and over to an agitated child. But the somebody also suggests Somebody Up There, whose love enfolds all human beings — even this oilsoaked crew.)

3. Do you find any similarity between "ESSO—SO—SO—SO" in "Filling Station" and "rainbow, rainbow, rainbow!" in "The Fish" (page 572)? (Both lines stand late in their poems

and sound similar; both express the speaker's glimpse of beauty—or at least, in "Filling Station," the only beauty the people can muster and the poet can perceive.)

Helen Vendler, discussing the poem in *Part of Nature, Part of Us* (Cambridge: Harvard UP, 1980), takes the closing statement to mean "God loves us all." But Irvin Ehrenpreis disagrees: "The '—so—so—so' of overlapping labels on stacked cans is supposed to comfort automobiles as if they were high-strung horses, i.e., like a mother, not a god." Doily and begonia indicate that some absent woman has tried to brighten up this gas station for her husband and her sons (review of Vendler's book in *The New York Review of Books*, 29 Apr. 1980).

Robert Pinsky has also written of "Filling Station" with high esteem. He calls the poem a kind of contest between "the meticulous vigor of the writer" and "the sloppy vigor of the family," both filling a dull moment and scene with "an unexpected, crazy, deceptively off-hand kind of elegance or ornament." He particularly admires the poet's choice of modifiers—including the direct, honest-seeming *dirty*. "Adjectives," he notes, "according to a sound rule of thumb for writing classes, do not make 'good descriptions.' By writing almost as though she were too plain and straightforward to have heard of such a rule, Bishop loads characterizations of herself and her subject into the *comfy* dog, the *dim* doily, the *hirsute* begonia; the quietest possible virtuoso strokes" (*The Situation of Poetry* [Princeton: Princeton UP, 1976] 75–77).

"I've sometimes thought 'Filling Station' would make a good exercise for acting students," observes critic and teacher David Walker, "given the number of different ways the first line—and much of the rest—might be stressed. Is the opening exclamation solemn and childlike, or prissy and fastidious, or enthusiastic? All we can identify with certainty, I think, is the quality of fascination, the intent gaze on the filling station's pure oiliness." Walker is reminded of Frost's "Design" (page 1493) in that both poets seek to discover "a meaningful pattern in apparently random details"—but while Frost points toward a sinister architecture in what he observes, Bishop finds beauty and harmony ("Elizabeth Bishop and the Ordinary," *Field*, Fall 1984).

Brad Leithauser has admired the poem's ingenious sound effects. At its end, "the cans of oil are arranged like cue cards to prompt that concluding sentence, the SO—SO—SO grading toward that 'Somebody loves us all.' Neatly, the message in the oil cans is reinforced by both the 'so' and the 'softly' in the fourth line from the end" ("The 'complete' Elizabeth Bishop," *New Criterion* Mar. 1983: 38).

Elizabeth Bishop, ONE ART, page 787

Like Thomas's "Do not go gentle," this villanelle manages to say something that matters while observing the rules of a tricky French courtly form. (For remarks on the villanelle and on writing it, see the entry on Thomas in this manual, page 55). A similar feat is performed in Bishop's ingenious recollection of childhood, "Sestina."

Question: What varieties of loss does the poet mention? (She goes from trivial loss—lost door keys—to lost time, to losing beloved places and homes, to loss of love.)

In recalling that she has lost "a continent," the poety may be speaking personally: she lived in Brazil for many years, but wrote this poem after returning to America.

William Blake, THE SICK ROSE, page 787

William Blake, THE TYGER, page 788

In "The Sick Rose," why is the worm, whose love is rape, *invisible*? Not just because it is hidden in the rose, but also because it is some supernatural dweller in night and storm. Perhaps the worm is unseen Time, that familiar destroyer—is the rose then mortal beauty? Those are usual guesses. For an unusual guess, see E. D. Hirsch, Jr., *Innocence and Experience*

(New Haven: Yale UP, 1964). "The rose's sickness, like syphilis, is the internal result of love enjoyed secretly and illicitly instead of purely and openly." In Hirsch's view, the poem is social criticism. Blake is satirizing the repressive order, whose hypocrisy and sham corrupt the woman who accepts it. Still, like all the best symbols, Blake's rose and worm give off hints endlessly, and no one interpretation covers all of them. We noted with interest that "The Sick Rose" is rightly included in *The Faber Book of Seduction* (London, 1988).

"The Tyger," from *Songs of Experience,* is a companion piece to "The Lamb" in *Songs of Innocence.* But while "The Lamb" poses a relatively easy question ("Little lamb, who made thee?") and soon answers it, "The Tyger" poses questions that remain unanswerable. Alert students may complain that some of Blake's questions have no verbs—what dread hand and what dread feet did *what?* While the incompleteness has been explained by some critics as reflecting the agitated tone of the poem, it may have been due to the poet's agitated habits of composition. Drafts of the poem in Blake's notebook show that, after writing the first three stanzas, he began the fourth stanza with the line "Could fetch it from the furnace deep," which would have completed the question in line 12. But then he deleted it, and wrote in stanza four almost as it stands now. (See Martin K. Nurmi, "Blake's Revisions of 'The Tyger,'" *PMLA* 71 [1956]: 669–85. Other useful discussions include that of Hirsch, who thinks the stars are the rebel angels who threw down their spears when they surrendered; and John E. Grant, "The Art and Argument of 'The Tyger'" in *Texas Studies in Literature and Language* 2 (1960): 38–60.

Louise Bogan, THE DREAM, page 789

In *Louise Bogan: A Portrait* (New York: Knopf, 1985), Elizabeth Frank discusses "The Dream," which Bogan wrote in her late thirties after having suffered a series of breakdowns. Bogan herself described the poem as "the actual transcript of 'a nightmare,' but there is a reconciliation involved with the fright and horror. It is through the possibility of such reconciliations that we, I believe, manage to live." Thus Frank sees the horse (the *night-mare*) as "the accumulated power of terror and rage Bogan had only recently confronted during her breakdowns," and the other woman in the poem, who comes to the rescue, as "Bogan's new self, the 'cured' woman, the emergent adult and artist." The glove, according to Frank, "becomes a symbol of challenge and submission, a token of the poet's strong feminine sexuality—and perhaps many other things as well. Put at the mercy of the beast, who is not only appeased, but in turn enchanted, the glove in the end represents the poet's whole self, triumphant over the lovingly submissive beast."

Mark Alexander Boyd, CUPID AND VENUS, page 789

Offer a prize to any student who, on a few days' notice, can read this poem aloud. (All it takes is nerve, and a passable Scottish burr.)

The allusions in lines 5–8 will need unraveling. Eros, god of love in Greek mythology (Cupid in Roman), is a youth, the son of Venus, and the lover of the maiden Psyche. Later tradition reduces him to a child and renders him blind or blindfolded. That Venus was born from the sea will be remembered from Botticelli's painting of the goddess on the half-shell, with which some students are probably acquainted.

A Scottish Petrarchan, Boyd shares (belatedly) the tradition that represents the lover as helpless and obsessed—visible in the Wyatt and Surrey versions of Petrarch's line on page 628. Ezra Pound has called this the most beautiful sonnet in the English language, and students may be asked what there is to admire in it.

Joseph Brodsky, BELFAST TUNE, page 790

"Belfast Tune" is like a broadside ballad in its relative simplicity, its metrical rimed stanza, its subject taken from recent events. In the conflict in Northern Ireland, where do the poet's sympathies lie? Does he take sides? (Unwilling to take sides in Russia, Brodsky doesn't take sides in Belfast either. As a poet, he has been deliberately apolitical—"defiantly so," as W. H. Auden remarked in a forward to Brodsky's *Selected Poems*.)

Students might like to know a little of the poet's personal history. A Russian Jew sentenced to internment at hard labor during a post-Stalinist purge of "social parasites," Brodsky remained unbowed. During his twenty-month ordeal he studied English and American poetry from one paperback anthology, with the aid of a Russian-English dictionary. In 1972 he was allowed—indeed, encouraged—to leave the Soviet Union, and came to America. He has taught at the University of Michigan, Mount Holyoke, and Boston University. Many who know Russian consider him the outstanding living poet in that language; for the past few years he has been writing in English as well. In 1987 he received the Nobel Prize for Literature.

In a review of *To Urania: Selected Poems 1965–1985* (Viking, 1988), from which "Belfast Tune" comes, Donald Davie takes Brodsky to task for not really knowing how English works. "The enjambment, the run over from one line into the next, is a more delicate instrument in English than in Russian precisely because it is potentially more disruptive." Davie calls Brodsky's use of enjambment in "Belfast Tune" "coarse and cavalier," and quoting the last two stanzas of the poem, indicates the runovers *say/ground, gray/bulb*, and *switch/hemispheres*. "In none of these cases does the whirl across the line-end, with the consequent jar or thud at the first pause in the next line, mirror a corresponding violence in feeling, in what is said. Accordingly, the verse-lines have a metrical and typographic but not a *musical* integrity" *(Times Literary Supplement,* December 23–29, 1988). Perhaps so, but we think the poem has a certain street-ballad-like charm to it.

Gwendolyn Brooks, A BLACK WEDDING SONG, page 790

A moving poem addressed to a bride and bridegroom, "A Black Wedding Song" deftly likens the married state to a missile launched on the wedding day. Why such a militaristic metaphor? Because "war comes in from the World / and puzzles a darling duet." Bride and groom must be armed to keep their marriage intact and alive, to withstand "the Assault that is promised."

What makes Brooks's poem specifically a *black* wedding song? Isn't every marriage threatened by assaults from the world? Since attending the Second Black Writer's Conference at Fisk University in 1967, Brooks has been persuaded that "black poets should write as blacks, about blacks, and address themselves to blacks." Perhaps she also feels that black marriages are especially endangered, for political and economic reasons.

Gwendolyn Brooks, THE RITES FOR COUSIN VIT, page 791

Cousin Vit, a vital, life-loving woman (a prostitute?), is dead. In this contemporary sonnet, the poet depicts her as too lively a presence to be confined by the casket, the "stuff and satin aiming to enfold her, / The lid's contrition nor the bolts before." To the poet, Cousin Vit appears to be still energetically moving about: walking, talking, drinking, dancing as of old. The wonderful *Is* that ends the poem emphatically sums up the dead woman's undying vigor.

Robert Browning, SOLILOQUY OF THE SPANISH CLOISTER, page 792

The "Soliloquy" is a poem especially valuable to combat the notion that poetry can deal only in love and gladness. Here, the subject is a hatred so intense that the speaker seems practically demented. In the last stanza, he almost would sell his soul to the Devil in order to blight a flowering shrub. A little background information on abbeys, their organization, and the strictness of their rules may help some class members. From internal evidence, it is hard to say whether this is a sixteenth-century cloister or a nineteenth-century one;

(text pages) 792–794

Barbary corsairs (line 31) plied their trade from about 1550 until 1816. The business about drinking in three sips (lines 37–39) may need explaining: evidently it refers to a symbolic observance, like crossing knife and fork.

It might be stressed that the person in this poem is not the poet: the tone isn't one of bitterness, but of merriment. Comedy is evident not only from the speaker's blindness to his own faults, but from the rollicking rhythm and multisyllable comic rimes (*abhorrence/Lawrence; horsehairs/corsair's; Galatians/damnations; rose-acacia/Plena gratia*).

Questions: With what sins does the speaker charge Brother Lawrence? (Pride, line 23—monogrammed tableware belonging to a monk!; lust, 25–32; and gluttony, 40.) What sins do we detect in the speaker himself? (Envy, clearly, and pride—see his holier-than-thou attitude in stanza 5. How persuasive are his claims to piety when we learn he secretly owns a pornographic novel?) "Soliloquy" abounds in ironies, and class members can spend a lively few minutes in pointing them out.

Thomas Carew, Ask me no more where Jove bestows, *page 793*

Carew's poem is an eloquent tribute to a lady whose complexion is so beautiful, the poet says, that it must emanate, with divine assistance, from the very roses. Her hair is made of sunbeams, her voice is like the nightingale's, and the stars twinkle in her eyes. Exaggerated? Yes. Still, what woman, even in the unromantic 1990s, wouldn't relish so lyrical a testimonial to her beauty?

You might want to ask your class to compare "Ask me no more where Jove bestows" with "My mistress' eyes are nothing like the sun" (page 756), where Shakespeare twits the Petrarchan conventions while of course relying as heavily as Carew does upon his readers' ability to appreciate a well-turned conceit.

Herbert J. C. Grierson, in his anthology *Metaphysical Lyrics and Poems of the Seventeenth Century* (Oxford UP, 1921), waxes poetic over the *dividing throat:* "One seems to hear and see Celia executing elaborate trills as Carew sits entranced."

There is a famous anecdote about Carew, which may or may not be true. One evening he was lighting King Charles I to the queen's chamber. Entering the room first, the poet saw the queen in the arms of Lord St. Albans. Before the king noticed that anything was amiss, Carew tactfully stumbled, thus extinguishing his candle long enough for the queen to adjust her position. His quick thinking endeared Carew to the queen from that day forward.

Fred Chappell, Skin Flick, *page 794*

Nastily playful though this poem may be, it makes a serious comment. Students will probably enjoy working through the details of Chappell's comparison between the porn movie's male audience and cowboys surrounded by Indians. (A theme: The more things change, the more they remain the same.) "No water for miles and miles" is a cliché from those grade-C western films now replaced by X-rated fare. Like cowboy movies of the 1940s and 1950s, skin flicks are loveless affairs: "No / Mushy love stuff for them." (Such mush was usually avoided in Hopalong Cassidy horse operas by having the hero ride off into the twilight whenever the heroine invited him to stay awhile and settle down.) In Chappell's poem, lack of water reflects lack of love as surely as it does in the anonymous "Western Wind" (page 778) or in Eliot's *Waste Land*.

The parallel appears in other ways. Naked bodies are a "different sort of cattle drive." Once Gene Autry was "in the saddle," nowadays it's hard-core porn stars John Holmes or Harry Reems. "The burning bush strikes dumb" points at least three ways: (1) the burning bush or wahoo tree of the West, possibly also sagebrush on the scorching plains; (2) luridly lit pubic hair; (3) the burning bush that astounded Moses in Exodus 3:2–4. In the last line, yet another cliché from western movies takes on new meaning. Like settlers ringed by hostile Indians, pornography addicts are worried and tense, a huddle of men on guard against the outside world.

(text pages) 794–795

Chappell, novelist and poet, shared the 1984 Bollingen prize for poetry with John Ashbery. Among his works in poetry is the tetralogy *Midquest* (Baton Rouge: Louisiana State UP, 1981). He teaches English at the University of North Carolina, Greensboro.

G. K. Chesterton, THE DONKEY, page 795

QUESTIONS FOR DISCUSSION:

1. Who is the speaker—some particular donkey? (No, the genetic donkey, looking back over the history of his kind.)

2. To what prehistoric era does Chesterton refer in lines 1–3? (To the original chaos out of which the world was made. The poet apparently imagines it in bizarre, dreamlike imagery: fish with wings, walking forests, fig-bearing thorn.) Chesterton was fascinated by the book of Genesis "because of its beginning in chaos," comments Garry Wills in his introduction to a reprint edition of Chesterton's novel of 1908, *The Man Who Was Thursday* (New York: Sheed, 1975). The novel hints at a playful God who enjoys returning things to chaos every now and then. Writing about the world of dream in a newspaper article in 1904, Chesterton remarked, "A world in which donkeys come in two is clearly very near to the wild ultimate world where donkeys are made."

3. Whose "ancient crooked will" is meant? The will of the devil in perversely designing the donkey, or the donkey's own venerable stubbornness? (We're not certain.)

4. What *fools* does the donkey chide in the last stanza? (Anybody who ever abused a donkey, or who thinks donkeys contemptible.)

5. Explain how the allusion in the last stanza is essential to the meaning of the poem.

6. What devices of sound contribute to the poem's effectiveness?

Samuel Taylor Coleridge, KUBLA KHAN, page 795

The circumstances of this poem's composition are almost as famous as the poem itself, and, for the convenience of instructors who wish to read to their students Coleridge's prefatory note, here it is:

> In the summer of the year 1797, the author, then in ill health, had retired to a lonely farmhouse between Porlock and Linton, on the Exmoor confines of Somerset and Devonshire. In consequence of a slight indisposition, an anodyne had been prescribed, from the effects of which he fell asleep in his chair at the moment that he was reading the following sentence, or words of the same substance, in *Purchas's Pilgrimage:* "Here the Khan Kubla commanded a palace to be built, and a stately garden thereunto. And thus ten miles of fertile ground were inclosed with a wall." The author continued for about three hours in a profound sleep, at least of the external sense, during which time he had the most vivid confidence that he could not have composed less than from two to three hundred lines; if that indeed can be called composition in which all the images rose up before him as *things*, with a parallel production of the correspondent expressions, without any sensation or consciousness of effort. On awaking he appeared to himself to have a distinct recollection of the whole, and taking his pen, ink, and paper, instantly and eagerly wrote down the lines that are here preserved. At this moment he was unfortunately called out by a person on business from Porlock, and detained by him above an hour, and on his return to his room, found, to his no small surprise and mortification, that though he still retained some vague and dim recollection of the general purport of the vision, yet, with the exception of some eight or ten scattered lines and images, all the rest had passed away like the images on the surface of a stream into which a stone has been cast, but, alas! without the after restoration of the latter!

(text pages) 795–799

It is clearly a vulgar error to think the poem a mere pipe dream, which anyone could have written with the aid of opium. The profound symbolism of "Kubla Khan" has continued to intrigue critics, most of whom find that the *pleasure-dome* suggests poetry, the *sacred river*, the flow of inspiration, or instinctual life. About the *ancestral voices* and the *caves of ice* there seems less agreement, and students might be invited to venture their guesses. For a valuable modern reading of the poem, see Humphry House, "Kubla Khan, Christable and Dejection" in *Coleridge* (London: Hart-Davis, 1953), also reprinted in *Romanticism and Consciousness*, ed. Harold Bloom (New York: Norton, 1970).

Some instructors may wish to bring in "The Rime of the Ancient Mariner" as well—in which case it may be a temptation to go on to Jung's theory of archetypes and to other dreamlike poems such as Yeats's "The Second Coming" (page 718). A fine topic for a term paper might be, after reading John Livingston Lowes's classic source study *The Road to Xanadu* (Boston: Houghton, 1927), to argue whether it is worth trying to find out everything that may have been going on in the back of a poet's mind, and to what extent such investigations can end in certainty.

Emily Dickinson, Because I could not stop for Death, page 797

Questions for Discussion:

1. What qualities does Dickinson attribute to Death? Why is Immortality going along on this carriage ride? (For the poet, death and immortality go together. Besides, Dickinson is amplifying her metaphor of Death as a gentleman taking a woman for a drive: Immortality, as would have been proper in Amherst, is their chaperone.)

2. Is the poem, as the poet wrote it, in some ways superior to the version first printed? Is *strove* perhaps a richer word than *played*? What is interesting in the phrase *Gazing Grain*? How can grain "gaze"? (It has kernels like eyes at the tips of its stalks. As the speaker dies, the natural world—like the fly in "I heard a Fly buzz" on page 708—is watching.) What is memorable in the rhythm and meaning of the line "The Dews drew quivering and chill?" (At *quivering*, the rhythm quivers loose from its iambic tetrameter. The image of cold dampness foreshadows the next stanza, with its images of the grave.)

3. What is the Carriage? What is the House?

4. Where is the speaker at the present moment of the poem? Why is time said to pass more quickly where she is now? (Eternity is timeless.)

5. What is the tone of the poem? (Complicated!—seriousness enlivened with delicate macabre humor? Surely she kids her own worldly busyness in the opening line.)

William Galperin reads the poem as a feminist affirmation. Not death, he finds, but immortality is Dickinson's subject. In the end the poet asserts a triumph possible only because she has renounced the proposal of Death, that threatening gentleman caller who might have married her ("Emily Dickinson's Marriage Hearse," *Denver Quarterly* Winter 1984: 62–73).

Emily Dickinson, I started Early—Took my Dog, page 798

It would be unfortunate if students were to regard this poem as nothing more than a sexual fantasy. Handled with frankness and tact, it can be an excellent class-awakener. The poet expresses feelings for the natural world so intense that, like a mystic choosing erotic imagery to speak of the Beatific Vision, she can report her walk on the beach only in the language of a ravishing. The humor of the poem also is essential: the basement-dwelling mermaids, the poet's self-picture as a mouse.

Emily Dickinson, My Life had stood—a Loaded Gun, page 799

This astonishing metaphysical poem (another hymnlike work in common meter) can be an excellent provoker of class debate. Before trying to fathom it, students might well examine its diction. *Sovreign Woods* ("sovereign" would be the more usual spelling) suggest an estate owned by a king. How do the Mountains *reply*? By echoing the gun's report. Apparently the *smile* is the flash from the gun's muzzle; and the *Vesuvian face*, a glimpse of the flaming crater of the volcano. The Eider-Duck, a sea duck, has particularly soft and silky down which is used in pillows and quilts. The gun's *Yellow Eye* seems, again, its flash: and the *emphatic Thumb* is presumably the impact of the bullet that flattens its victim. (Some will say the thumb is a trigger finger, but you don't pull a trigger with your thumb.)

Argument over the meaning of the poem will probably divide the class into two camps. One will see the poem, like "Because I could not stop for Death," as an account of resurrection, with the Owner being God or Christ, who carries away the speaker, life and all, to the Happy Hunting Grounds of Paradise. Personally, we incline toward the other camp. In that view the Owner seems a mere mortal, perhaps a lover. The last stanza reveals that he can die. So taken, the last two lines make more sense. Not having the power to die, the speaker feels something lacking in her. She doesn't wish to outlive her huntsman and be a lonely killer.

Philip Larkin admits the possibility of both views: "This is romantic love in a nutshell, but who is its object? A religious poet—and Emily was this sometimes—might even have meant God" (*Required Writing* [New York: Farrar, 1984] 193).

Lately a third camp has appeared, proclaiming a feminist interpretation. The poem, as summed up by Adalaide Morris, "tells about a life packed with a potential that the self was not empowered to activate." From this point of view, the poem is overtly political, exhilarating to teach because it recognizes long suppressed animosities ("Dick, Jane, and American Literature: Fighting with Canons," *College English* 47 [1985]: 477).

But the poem remains tantalizingly ambiguous. You won't know until you go into class what a discussion may reveal.

Deborah Digges, FOR THE DAUGHTERS OF HANNAH *Bible Class* OF TIPTON, MISSOURI'S WOMEN'S PRISON: MOTHER'S DAY, 1959, page 799

Why did the speaker as a child fear the prisoners? (She has this notion that she'll turn into a criminal if she merely looks at them—a horrific transformation, perhaps like Lot's wife's turning into a pillar of salt.)

What did she fear within herself? (Sins that bloomed like bruises—perhaps involuntary sexual dreams?)

Now, in maturity, why does she regard the prisoners with new compassion and understanding? (She, too, is a mother now, and can feel for the prisoners' numb anguish in being deprived of their own children.)

What do you make of the last lines? (They complete a metaphor. In lines 16–18 the poet likens the prisoners' numb, frozen, pent-up motherlove to the ice that clogged the roads and didn't melt until spring. Does the speaker now feel a similar emotional thaw? It's hard to put such a symbol into other words. But here's one attempt: Like the ice now melting from the eaves, the child-speaker's old fear has dissolved, replaced by a bright memory, or perhaps by a fresh joy of her own.)

Digges's first book, *Vesper Sparrows* (Atheneum, 1986), in which this is the opening poem, won the Delmore Schwartz Award. A second collection, *Late in the Millenium*, was published by Knopf in 1989. Born and raised in Missouri, she currently teaches at Tufts University.

John Donne, THE FLEA, page 800

This outrageous poem is a good class-rouser on a dull day, but we don't urge you to use it unless the class seems friendly. (Some women students tend to be offended by Donne's levity; men tend to be put off by his ingenuity.)

A little familiarity with a seventeenth-century medical notion may help make Donne's metaphor clear. Conception, it was thought, took place when the blood of man and woman

(text pages) 800–802

mingled during intercourse. That is why Donne declares in line 11 that "we almost, yea more than married are." Bitten by the flea containing his blood, the woman may already be pregnant.

Instructors fond of Donne's knotty poems will be grateful for Theodore Redpath's valuable crib-book *The Songs and Sonnets of John Donne: An Editio Minor* (London: Methuen, 1956; also New York: University Paperbacks, 1967). Redpath works through the poems line by line, explicating difficulties. He explains line 18: The woman would commit "three sins in killing three" in that she'd commit murder in killing him, suicide in killing herself, and sacrilege in killing the flea. Why sacrilege? Because she would be attacking a "marriage temple."

Patricia Meyer Spacks has treated the poem to scrutiny in *College English* 29 (1968): 593–94.

John Donne, A VALEDICTION: FORBIDDING MOURNING, *page 801*

In his *Life of Donne*, Izaak Walton tells us that Donne wrote this poem for his wife in 1611, when he was about to depart on a diplomatic mission to France.

Much of the meaning of the poem depends upon the metaphor of the compasses in the last three stanzas. There is probably no better way to make sure students understand it clearly than to bring in a draftsman's compass—even the Woolworth's variety—and to demonstrate the metaphor with it. There'll always be someone who thinks Donne means the kind of compass that indicates north.

QUESTIONS FOR DISCUSSION:

1. What is a *valediction* anyway? (What is a high school "valedictorian"?)

2. Why does the speaker forbid mourning? Do lines 1–4 mean that he is dying? Explain this metaphor about the passing away of virtuous men. (As saints take leave of this world—so sweetly and calmly that one hardly knows they're gone—let us take leave of each other.)

3. In lines 7–8, what is suggested by the words with religious denotations? *Profanation* (the desecration of a sacred thing), *the laity.* What is the idea? (Love seems to the speaker a holy mystery. He and his wife are its priests or ministers.)

4. Explain the reference to astronomy in the third stanza. (Earthquakes shake, rattle, and roll; Ptolemaic spheres revolve gently and harmlessly. This takes us to the notion of *sublunary* lovers in stanza 4. In the medieval cosmos, the heavenly bodies are fixed and permanent, while everything under the moon is subject to change.)

5. Paraphrase stanza 4. (Unlike common lovers, bound to their earthly passions, we have less need of those things that serve sensual love: namely, bodies.)

6. Why is beaten gold an appropriate image in the sixth stanza? What connotations does gold have? (Refined, precious, durable, capable of being extended without breaking.)

7. Comment on the word *hearkens*, line 31. (As a draftsman's compass will illustrate, the fixed central foot leans forward when the compass is extended, as if, in Donne's metaphor, eager for the other mate's return.)

Rita Dove, DAYSTAR, *page 802*

What suggestions can be found in the title of Dove's poem? The word is a trove of connotations. You may wish to make sure at the start that your students are familiar with the dictionary meaning of *daystar:* the morning star or, poetically, the sun. From the poem's last line, we infer that it was the sun the poet had in mind. Why *daystar* instead of *sun?* Does the word suggest to the poet, as to us, joy and happiness? Also, a star is a "self-luminous, self-containing mass of gas in which the energy generated by nuclear reactions in the interior is balanced by the outflow of energy to the surface and the inward-directed gravitational forces are balanced by the outward-directed gas and radiation pressures." Perhaps the poet is suggesting that a similar state of balance is what the wife and mother in the poem seeks—and finds—when she sits behind the garage, "building a palace," while her children are taking their naps. The chance to daydream in solitude sustains her not only through her toils but "that night when Thomas rolled over and / lurched into her."

Thomas and Beulah, in which "Daystar" appears, won the 1987 Pulitzer prize for poetry. The book contains a series of poems about the lives and concerns of an ordinary black couple.

John Dryden, To the Memory of Mr. Oldham, page 803

With the aid of Dryden's great short poem (and the selections in the book from Swift, Pope, and Johnson), one at least can acquaint students with a little of neoclassical poetry. One can point out, too, that such poetry is not quite dead in America in our own day, as may be seen from the poems of Yvor Winters (page 895) and J. V. Cunningham (pages 543, 672). The directness and plainness of Dryden's poem are clear from its very opening, and in teaching it one can question the assumption that neoclassical poetry is written only in bookish and Latinate words.

In teaching Dryden's poem, one can also mention (and define) the elegy, and can refer students to other famous elegies in the text: those of Milton (page 850) and Gray (page 760). If one is going to teach "Lycidas," Dryden's succinct poem might well serve as an introduction. It may be readily compared with "Lycidas" in that it mourns the death of a poet, expresses abhorrence for knaves, and observes a few classical conventions.

A. E. Housman's "To an Athlete Dying Young" (page 832) also may be likened to Dryden's poem in that both poets favor classical conventions: footraces with laurels as crowns and the dead hero's descent into the underworld. Both poets find that premature death can confer benefits. What would Oldham have gained had he survived? More polish as a poet, yet he would have lost much of his force. In reading Housman's poem, students can be helped to recognize its metaphors: the comparison in the first two stanzas of victor's chair and dead lad's coffin, the comparison in line 5 of all human life to a footrace, with death at the finish line. Students might be asked if they know of any living proof of Housman's observation that sometimes the name dies before the man (a truth often shown by the wistfulness of old football players at alumni weekends).

T. S. Eliot, Journey of the Magi, page 804

The speaker is a very old man ("All this was a long time ago . . . "), looking forward to his death. As his mind roves back over the past, it is mainly the discomforts and frustrations of his journey that he remembers, and when he comes to the part we have been waiting for, his account of the Nativity, he seems still mystified, as though uncertainly trying to figure out what had happened—"There was a Birth, certainly." Apparently the whole experience was so devastating that he prefers to omit all further details. His plight was to recognize Christ as God and yet to be unable to accept Christ as his savior. Being a king, he did not renounce his people but they henceforth seemed alien to him, clutching their discredited gods like useless dolls.

The passage beginning "Then at dawn" (lines 21–28) is full of foreshadowings, both hopeful and sinister. Besides the symbolic white horse, the vine leaves suggest Christ who said to his disciples, "I am the vine, ye are the branches" (John 15:5). The tavern customers suggest the Roman soldiers who will drink and cast dice at the cross.

Although Eliot's dissatisfied Magus isn't one of the kings portrayed by Yeats in "The Magi" (page 902)—being dissatisfied for different reasons—it is curious that Eliot may have taken the dramatic situation of his poem from one of Yeats's stories. In "The Adoration of the Magi" in Yeats's prose collection *Mythologies* (reprinted in 1925, two years before Eliot first published his poem), three old men call on the storyteller, and, drawing close to his fire, insist on telling him of a journey they had made when young, and of a vision of Bethlehem. Like Eliot's speaker, who repeats "set down / This set down / This," they demand that their story be taken down word for word.

Among the useful discussions of Eliot's poem are Elizabeth Drew's in *T. S. Eliot: The Design of His Poetry* (New York: Scribner's, 1949) 118–22, and Grover Smith's in *T. S. Eliot's Poetry and Plays* (Chicago: U of Chicago P, 1960) 121–25. More recently, Daniel A. Harris has characterized the Magus as a primitive Christian with a "baffled consciousness of mystery." See his article "Language, History, and Text in Eliot's 'Journey of the Magi,'" *PMLA* 95 (1980): 838–56. But Harris's opinions are questioned by William Skaff in a letter in *PMLA* 96 (1981): 420–22: "In 'Journey' Eliot adopts the dramatic mask of the Magus in order to express his own struggles with literal belief, his real 'religious position of 1927.'"

T. S. Eliot, THE LOVE SONG OF J. ALFRED PRUFROCK, page 806

Teaching any basic course in literature, an instructor would have to be desperate for time not to devote to "Prufrock" at least a class or two. Eliot's early masterpiece can open such diverse matters as theme, tone, irony, persona, imagery, figures of speech, allusion, symbolism, and the difference between saying and suggesting. Most students will enjoy it and remember it.

QUESTIONS FOR DISCUSSION:

1. Why the epigraph from Dante? What expectations does it arouse? (Perhaps that this "song" will be the private confession of someone who thinks himself trapped and unredeemable, and thinks it of his hearer, too.)

2. What facts about J. Alfred can we be sure of? His age, his manner of dress, his social circles? What does his name suggest? Can you detect any puns in it? (A prude in a frock—a formal coat.)

3. What do you make of the simile in lines 2–3? What does it tell us about this particular evening? (*Etherized* suggests fog, also submission, waiting for something grim to happen.) What does it tell you about Prufrock's way of seeing things? ("A little sick," some students may say, and with reason.)

4. What gnaws at Prufrock? (Not just his sense of growing old, not just his inability to act. He suffers from Prufrock's Complaint: dissociation of sensibility. In line 105, unable to join thought and feeling, he sees his own nerves existing at one remove from him, as if thrown on a screen by a projector.)

5. Who are "you and I" in the opening line? Who are "we" at the end? (Some possibilities: Prufrock and the woman he is attending. Prufrock and the reader. Prufrock and Prufrock—he's talking to himself, "you" being the repressive self, "I" being the timid or repressed self. Prufrock and the other eggheads of the Western world—in this view, the poem is Eliot's satire on the intelligentsia.)

6. What symbols do you find and what do they suggest? Notice those that relate to the sea, even *oyster-shells* (line 7). (XJK points out blatantly that water has connotations of sexual fulfillment, and quotes "Western Wind" [page 778]. Eliot hints that unlike Prufrock, the vulgar types who inhabit cheap hotels and fish shops have a love life.)

7. Try to explain the last three lines.

8. Now summarize the story of the poem. What parts does it fall into? (Part one: Prufrock prepares to try to ask the overwhelming question. Then in lines 84–86 we learn that he has failed to ask it. In 87–110 he tries to justify himself for chickening out. From 111 to the end, he sums up his static present and hollow future.)

That Eliot may have taken the bones of his plot from Henry James's story "Crapy Cornelia" (1909) is Grover Smith's convincing theory. "This is the story of White-Mason, a middle-aged bachelor of nostalgic temperament, who visits a young Mrs. Worthington to propose marriage but reconsiders owing to the difference in their worlds" (*T. S. Eliot's Poetry and Plays* [Chicago: U of Chicago P, 1960] 15).

"The meter of 'Prufrock' is peculiar," observes John Heath-Stubbs. "It is not simply free verse, as in [Eliot's] earlier Laforgueian pieces, but in its lines of irregular length, many but not all of which rhyme, suggests a free version of the Dantesque Canzone." This suggestion, and the poem's epigraph from the *Inferno*, point to Eliot's growing preoccupation with Dante ("Structure and Source in Eliot's Major Poetry," *Agenda*, Spring-Summer 1985: 24).

T. S. Eliot Reading His Poetry (Caedmon recording TC 1045) includes the poet's rendition of "Prufrock."

Louise Erdrich, INDIAN BOARDING SCHOOL: THE RUNAWAYS, page 809

For discussion: Have you ever been somewhere you wanted to run away home from? If so, how does your memory of that experience compare with that of the speaker in this poem?

This poem is long on wounds: the railroad tracks are scars (6), the runaways' old welts are like roads (15), the names they wrote in wet cement and the leaves they pressed into the sidewalk before it dried recall "delicate old injuries" (22–24). All these things carry powerful connotations of being wounded, mistreated, beaten down—like the runaways themselves.

Jacklight (New York: Holt, 1984), the collection in which this poem appears, contains several other realistic poems of Indian life. Since the success of *Love Medicine* (1984), *The Beet Queen* (1986), and *Tracks* (1988), Louise Erdrich is best known as a novelist, but we think her poetry warrants attention too.

Erdrich, born in Little Falls, Minnesota, grew up in Wahpeton, North Dakota, and now lives in New Hampshire with her husband, Michael Dorris, also a novelist and a nonfiction writer.

Robert Frost, AWAY! page 810

Robert Frost, BIRCHES, page 811

These two poems, one written early in Frost's career and one written late, make for a surprising comparison. "Birches" is the poem most often studied; we'll consider it first here.

"Birches," according to Lawrance Thompson, was written during a spell of homesickness in 1913–14, when Frost and his family were living in Beaconsfield, Buckinghamshire, England (*Robert Frost: The Years of Triumph*, Holt, 1970; 37, 541).

Students may be led to see the poem as much more than a nostalgic picture of boyhood play. From line 43 on, the poem develops a flamboyant metaphor. Richard Poirier has given us a good summary of the poem's theme: "While there are times when the speaker of ['Birches'] would 'like to get away from earth awhile,' his aspiration for escape to something 'larger' is safely controlled by the recognition that birch trees will only bear so much climbing before returning you, under the pressure of human weight, back home" (*Robert Frost: The Work of Knowing* [Oxford UP, 1977] 172).

One line in "Birches" meant most to Frost, the line about feeling lost in the woods, facing too many decisions about which way to go. He pointed it out to audiences on several occasions: "It's when I'm weary of considerations" (line 43). Reading the poem at Bread Loaf in July, 1954, he remarked of the line, "That's when you get older. It didn't mean so much to me when I wrote it as it does now" (*Robert Frost: A Living Voice*, ed. Reginald Cook [U of Mass P, 1974] 51). Radcliffe Squires has written interestingly of the birch tree as a path toward heaven fraught with risk, suspense, even a kind of terror. The climbing boy performs his act of birch-bending gracefully, but in doing so goes almost too far, like one filling a cup "even above the brim" (*The Major Themes of Robert Frost* [U of Michigan P, 1963] 55–56).

Sidelights on the poem: Frost wrote to his friend, Charles Madison, in 1950, " 'Birches' is two fragments soldered together so long ago I have forgotten where the joint is." Can anybody find it? . . . A particular word he congratulated himself on finding was *crazes* in line 9: "cracks and crazes their enamel" (Cook, 230). Frost's concern for scientific accuracy is well known. He sought evidence to confirm his claim that birches bend to left and right. "With disarming slyness, he said: 'I never go down the shoreline [from Boston] to New York without watching the birches to see if they live up to what I say about them in the poem.' His birches, he insisted, were *not* the white mountain or paper birch of northern New England (*Betula papyrifera*); they were the gray birch (*Betula populifolia*)" (Cook, 232).

"Away!", which we think the best poem in Frost's final collection *In the Clearing* (1962), is wonderfully dextrous and shows the venerable poet still in command of a highly demanding form. Try to imitate perfectly rimed *abab* quatrains like these! (And try to write and apparently think in them with seeming effortlessness.)

But the poem is no hollow *tour de force;* Frost deals with a theme of lifelong importance to him. In his deliberate bepiddlement of the story of the Fall of Man and his directive, "Forget the myth," Frost seems to come out firmly on the side of atheism or agnosticism. But, as usual, he hedges: "Unless I'm wrong." The really interesting (and outrageous) statement comes in the last stanza: if he doesn't like the knowledge that death reveals, he'll refuse to stay dead and come back to earth again. (Frost died the year after this poem was published!) It is this return to earth that recalls "Birches." As that poem puts this wish: "I'd like to get away from earth awhile / And then come back to it and begin over." Can students find a connection between these two seemingly dissimilar poems?

Robert Frost, STOPPING BY WOODS ON A SNOWY EVENING, page 812

Students will think they know this poem from their elementary school textbooks, in which it is usually illustrated as though it were about a little horse, but they may need to have its darker suggestions underlined for them. Although one can present a powerful case for seeing Frost as a spokesman for the death wish, quoting other Frost poems such as "Come In," "To Earthward," and "Into My Own," we think it best to concentrate on this familiar poem and to draw the class to state what it implies. The last stanza holds the gist of it. What would he do if he *didn't* keep his promises? There is sense, however, in an objection a student once made: maybe he'd just stay admiring the snow for another fifteen minutes and be late for milking. "People are always trying to find a death wish in that poem," Frost told an audience at the Bread Loaf Writers' Conference in 1960. "But there's a life wish there—he goes on, doesn't he?"

Ask students if they see anything unusual about the rime scheme of the poem (rimes linking the stanzas as in terza rima or as in Shelley's "Ode to the West Wind"), and then ask what problem this rime scheme created for the poet as the poem neared its end. How else would Frost have ended it if he hadn't hit upon that magnificent repetition? In 1950 Frost wrote to a friend, "I might confess the trade secret that I wrote the third line of the last stanza of Stopping by Woods in such a way as to call for another stanza when I didn't want another stanza and didn't have another stanza in me, but with great presence of mind and a sense of what a good boy I was I instantly struck the line out and made my exit with a repeat end" (qtd. in Lawrance Thompson, *Robert Frost: The Years of Triumph* [New York: Holt, 1970] 597–98). On another occasion Frost declared that to have a line in the last stanza that didn't rime with anything would have seemed a flaw. "I considered for a moment winding up with a three line stanza. The repetend was the only logical way to end such a poem" (letter of 1923 to Sylvester Baxter, given by R. C. Townsend, *New England Quarterly* 36 [June 1963]: 243.)

(text pages) 812–814

That this famous poem may be sung to the old show tune "Hernando's Hideaway" (from *The Pajama Game*) was discovered by college students working as waiters at the Bread Loaf Writers' Conference in 1960.

Paper topic: Read Lionel Trilling's speech at Frost's eight-fifth birthday dinner, in which Trilling maintained, "I think of Robert Frost as a terrifying poet" ("A Speech on Robert Frost: A Cultural Episode," *Partisan Review* 26 [Summer 1959]: 445–52; also reprinted in *Robert Frost: A Collection of Critical Essays*, ed. James M. Cox [Englewood Cliffs: Prentice, 1962]). Referring to "Stopping by Woods" and other Frost poems, state to what extent you agree with Trilling's view or disagree with it.

Frost reads the poem on *An Album of Modern Poetry* (Library of Congress, PL 20) and on *Robert Frost Reading His Own Poems*, record no. 1 (EL LCB 1941, obtainable from the National Council of Teachers of English, 1211 Kenyon Road, Urbana, IL 61801). Both recordings also include "Fire and Ice."

Allen Ginsberg, A SUPERMARKET IN CALIFORNIA, page 813

A comparison of this poem with Walt Whitman's "I Saw in Louisiana a Live-Oak Growing" (page 890) and "To a Locomotive in Winter" (page 512) demonstrates the extent to which Ginsberg, in his tribute to Whitman, uses very Whitmanlike "enumerations." Ginsberg's long sentences, use of free verse, parentheses, and fulsome phrases ("childless, lonely old grubber," "lonely old courage-teacher," etc.) are further indications that he is paying tribute to Whitman in part by echoing his style.

There is in "A Supermarket in California" as well a quality of surrealism that is Ginsberg's own. The existence of a "Neon fruit supermarket," the juxtaposition of past and present, the inclusion of Spanish poet Garcia Lorca (like Ginsberg and Whitman, a homosexual) "down by the watermelons," and the references to Charon and to the River Lethe all hover at the edges of dream.

Questions for discussion: What does Ginsberg mean when he speaks of "the lost America of love"? What does the poem say about loneliness? about death? (Whitman's death, in the poem, is as lonely a journey as Ginsberg imagines his life to have been.)

Dana Gioia, CALIFORNIA HILLS IN AUGUST, page 814

For this manual, Gioia has provided this note on his poem:

> "California Hills in August" was conceived as a defense of the special beauty of the dry landscape of California and the American Southwest. While the poem admits that an outsider might initially find this environment harsh and even hostile in the hot summer months, it asks that the landscape be seen on its own terms and not judged as some deficient form of conventionally verdant summer scenery.
>
> Strangely, I did not write this poem until I had been living in the Northeast for several years. It was only then that I realized how foreign the sparse August hills of California seemed to most Americans who come from areas which are lushly green and overgrown during the summer. Likewise, living back East, I finally understood how much the symbolic framework of traditional English poetry depended on the particular kind of climate shared by England and Eastern America. Until then the seasonal patterns of this poetry with its snowy winters and green summers had always seemed remote and artificial to a Westerner like me, raised in a climate which endowed a radically different sense of Nature's cycles. I wondered why Western writers had not played with that paradox more forcefully in their poetry.
>
> These thoughts must have been in the back of my mind when suddenly I returned home for a family emergency. Two sleepless days later, after it had been resolved, I took a noontime walk through some nearby hills. In the strangely heightened

consciousness of physical exhaustion and dislocating travel, I found myself overwhelmed by the delicate, sun-bleached beauty of the place. This was a landscape which I had left long enough ago both to hunger for as a native and to recoil from as an outsider. Sorting out these contradictions as I walked home, I heard myself say the first few lines of the poem. Writing out the opening stanzas later that day, I realized how image by image this poem had been forming for a long time in my unconscious.

The poem divides into two uneven parts. The first four stanzas view the landscape through the eyes of an imaginary Easterner who sees only the emptiness, deprivation, and savagery of the place. But then in the final stanza the speaker suddenly upsets everything said up to this point by suggesting how a native might see the same place from a different perspective. Isn't one of the purposes of poetry to make us see the unexpected beauty of some person, place or thing we previously took for granted or dismissed? That transformation was what I hoped to accomplish in "California Hills in August."

Gioia (the name is pronounced "Joy-a") is a business executive who finds time to write literary criticism as well as poetry. Recently he has edited the short stories of Weldon Kees and (with William Jay Smith) *Poems from Italy.* "California Hills in August" was first published in *The New Yorker,* and appears in his first collection, *Daily Horoscope* (Saint Paul: Graywolf, 1986).

H.D., HELEN, page 814

What a cold, hate-inspiring queen H.D. portrays! She makes a sharp contrast with the loveable image of Helen as a child in Yeats's "Long-legged Fly."

For a few notes on recent scholarly activity, see the entry on H.D.'s "Heat" in this manual.

Donald Hall, NAMES OF HORSES, page 815

Hall is apparently eulogizing not one horse but a long succession of them, each taking on its predecessors' duties through the years. The poem enumerates the everyday chores the horses had to do, "generation on generation," Sundays included. The "man, who fed you and kept you, and harnessed you every morning" represents not a single farmer, apparently, but all those on this New Hampshire farm who cared for and finally buried their horses in the time-honored way "for a hundred and fifty years." Like all dead animals (including people), the horses when they die become "soil makers," useful even in their graves.

The wonderful list in the poem's last line delivers what the title has promised, names of horses. The last one, Lady Ghost, is the most connotative. You might wish to have students explore its suggestions.

First published in *The New Yorker,* this poem later appeared in Hall's seventh book of poems, *Kicking the Leaves* (New York: Harper, 1978). The farm in New Hampshire where Hall lives was once a working farm run by his grandparents and, before them, his great-grandparents. Hall has written a prose memoir, *String Too Short to Be Saved* (Boston: Godine, 1981), about the boyhood summers he spent there with his grandmother and grandfather. "Names of Horses," too, seems to depend heavily on the poet's fond memories of the farm he still loves.

Thomas Hardy, THE CONVERGENCE OF THE TWAIN, page 816

The discovery in September 1985 of the well-preserved wreck of the *Titanic* in the North Atlantic, and the subsequent squabble over possession of it, has given this old favorite poem a certain immediacy. Most students will be familiar with the history of this great disaster from recent news reports (or from the popular book and film *Raise the Titanic!*). Still, a few facts may need to be recalled. The fateful day was April 15, 1912. The pride of the

British White Star lines, the *Titanic* was the world's largest ship in its day, celebrated for luxurious trappings (including Turkish baths and a fully equipped gym). Many of the unlucky passengers were wealthy and famous. One reason the *Titanic* sank with such cost of life was that the builders, smugly assuming the ship to be unsinkable, had provided lifeboats for fewer than half the 2,200 passengers. (Only 705 survived.) Hardy wrote the poem for the souvenir program of a benefit show for the *Titanic* Disaster Fund (to aid survivors and the bereaved) given at Covent Garden, May 14, 1912.

Hardy has been seen as an enemy of science and industrialism, those spoilers of rural England, but Donald Davie argues that "The Convergence of the Twain" shows no such animosity. The poem censures vanity and luxury, "but not the technology which built the great ship and navigated her" (*Thomas Hardy and British Poetry* [Oxford: Oxford UP, 1972]).

Although Hardy personally knew two victims of the disaster, the "Convergence," as J. O. Bailey points out, is not a personal lament; indeed, the drowned are hardly mentioned. The poem is a philosophic argument, with the Immanent Will punishing man for pride: "It acts like the Greek concept of Fate that rebukes *hubris*" (*The Poetry of Thomas Hardy* [Chapel Hill: U of North Carolina P, 1970]). Fate, however, seems personified in the poem as the Spinner of the Years, a mere agent of the Will.

Students can concentrate profitably on the poet's choice of words: those that suggest the exotic unnaturalness of the *Titanic*'s furnishings (*salamandrine, opulent, jewels . . . to ravish the sensuous mind, gilded*). Diction will also point to the metaphor of the marriage between ship and iceberg: the *intimate welding* and the *consummation*. The late Allen Tate was fond of reading this poem aloud to his friends, with mingled affection and contempt, and remarking (according to Robert Kent) that it held "too many dead words, dead then as now, and all the more obtuse for having been long dead in Shelley. 'Stilly,' for example." From Hardy's original printed version of the poem, as given in *The Variorum Edition of the Complete Poems of Thomas Hardy*, ed. James Gibson (New York: Macmillan, 1979), it appears that he originally cast line 6: "The cold, calm currents strike their rhythmic tidal lyres." Isn't *thrid* an improvement, even though it is stiltedly archaic?

Thomas Hardy, DURING WIND AND RAIN, page 818

A view of the passage of time and the coming of death to ordinary people, this great lyric invites comparison with E. E. Cummings's "anyone lived in a pretty how town." It differs, though, in sticking to conventional language and in utterly lacking sentiment. Is there a more blood-chilling line in poetry than "And the rotten rose is ripped from the wall," a powerful example of alliteration that means business?

In *A Commentary on the Poems of Thomas Hardy* (New York: Harper, 1977), F. B. Pinion sees this poem as inspired by Hardy's reading of his wife's reminiscences. The scenes in the poem, he says, depict "two of the homes and gardens of the Gifford family, which Emma remembered from her girlhood at Plymouth. Each of the recollections, bright and happy in itself, is seen through the regret which years and death have brought."

Donald Davie, in *Thomas Hardy and British Poetry* (New York: Oxford UP, 1972), calling this poem "one of Hardy's greatest achievements," admires in it "the cunning variations of the accomplished metrist."

In "During Wind and Rain" the variation that is decisive is on the seventh and last line of each stanza:

"How the sick leaves reel down in throngs!"
"See, the white storm-birds wing across!"
"And the rotten rose is ripped from the wall"
"Down their carved names the rain-drop ploughs."

Rhythmically so various, the lines are metrically identical. Though anapests are twice substituted for iambs in the line about the rose, the expectation of symmetry proves that line too is iambic tetrameter like the others. . . . The effect is that when we reach the refrain at the end of the third stanza, though symmetry is maintained between that stanza and the first two, it is so masked by the rhythmical variation that, instead of checking back to register how this stanza reproduces the earlier ones, we are propelled forward to see what will happen in the last.

Thomas Hardy, FOUR SATIRES OF CIRCUMSTANCE, page 819

This is a selection from the fifteen "Satires of Circumstance," a sequence that (at the publisher's insistence) lent its title to Hardy's poetry volume of 1914. This major collection appeared in a momentous year for the poet: after his second marriage and shortly after the outbreak of World War I. Hardy himself had voiced doubts about putting these bitter items in a book together with more recent, poignant elegies for his first wife. He called them "caustically humorous productions which had been issued with a light heart before the war"; and later remarked to Edmond Gosse: "The scales had not fallen from my eyes when I wrote them, and when I reprinted them they had" (letter of 16 Apr. 1918).

Robert Gittings, a recent biographer, agrees with reviewers of 1914 that the "Satires" would have been better off as short stories (*Thomas Hardy's Later Years* [Boston: Atlantic-Little, 1978] 161–63). Yet the astute John Crowe Ransom has been fond of these caustic items. "They are satires rather than proper tragedies," he observes, "being poems in which the victims are not entitled to our sympathy. The joke is upon persons who have to be punished because they were foolish; because they were more innocent than anybody can afford to be in this world." The "Satires" show Hardy in a mood of ferocity that enlarges his range, "though the gentle reader may not like him any the better" (Introduction, *Selected Poems of Thomas Hardy* [New York: Collier, 1966]).

We admire Hardy's extreme economy in telling these stories, in centering on moral crises so painful that, unless we laugh, we must cringe. Together, the "Satires" form a small gallery of various kinds of irony.

J. O. Bailey has illuminated the title "Satires of Circumstance." Earlier, in a novel of 1876, Hardy had used the phrase in describing a man who "sometimes had philosophy enough to appreciate a satire of circumstance, because nobody intended it" (*The Hand of Ethelberta*, chapter 12). Most of the poems in this series, Bailey notes, present ironic situations that no one intended. Some, like "In Church," depend on things accidentally noticed or overheard (*The Poetry of Thomas Hardy: A Handbook and Commentary* [Chapel Hill: U of North Carolina P, 1970] 334).

"In Church": Irony lies in a contrast between the pulpit personality of the preacher and the vain, self-satisfied actor he is when behind the scenes. We feel for his disillusioned little pupil: a terrible contrast is implied between her before and after views of her idol.

"In the Room of the Bride-elect": Ironies inhere in a contrast between what the parents wanted and what the bride wanted; between the bride's former mind and her mind now; between what the bridegroom knows and what we know.

"In the Cemetery": There's an ironic discrepancy between how the dead should be treated and this shuffling-about of their remains; "In tone a poem of cynical humor, but Hardy's compassion is implied" (Bailey 337). The custodian's expressed view is certainly not the poet's. Ironically, this poem recalls a classic British folk ballad, "They're Digging Up Father's Grave to Build a Sewer."

"In the Nuptial Chamber": An ironic contrast lurks between the happy serenade and this horrific revelation. What a remarkably forthright bride! Wonderful detail: how, in line 2, the music turns her into a sinister ghost. As a child, Hardy had delighted in a country band organized by his father and grandfather; no doubt this combo serenaded many newlyweds. Compare the ironic situation of this poem with that in James Joyce's great story "The Dead" (in *Dubliners*).

Robert Hayden, THOSE WINTER SUNDAYS, page 821

QUESTIONS FOR DISCUSSION:

1. Is the speaker a boy or a man? How do you know?
2. Summarize the poem's theme—its main idea.
3. What do you understand from "the chronic angers of that house"?
4. How does this poet's choice of words differ from that of most writers of prose? Suggestion: Read the opening five lines aloud.

This brief poem, simple in the word's best sense, has a depth that rewards close reading. That the speaker is a boy is a safe inference: most grown-ups do not lie abed while their fathers polish their shoes. Besides, evidently years have intervened between the speaker today and his previous self—the observing child. Now the speaker understands his father better, looks back on himself, and asks, "What did I know?"

The poem states its theme in its wonderful last line (worth quoting to anyone who distrusts abstract words in poetry). Students can miss Hayden's point unless they understand its vocabulary. *Austere* can mean stern, forbidding, somber, but it can also mean (as it does here) ascetic, disciplined, self-denying. To rise in the freezing house takes steely self-discipline. That the father's life is built on austerity we get from his labor-worn hands. What is an *office*? A duty, task, or ceremony that someone assumes (or has conferred on him): the tasks of shining shoes, of stirring banked fires in a furnace (or a coal-burning stove?). James Wright, a keen admirer of Hayden's poem, has spoken of it in an interview:

> The word *offices* is the great word here. *Office,* they say in French. It is a religious service after dark. Its formality, its combination of distance and immediacy, is appropriate. In my experience uneducated people and people who are driven by brute circumstance to work terribly hard for a living, the living of their families, are very big on formality. (*The Pure Clear Word: Essays on the poetry of James Wright,* ed. Dave Smith [Urbana: U of Illinois P, 1982] 10.)

Perhaps the "chronic angers" belong to the father: the boy gets up slowly and fearfully as though in dread of a tongue-lashing. Yet this reading does not seem quite in keeping with the character of the father as he emerges: stoic, patient, long-suffering, loving. Hayden does not invest these angers in the father exclusively. Perhaps any tenant of this bitterly cold house has reason to dread getting up in it.

When read aloud, the opening stanza reveals strong patterns of sound: the internal alliteration of the *k*-sound in *blueblack, cracked, ached, weekday, banked, thanked* (and, in the next stanza, in *wake, breaking, chronic*)—staccato bursts of hard consonants. Rather than using exact rime at the ends of lines, Hayden strengthens lines by using it internally: *banked/thanked* (line 5), *wake/breaking* (6); perhaps off-rime, too: *labor/weather* (4), *rise/dress* (8). Alliteration and assonance occur in *clothes . . . cold* (2), *weekday weather* (4). If you assign this poem early in your investigation of poetry, probably it matters more that students hear and respond to the rich interplay of repeated sounds than that they be able to give these devices labels.

"Those Winter Sundays" is the most often reprinted poem of Robert Hayden. A black poet who grew up in Detroit and who for many years was an English professor at the University of Michigan, he has written other poems apparently drawn from childhood and memory, among these "Obituary," another moving tribute to his father. Hayden's posthumous *Collected Poems* (New York: Norton/Liveright, 1985) belongs, we think, in every library.

Seamus Heaney, DIGGING, page 821

When Irish poet Seamus Heaney went to Lewiston, Maine, in 1986 to receive an honorary degree at Bates College, he read "Digging" aloud to the assembled graduates, parents, and friends. It seemed an appropriate choice. Some of the Maine students in his audience must have found expressed in Heaney's poem their own admiration for hard-working forebears whose course through life they had decided not to follow. Is the speaker in the poem uneasy about choosing instead to be a poet? It is clear that he admires the skill and strength his father and grandfather displayed in their work. But the poem ends on a positive note. The poet accepts himself for what he is.

In "Feeling into Words," an essay in his *Preoccupations* (New York: Farrar, 1980), Heaney likens the writing of poetry to digging up archaeological finds. Apparently it is a matter of digging a spade into one's past and unearthing something forgotten. "Digging," written in 1964, was his earliest poem in which it seemed that his feelings had found words. "The pen/spade analogy," he adds, "was the simple heart of the matter and *that* was simply a matter of almost proverbial common sense." As a schoolboy, he was often told to keep studying "because 'learning's easily carried' and 'the pen is lighter than the spade.'"

Seamus Heaney, MOTHER OF THE GROOM, page 822

This tender poem enters the thoughts of someone often forgotten: the mother of the groom. Floating through her mind at the wedding is a vivid memory of her son as a little boy. Now her lap is "voided." Though she "hears a daughter welcomed," she is not yet ready to take part in the welcoming.

> It's as if he kicked when lifted
> And slipped her soapy hold.

The third stanza shifts to another theme: the permanence of marriage—a generation-old Irish Catholic marriage, at any rate—symbolized by the mother's wedding ring, "bedded forever now / In her clapping hand."

Is the poem sentimental? We think not. It sticks to just a few closely observed details, understated but implying volumes.

Anthony Hecht, THE VOW, page 822

The fear that too powerful a blend of Jewish and Irish spirits caused the miscarriage is foreshadowed in the opening lines: "The mirth of tabrets ceaseth," as does the joy of the Irish minstrel harp. (*Ceaseth* gives the line an Old Testament ring.) Apparently the poem reflects part of the poet's life: Hecht and his first wife, Patricia Harris, were married in 1954; they had two sons. "The Vow" was first printed in 1957.

Does the speaker accept the dream-child's tragic view that the best of all fates is not to be born? Evidently not, or he wouldn't make the vow. The *bone gates* (line 13) are not only the classical gates of horn, but literally the mother's pelvic girdle. Stanza 3 seems a weak one in an otherwise powerful poem. In rhetoric, diction ("Mother, . . . "), and imagery, it recalls the earlier Robert Lowell (see for instance "Christmas Eve Under Hooker's Statue" in *Lord Weary's Castle*).

In the last stanza, the metallurgical metaphor will probably need explaining. Gentile and Jewish parents will be *tried* (refined, perfected) by the flames of their love. Possibly Hecht has in mind amalgamation—the only gold-refining process that uses a furnace—in which mercury and crude gold unite, then separate under high heat to produce pure gold.

Hecht has recorded "The Vow" for *The Spoken Arts Treasury of 100 Modern American Poets*, vol. 15 (SA 1054).

George Herbert, LOVE, page 824

Herbert's poem is often read as an account of a person's reception into the Church; the eaten *meat*, as the Eucharist. Herbert's extended conceits or metaphors are also evident in "The Pulley" (page 597) and "Redemption" (page 707).

(text pages) 824–828

For discussion: compare "Love" with another seventeenth-century devotional poem. Donne's "Batter my heart" (page 540). What is the tone of each poem? Herbert may seem less intense, almost reticent by comparison. Douglas Bush comments, "Herbert does not attempt the high pitch of Donne's 'Divine Poems.' His great effects are all the greater for rising out of a homely, colloquial quietness of tone; and peace brings quiet endings—'So I did sit and eat'" (*English Literature in the Earlier Seventeenth Century* [New York: Oxford UP, 1945] 139).

Herbert, by the way, is an Anglican saint—the only one who does not also appear in the Roman Catholic calendar.

Robert Herrick, THE BAD SEASON MAKES THE POET SAD, page 825

A poem of interest for several reasons. For one, it makes the ability to write poetry depend on congenial politics. As a priest of the Church of England whom the Puritans ousted from his sinecure, Herrick, of course, was a staunch Royalist. For another, it is a kind of sonnet written in couplets.

The "many fresh and fragrant mistresses"? Perhaps Herrick refers to his poems, perhaps to the Muses, perhaps to the many lovely women his poems portray.

Like so many of Herrick's poems, this one is rich in classical lore. *Tyrian dews* echoes Tibullus; the last line, Horace. In line 7, there's an allusion to the myth of the golden age (the degeneration of the ages of man through gold, to silver, and finally to iron).

The last line strikes us as a quite wonderful description of the poetic experience. Even if it is borrowed from the Latin poet, it still seems fresh and fragrant.

Robert Herrick, TO THE VIRGINS, TO MAKE MUCH OF TIME, page 825

Roses would have suited Herrick's iambic meter—why is *rose-buds* richer? Rosebuds are flowers not yet mature, and therefore suggest virgins, not matrons. There may be a sexual hint besides: rosebuds more resemble private parts than roses. But in this poem, time flies; the rosebuds of line 1 bloom in line 3. *Rose-buds* is also rhymically stronger than roses, as Austin Warren has pointed out: it has a secondary stress as well as a primary. Warren has recalled that when he first read the poem in college in 1917, he misread *rose-buds* as *roses*, kept misreading it ever after, and only a half-century later realized his mistake and found a new poem in front of him. "In untutored youth, the sentiment and the rhythm suffice: the exactness of the language goes unnoticed. And in later life a remembered favorite escapes exact attention because we think we know it so well" ("Herrick Revisited," *Michigan Quarterly Review* 15 [Summer 1976]: 245–67).

Question for discussion: What do you think of Herrick's advice? Are there any perils in it?

Garrett Hongo, THE CADENCE OF SILK, page 826

Here's one to convince basketball watchers, especially Lakers fans, that the subjects of poetry don't have to be arcane. Not only a poem about sports, "The Cadence of Silk" may refer also to writing poetry: watching the Sonics taught the speaker about cadence, rhythm, and rhetoric (lines 6–14). Appropriately, when the final ball swishes through the net, the poem ends too.

Worth noticing: Hongo's colorful imagery and similes ("sleek as arctic seals," "like a waiter balancing a tray").

Garrett Hongo, THE HONGO STORE / 29 MILES VOLCANO / HILO, HAWAII, page 828

"The Hongo Store," apparently based on family lore as well as an old photograph, is gently humorous in tone. The father's saying "Be quiet" to the crying baby, the fact that the family car is an Edsel, the father's panicky defiance of the advice on the radio, and his broken

(text pages) 829–830

English, all evoke gentle smiles. Yet an air of genuine thankfulness and relief pervades the final stanza. The family's only damage from the volcano's rumblings is a broken store window.

Among Hongo's strengths as a poet are his verbs. The volcano and the bells *thud*, the baby *squalls*, the radio *squeals*, the car *grinds*, the news *barks* and *crackles*. Such sensuous language makes it easy for the reader to take in Hongo's world.

Gerard Manley Hopkins, SPRING AND FALL, page 829

Hopkins's tightly wrought syntax may need a little unraveling. Students may be asked to reword lines 3–4 in a more usual sequence ("Can you, with your fresh thoughts, care for leaves like the things of man?") and then to put the statement into more usual words. (An attempt: "Do you, young and innocent as you are, feel as sorry for falling leaves as for dying people?") Lines 12–13 may need a similar going-over and rough paraphrase. ("Neither any human mouth nor any human mind has previously formed the truth that the heart and spirit have intuited.") "Sorrow's springs are the same"—that is, all human sorrows have the same cause: the fact that all things pass away. A world of constant change is "the blight man was born for": an earth subject to death, having fallen from its original state of a changeless Eden. The difficulties of a Hopkins poem result from a swiftly thoughtful mind trying to jam all possible meaning into a brief space (and into words that are musical).

Wanwood is evidently a term the poet coined for pale autumn woods. W. H. Gardner, editor of Hopkins's poems, finds in it also the suggestion of "wormwood"—bitter gall, also wood that is worm-eaten. The term *leafmeal* reminds him of "piecemeal," and he paraphrases line 8: "One by one the leaves fall, and then rot into mealy fragments."

John Crowe Ransom's "Janet Waking" is another poem in which a sophisticated poet contemplates a grieving child. How do the poems differ? Ransom tries to convey the intensity of Janet's grief over her dead hen; Hopkins is content to talk to Margaret, like a priest trying to console her, and to philosophize.

Hopkins's great lyric ought to survive George Starbuck's brilliant travesty (page 739). The two are worth comparing in case students assume that what matters in poetry is the message alone, that particular words have no consequence. Starbuck obviously is having a good time translating "Spring and Fall" into the American vernacular. His parody may be useful as a way into a number of crucial matters: the diction of a poem and how language indicates it. Who or what is the butt of Starbuck's ridicule? Is it Hopkins and his poem; or the speaker himself, his crudeness, his hard-boiled simplemindedness? (Howard Moss performs a comparable reduction of a sonnet of Shakespeare's on page 586.)

Gerard Manley Hopkins, THOU ART INDEED JUST, LORD, IF I CONTEND, page 830

Like Milton in his sonnet on page 850, Hopkins begins with a complaint. Unlike Milton he seems, at last, far from being reconciled to the Lord's will; and tension between his faith and the testimony of his experience threaten to (but do not quite) overpower him. In the octet, to be sure, Hopkins is not merely speaking for himself: he is paraphrasing Jeremiah, who also complains that God lets scoundrels prosper. One of Hopkins's so-called "Terrible Sonnets," the poem ends not with any sense of deliverance, but with a prayer.

In line 1, why is the Lord said to be just? Well, at least he gives the poor suffering mortal a hearing. "When I have a friend like you, who needs an enemy?" seems the near-despairing thrust of lines 5–9. Although in the sestet the speaker sees the possibility of change, drawing hope from the fact that nature is renewed in spring, nature only reminds him once more of his own impotence and sterility. Birds build nests, but he can't.

In this as in other Hopkins poems, close reading is essential, but will reveal a few vexing difficulties. In line 2, "but, sir, so what I plead is just," the *so what* can give trouble—more so than the way Hopkins worded the line originally: "but, sir, what I shall speak is just." Syntax is inverted in lines 12–13 to stress the *no* and the two *nots*.

Norman MacKenzie has suggested that this poem could have been entitled "Bitterness in Spring," and thinks the poet's envy of the *fresh wind* (which shakes those lucky birds) suggests lack of poetic inspiration. (You might contrast this tormented poem with Hopkins's rapturous "Pied Beauty" [page 576] and with "The Windhover" [next], in which the *sheer plod* of the religious life is triumphantly justified.) It is ironic, notes MacKenzie, "that one of [Hopkins's] most enduring poems should be based on the conviction that no work of his would ever endure" (*A Reader's Guide to Gerard Manley Hopkins* [Ithaca: Cornell UP, 1981] 204).

Gerard Manley Hopkins, THE WINDHOVER, page 830

"The best thing I ever wrote," said Hopkins. If your students have enjoyed "Pied Beauty" (page 576) or "God's Grandeur" (page 635) without too much difficulty, then why not try "The Windhover," despite its famous ambiguities? Some students may go afield in reading the opening line, and may take *I caught* to mean that the poet trapped the bird; but they can be told that Hopkins, a great condenser, probably means "I caught a glimpse of."

Dispute over the poem often revolves around whether or not the windhover is Christ and around the meaning of *Buckle!* Most commentators seem to agree that the bird is indeed Christ, or else that Christ is like the bird. (Yvor Winters, who thought the poem "minor and imperfect," once complained, "To describe a bird, however beautifully, and to imply that Christ is like him but greater, is to do very little toward indicating the greatness of Christ.") Some read *Buckle!* as a plea to the bird to descend to earth; others, as a plea to all the qualities and things mentioned in line 9 *(Brute beauty, valor, act)* to buckle themselves together into one. Still others find the statement ending in *Buckle!* no plea at all, but just an emphatic observation of what the poet beholds. If Christ is the windhover (other arguments run), in what sense can he be said to buckle? Two of the answers: (1) in buckling on human nature and becoming man, as a knight buckles on armor; (2) in having his body broken on the cross. Students can be asked to seek all the words in the poem with connotations of royalty or chivalry—suggestive, perhaps, of Christ as King and Christ as noble knight or chevalier. Why the *sheer plod?* Hopkins reflects (it would seem) that if men will only buckle down to their lowly duties they will become more Christlike, and their spiritual plowshares will shine instead of collecting rust. Hopkins preached a sermon that expressed a similar idea: "Through poverty, through labor, through crucifixion His majesty of nature more shines." The *embers*, we think, are a metaphor: moist clods thrown by the plow going down the sillion. Hopkins likes to compare things to hearth fire: for instance, the "fresh-firecoal chestnut-falls" in "Pied Beauty."

For detailed criticism, one might start with Norman H. MacKenzie, *A Reader's Guide to Gerard Manley Hopkins* (Ithaca: Cornell UP, 1981). MacKenzie provides facts from ornithology and his own kestrel-watching: no other birds are so expert in hovering, body horizontal, tail and head pointing down as they study the ground for prey. To hang stationary in the air over one spot, they must fly into the wind "with rapidly quivering (*wimpling*, line 4) wings, missing a few beats as gusts die, accelerating as they freshen"—responding to variations in the wind with nearly computer speed. Once in about every eight hovers, the kestrel will dive, not inertly but with wings held tense and high—it doesn't "buckle" in the sense of *collapse*. If it finds no victim, the bird swings and banks and takes an upward "stride," to hover once more. Hopkins's "how he rung upon the rein" doesn't mean that the kestrel climbs in a spiral. No gyring Yeats-bird, he.

Despite his fondness for Old and Middle English, Hopkins luckily refrained from calling the windhover by its obsolete name: *fuckwind* or *windfucker*. (No, that *f* is not a long *s*.) Thomas Nashe in *Lenten Stuffe* (1599) speaks of the "Kistrilles or windfuckers that filling themselves with winde, fly against the winde evermore." See *windfucker* in the *Oxford English Dictionary*. (For this dumbfounding discovery, thanks to David Lynch, who copyedited *Literature*, 4th ed.)

(text pages) 831–835

A. E. Housman, LOVELIEST OF TREES, THE CHERRY NOW, page 831

What is Housman's theme? Good old *carpe diem*. If you ask students to paraphrase this poem (it's not hard), a paraphrase might add, to catch the deeper implication, "Life is brief—time flies—I'd better enjoy beauty now."

Not part of the rough poem Housman began with, the second stanza was added last. Lines 9–10 originally read: "And since to look at things you love / Fifty times is not enough." What can be said for Housman's additions and changes? (These and other manuscript variations are given by Tom Burns Haber in *The Making of "A Shropshire Lad"* [Seattle: U of Washington P, 1966].)

A. E. Housman, TO AN ATHLETE DYING YOUNG, page 832

For a comment on this poem, see the note in this manual on Dryden's "To the Memory of Mr. Oldham," page 803 in the text.

Langston Hughes, DREAM DEFERRED, page 833

Simile by simile, Hughes shows different attitudes, including violent protest, that blacks might possibly take toward the long deferral of their dream of equality. Students might be asked what meaning they find in each comparison. The sugared crust (line 7) is probably the smiling face obligingly worn by Uncle Toms.

The angry, sardonic tone of the poem is clearly different from the sorrowful tone of Dudley Randall's "Ballad of Birmingham" (page 615).

Hughes's poem supplied the title for Lorraine Hansberry's long-running Broadway play *A Raisin in the Sun* (1958), in which the Youngers, a family descended from five generations of slaves, come to a Chicago ghetto in hopes of fulfilling their dream.

Donald Ritzhein has written a moving account of what the poem has meant to him, starting when his mother cut it out of a newspaper and pasted it to his bedroom door. "By the time I got to high school . . . I still didn't know a lot about the misery of deferred dreams . . . I knew a little more about them when I heard Martin Luther King, Jr., talk about dreams in Washington. I finally felt a little of what it's like to defer dreams when John F. Kennedy was killed" ("Langston Hughes: A Look Backwards and Forwards," *Steppingstones*, a little magazine published in Harlem, Winter 1984: 55–56). Have you any student who would care to write about what the poem has meant to her or him?

Langston Hughes, ISLAND, page 834

In both "Island" and Yeats's "The Lake Isle of Innisfree," the unhappy speaker dreams of an island where he yearns to take refuge. In the Yeats poem, the island is a real place with a name. In Hughes's poem the island seems rather to be a metaphor like the wave he hopes will take him there. The wave represents a sorrow so overwhelming that the speaker fears it will drown him. The island, where the "sands are fair," appears to promise surcease.

Randall Jarrell, THE DEATH OF THE BALL TURRET GUNNER, page 834

The speaker seems an unknown citizen like Auden's (page 523). Jarrell's laconic war poem is complex in its metaphors. The womb is sleep; the outside world, waking; and the speaker has passed from one womb to another—from his mother into the belly of a bomber. His existence inside the ball turret was only a dream, and in truth he has had no mature life between his childhood and his death. Waking from the dream, he wakes only to nightmare. In another irony, the matter-of-fact battle report language of the last line contrasts horribly with what is said in it. How can the dead gunner address us? Clearly the poet had written his epitaph for him—and has done so as Jarrell said he wrote "The Woman at the Washington Zoo," "acting as next friend."

Robinson Jeffers, TO THE STONE-CUTTERS, page 835

If students will compare this poem with Shakespeare's sonnet "Not marble nor the gilded monuments," they'll be struck by a sharp difference of view. Shakespeare, evidently, is the optimist. For him, a poem can confer immortality and inspire love until doomsday. For Jeffers, poems will merely bring a little respite to "pained thoughts"—much as a spoonful of honey helps a hangover, according to one popular belief.

For a short assignment: Write a paraphrase of each of these two poems. Bring your work to class, ready to read it aloud. We'll discuss the poems, taking off from your work on them.

For another pessimistic view of monuments, direct students to Shelley's "Ozymandias." You might ask them to compare its theme with those of Shakespeare and Jeffers.

A sidelight on this poem: Any doubt he may have had that stone monuments last long didn't prevent Jeffers from building, with his own hands, a stone tower next to his home at the ocean's edge in Carmel, California.

Elizabeth Jennings, DELAY, page 835

QUESTIONS FOR DISCUSSION:

1. What is the poet comparing in this metaphor?
2. Why would the metaphor not have been available to a poet in Shakespeare's time?
3. What is the tone of the poem? How does the speaker feel about the *radiance of that star*—simply joyous and glad, as we might expect?
4. What connotations has the word *impulse* (line 6)? What is the implied metaphor here?

This poem, wistful in tone, is one sustained metaphysical conceit. To help students see it, members of the class can be asked to provide a few facts from modern astronomy: what a light year is, the distances of stars from earth (Alpha Centauri, the nearest, takes nearly four years and three months to transmit its light). Why does the poet say that the star's radiance *leans*? Possibly because the starlight strikes the earth on a slant; also because, in order to be perceived, the light depends or relies upon an observer.

Jennings, who seems to us one of the half-dozen finest living English poets, commands splendid effects of meter. One occurs in line 6: ordinarily, a reader would unhesitantly stress the first syllable of IM-*pulse,* but the iambic pattern sets up a powerful expectation. One almost wants to say *im*-PULSE, impulsively. One hesitates, and the word takes deserved emphasis. The implied metaphor is that a burst of feeling is like an electrical impulse: a sudden flow of current in one direction. The light of certain stars does fluctuate, incidentally; astronomers measure the distances of remote stars (those more than 300 light-years away) by measuring their fluctuations.

Jennings's *Collected Poems 1953–1985* (New York: Carcanet, 1986) is full of rewards.

Ben Jonson, ON MY FIRST SON, page 836

This heartbreaking poem from Jonson's *Epigrammes,* requested by several instructors, repays close reading. What is "the state [man] should envy?" Death. Why the dead child should be envied is made clear in the lines that immediately follow (7–8). The final couplet is difficult in its syntax, and contains a pun on *like* in a sense now obsolete. The speaker vows, or prays (*vow,* along with *votive,* comes from the Greek *euchesthai:* "to pray"), that anyone whom he loves may not live too long. The seriousness of Jonson's wit is shown in this colossal pun: *like* meaning "thrive, do well, get on" as well as "to be fond." See *like* in the *OED* for other illustrations:

> Shallow to Falstaff: "By my troth, you *like* well and bear your years very well" (*Henry IV,* P.2, 3.2.92).
>
> "Trees generally do *like* best that stand to the Northeast wind" (Holland's *Pliny,* 1600).

(text pages) 836–838

"Poems Arranged by Subject and Theme" in this manual lists the book's eleven other poems about fathers and children. In this Anthology section, see especially the father-and-son poems by Robert Phillips, Stephen Shu-ning Liu, and James Wright ("Autumn Begins in Martins Ferry, Ohio").

Donald Justice, ON THE DEATH OF FRIENDS IN CHILDHOOD, page 836

There is more emotional distance, less grief in this poem than in Ben Jonson's. Neither does the speaker in "On the Death of Friends in Childhood" seem to be mourning one specific loss. The "Friends" he mentions suggest friends in general, perhaps other people's as well as his own. Yet, though time has softened the impact of long-ago losses, the narrator urges that we remember dead childhood friends and what was shared with them.

John Keats, ON FIRST LOOKING INTO CHAPMAN'S HOMER, page 837

QUESTIONS FOR DISCUSSION:

1. What are the *realms of gold*? Can the phrase have anything to do with the fact that early Spanish explorers were looking for El Dorado, a legendary city of treasure in South America?
2. Does Keats's boner about Cortez mar the poem?
3. Did you ever read anything that made you feel like a stout Cortez? If so, what?

John Keats, WHEN I HAVE FEARS THAT I MAY CEASE TO BE, page 838

Students will see right away that the poem expresses fear of death, but don't let them stop there: there's more to it. Why does the poet fear death? Because it will end his writing and his loving. The poem states what both loving and writing poetry have in common: both are magical and miraculous acts when they are spontaneous. Besides favoring "unreflecting love" for its "fairy power," Keats would write "with the magic hand of chance." And—if you care to open up a profundity—what might the poet mean by those "huge cloudy symbols of a high romance"? Literal cloud shapes that look like Tristram and Isolde's beaker of love-potion, or what?

Note that this poem addresses not Keats's beloved Fanny Brawne, but "the memory of the mysterious lady seen in adolescence one brief moment at Vauxhall long ago in the summer of 1814," according to Robert Gittings in *John Keats* (Boston: Atlantic, 1968) 188. The poem is about a "creature of an hour." (Fanny, of course, occupied not one hour but many.)

Gittings has found in the poem echoes of two sonnets of Shakespeare, both about devouring time: #60 ("Like as the waves make towards the pebbled shore, / So do our minutes hasten to their end") and #64 ("When I have seen by Time's fell hand defaced"). In the copy of the *Sonnets* that Keats co-owned with his friend Reynolds, these two were the most heavily marked.

This poem has had a hefty impact on later poets, notably John Berryman, who took from it the title for an autobiographical collection of his own poems on ambition and desire: *Love and Fame* (1972).

John Keats, TO AUTUMN, page 838

Although "To Autumn" was to prove the last of the poet's greatest lyrics, we have no evidence that Keats (full of plans and projects at the time) was consciously taking leave of the world. On September 21, 1819, three days after writing the poem, Keats in a letter to his friend John Hamilton Reynolds spoke of his delight in the season: "I never lik'd stubble fields so much as now—Aye better than the chilly green of the Spring. Somehow *a stubble plain looks warm — in the same way that some pictures look warm — this strikes me so much in my Sunday's walk that I composed upon it."*

QUESTIONS FOR DISCUSSION:

1. In the opening stanza, what aspects of autumn receive most emphasis? To what senses do the images appeal?

2. In the first two stanzas, autumn is several times personified (lines 2–3, 12–15, 16–18, 19–20, 21–22). Who are its different persons? (Conspiring crony, careless landowner, reaper, gleaner, cider presser.)

3. In the third stanza, how does the tone change? Has there been any progression in scene or in idea throughout the poem? (Tone: calm serenity. In the first stanza, autumn is being prepared for; in the second, busily enjoyed; in the third, calmly and serenely contemplated. There is another stanza-by-stanza progression: from morning to noon to oncoming night. Like the *soft-dying day*, the light wind sometimes dies. The gnats in *wailful choir* also have funeral, mourning suggestions, but the stanza as a whole cannot be called gloomy.)

4. What words in stanza 3 convey sounds? (*Songs, music, wailful choir, mourn, loud bleat, sing, treble, whistles, twitter.* What an abundance of verbs! The lines convey a sense of active music making.)

5. Do you see any case for reading the poem as a statement of the poet's acceptance of the facts that beauty on earth is transitory and death is inevitable? (Surely such themes are present; the poem does not have to be taken to mean that the poet knows he himself will soon perish.)

For an unusually grim reading of the poem, see Annabel M. Patterson, "'How to load . . . and bend': Syntax and Interpretation in Keats's *To Autumn*," *PMLA* 94 (1979): 449–58. Finding that the poem "undermines" our traditional notion of Autumn, Patterson argues that Keats subversively portrays the goddess as deceptive, careless, and demanding. Her proffered ripeness leads only to *last oozings* and *stubble-plains* — dead ends not to be desired. In the poet's view (as she interprets it), "Nature is amoral and not to be depended upon." Try this argument on the class. Do students agree? Whether or not they side with Patterson, they will have to examine the poem closely in order to comment.

For a good discussion of ways to approach Keats's poem with a class, see Bruce E. Miller, *Teaching the Art of Literature* (Urbana: NCTE, 1980) 75–84. Miller points out that not all students will know how cider is made, and suggests asking someone to explain Keats's reference to the cider-press in stanza 2. He recommends, too, borrowing from your nearest art department some reproductions of landscape paintings: "Constable's work, which was contemporary with Keats, to my mind almost catches the spirit to 'To Autumn,' but it is a little more literal and photographic. . . . "

With this rich poem, you might well start a discussion of imagery, or review this topic if students have met it earlier. "The students," remarks Miller, "need not ask themselves as they read, 'What does it mean?' Rather they should continually ask, 'What do I see?' 'What do I hear and touch?' 'What do I feel?'" (Not that in reading the poem they're actually seeing or touching anything except printed books, of course.)

Philip Larkin, HOME IS SO SAD, page 839

Larkin's considerable achievement in "Home is so Sad" is that he so beautifully captures the ring of ordinary speech within the confines of a tight *a b a b a* rime scheme and iambic pentameter. Note the slant rimes in the second stanza: *as, was,* and *vase.*

In an interview, Larkin has recalled a letter he received from a middle-aged mother who had read his poem: "She wrote to say her children had grown up and gone, and she

(text pages) 839–841

felt precisely this emotion I was trying to express in the poem" ("Speaking of Writing XIII: Philip Larkin," [London] *Times* 20 Feb. 1964: 16). Bruce Martin finds the poem written not from a mother's point of view, but from that of a son who used to live in this house himself. The speaker projects his own sadness into it: it has remained pathetically changeless. What has changed is himself and others who once lived here (*Phillip Larkin* [Boston: Twayne, 1978] 52).

DMK, taking a different view, thinks it quite possible to read the poem as though (as in Larkin's well-known "Mr. Bleaney" from the same collection, *The Whitsun Weddings*) the former inhabitants of this house have died. The speaker is left unidentified, an impersonal seeing-eye; and so your students, too, will probably come up with differing interpretations.

Is Larkin's poem sentimental? Hard-eyed and exact in its observations, aided by the colloquialism of line 7, "A joyous shot at how things ought to be," it successfully skirts the danger. Students might like to discuss the three words that follow: "Long fallen wide." Does Larkin mean that the home was an unhappy one, or merely ordinary in its deviation from the ideal? Or is it that the arrow—the "joyous shot"—has fallen merely in the sense that, meant to be full of life, the home is now "bereft / Of anyone to please"?

Philip Larkin, POETRY OF DEPARTURES, page 840

The speaker wonders why he doesn't break with his dull, tame life, just walk out, chuck everything, launch into the Romantic unknown like a highwayman. What an appealing notion! Question: Have you ever yearned to do that very thing? (Who hasn't?)

Well, why doesn't he take off? Because he sees all too clearly and painfully that such a grandiose gesture would be ridiculous. Commenting on lines 25–27, in which the speaker imagines himself swaggering nut-strewn roads and crouching in the fo'c'sle, David Timms finds him "unconvinced by such daydreams, though sympathetic to the dreamers, for he is one himself." And so he dismisses his Romantic urge as too studied and belabored, ultimately false. Still, just because he sees through those dreams, he won't let himself feel superior. The dreams may be artificial, but so is his own tame life, his room, his specially chosen junk. As he is well aware, at the end, his greatest danger is to be trapped in owning things and neatly arranging them: books, china, all fixed on shelves in an order "reprehensibly perfect." (We have somewhat expanded on Timms' paraphrase, from *Philip Larkin* [Edinburgh: Oliver & Boyd, 1973] 87.)

William H. Pritchard has remarked that Larkin himself expunged Romantic possibilities from his life, the better to entertain them in his writing ("Larkin's Presence," in *Philip Larkin: The Man and His Work*, ed. Dale Salwak [U of Iowa Press, 1989] 74–75).

Irving Layton, THE BULL CALF, page 840

Sentimental poets frequently shed tears over concrete objects, while (in their imagery and diction) failing to open their eyes to the physical world. Such is not the case in Layton's "The Bull Calf," in which the poet tells us that he weeps only after having portrayed the dead calf in exact detail ("one foreleg over the other").

Layton's poem develops a series of contrasts. In the first section, the calf's look of nobility ("the promise of sovereignty," "Richard II") is set against his immaturity. The "fierce sunlight," in an implied metaphor, is compared to the calf's mother: taking in maize, licking her baby. In line 14, the "empty sky," suggestive of the calf's coming death, seems the turning point of the poem. In the remainder, the calf, which had been portrayed at first as full of life and pride, becomes an inanimate object, "a block of wood," a numb mass that emits ugly sounds when handled ("a sepulchral gurgle"). But in the closing lines, introducing still another contrast, Layton seems to show the calf as a living sleeper, or perhaps a statue or finished work of art.

(text pages) 841–843

Probably the best-known living poet in Canada, Layton was born in Rumania and came to Montreal early in life. Standing outside both British and French communities, Layton's perspective often has been that of an outsider, a Jew, a satirist, and a revolutionary. *The Selected Poems of Irving Layton,* with an introduction by Hugh Kenner (New York: New Directions, 1977), is an attempt to widen his audience south of the border.

Denise Levertov, DIVORCING, *page 842*

Levertov's poem more closely resembles Grace Schulman's "Hemispheres" than Margaret Atwood's "You fit into me." Atwood's is a hate poem; Schulman's is a love poem; curiously, so is Levertov's, in a way, even though she writes about divorce. Both poems contain vivid images of closeness between a man and a woman. To Schulman this merging of two bodies is joyous and sustaining. In "Divorcing," ironically enough, it is the closeness of marriage, "our fragrant yoke," that seems partly responsible for the divorce. It "could choke us. . . . We were Siamese twins." Clearly a feeling of claustrophobia accompanied the sweetness of the garland that "was twined round our two necks." The garland had thorns as well as flowers and leaves. It imparted a "scratchy grace." Like many liberated couples during the seventies, husband and wife felt that marriage was too great a threat to each partner's autonomy. They needed to see if they could "survive, severed."

Philip Levine, ANIMALS ARE PASSING FROM OUR LIVES, *page 842*

Remarkable about this poem is its point of view—a pig's—and its insouciant tone, considering the circumstances. This pig is on its way to market, and to death. An intelligent animal with a startlingly human persona (even jauntily condescending to the boy who, it says, thinks it will act "like a beast" [line 22]), it knows exactly the kind of death that awaits, and takes pride in facing its fate with courage. Its English is colloquial ("squeal / and shit like a new housewife / discovering television," "Not this pig"), its manner breezy.

Levine's title seems ironic. It leads us to expect a self-consciously "significant" poem about endangered species and the urgent need to save them. But in fact, it surprises us by focusing instead on meat animals, killed in great numbers every day without our thinking twice. Like Simic in "Butcher Shop," Levine sets forth the unpalatable details of the slaughter and the discomfort sensitive people feel when they reflect on the bloodshed necessary to provide meat for the table. Levine, describing the dreams in which the snouts of slaughtered pigs "drool on the marble," calls attention as well to the hypocrisy of "the consumers / who won't meet their steady eyes / for fear they could see." Given the title and the opening lines, who could predict where this poem was going?

Stephen Shu-ning Liu, MY FATHER'S MARTIAL ART, *page 843*

Born in China, Stephen Shu-ning Liu has lived in this country since 1952 and currently teaches at Clark County Community College in Las Vegas. In *Breaking Silence, An Anthology of Contemporary Asian American Poets* edited by Joseph Bruchac (New York: Greenfield Review, 1983) he remarks:

> My philosophy in writing poetry is that poetic language should be simple, clear and direct. Like fresh air and wholesome bread, poetry is for the crowd; and the poet, since he is just another human being, does not necessarily have a tattoo or a weird hairdo. A poet should work alone and leave group exercise to the football players.

In "My Father's Martial Art," the father has apparently died ("the smog / between us deepens into a funeral pyre"). His son remembers him and his skill with the martial arts, learned during a three-year stay in a monastery. The poignant final stanza makes evident the speaker's love and longing for his absent father. "But don't retreat into night, my father" calls to

(text pages) 843–845

mind another elegy for a father, Dylan Thomas's famous "Do not go gentle into that good night" (page 674).

Dream Journeys to China, an English-Chinese bilingual edition of his poems (Beijing: New World Press, 1982) is available from the China Publications Center, P.O. Box 399, Beijing, People's Republic of China.

Robert Lowell, SKUNK HOUR, page 844

Students should have no trouble in coming up with the usual connotations of *skunk*, but they may need help in seeing that the title is a concise expression of Lowell's theme. This is an evil-smelling hour in the speaker's life; and yet, paradoxically, it is the skunks themselves who affirm that life ought to go on. After the procession of dying and decadent people and objects in the first four stanzas, the mother skunk and her kittens form a triumph: bold, fecund, hungry, impossible to scare. Although they too are outcasts (surrounded by their aroma as the poet is surrounded by his madness and isolation?), they stick up for their right to survival.

The poem is rich in visual imagery. In the mind's eye, there are resemblances between the things contained in stanza 5 (the Ford car and the hill's skull), also the objects set in fixed rows (love-cars, tombstones, beached hulls). Water and the sea (by their decline or absence) are to this poem what they are to Eliot's *Waste Land*. Even the Church is "chalk-dry"; its spire has become a spar like that of a stranded vessel.

This poem is intensively analyzed in *The Contemporary Poet as Artist and Critic: Eight Symposia*, ed. Anthony Ostroff (Boston: Little, 1964). Richard Wilbur, John Frederick Nims, and John Berryman comment on the poem, after which Lowell comments on their comments. Lowell calls the opening of the poem "a dawdling, more or less amiable picture of a declining Maine sea town.... Sterility howls through the scenery, but I try to give a tone of tolerance, humor, and randomness to the sad prospect." He sees the skunk hour itself as a sort of dark night of the soul and refers readers to the poem by Saint John of the Cross. Lowell's night, however, is "secular, puritan, and agnostical." Lowell notes that the phrase *red fox strain* was intended only to describe the color of vegetation in the fall on Blue Hill, a mountain in Maine.

Elizabeth Hardwick, Lowell's wife when "Skunk Hour" was written, has affirmed that all the characters in the poem were actual—"were living, more or less as he sees them, in Castine [Maine] that summer. The details, not the feeling, were rather alarmingly precise, I thought. But fortunately it was not read in town for some time" (quoted by Ian Hamilton, *Robert Lowell: A Biography* [New York: Random, 1982] 267).

Sandra M. Gilbert, who sees the poem as "richly magical," reads it for its embodiment of myth. She explores it as a vision of Hell, pointing out that its events happen not on Halloween, but "somewhere in Hallowe'en's ritually black and orange vicinity." (The fairy decorator's shop is "sacramentally orange.") The summer millionaire has departed in fall, like a vegetation deity—Osiris or Attis. Nautilus Island's witchlike hermit heiress is Circe, Hecate, Ishtar, Venus, "the goddess of love turned goddess of death in an All Soul's Night world" ("Mephistopheles in Maine: Rereading Lowell's 'Skunk Hour,'" *A Book of Rereadings*, ed. Greg Kuzma [Lincoln, Neb.: Pebble and Best Cellar, 1979] 254–64).

Archibald MacLeish, THE END OF THE WORLD, page 845

QUESTIONS FOR DISCUSSION:

1. Where does the action of this poem take place?

2. To see for yourself how the sonnet is organized, sum up what happens in the octave. Then sum up what happens in the sestet.

3. How does the tone of the octave contrast with that of the sestet? (If you need to review *tone*, see pages 509–515.) Comment in particular on the clause in line 8: *the top blew off*. How do those words make you feel? Grim? Horrified? Or what?

4. Now read the closing couplet aloud. Try to describe (and account for) its effectiveness. Suppose MacLeish had wanted to write an Italian sonnet; he might have arranged the lines in the sestet like this—

> And there, there overhead, there, there hung over
> Those thousands of white faces, those dazed eyes,
> There in the sudden blackness the black pall,
> There in the starless dark the poise, the hover,
> There with vast wings across the canceled skies
> Of nothing, nothing, nothing—nothing at all.
> Would that have been as effective?

What is the tone of this sonnet? Not at all grim, though the poet speaks of the most horrific event imaginable. Our pleasure in the poem comes from many elements besides its subject: its sounds (including rimes), its rhythm (metrical lines of even length), its portrait of a circus frozen in a split second, and its colossal pun in *the top blew off* (*top* meaning the "big top" or circus tent, perhaps also the lid of the enormous pot of all Creation). Whose are the "vast wings"? Nothingness's.

The octave, with its casual description of routine merriment, stands in contrast to the sestet, which strikes a note of awe. Perhaps the pale faces and dazed eyes of the circus spectators reflect the attitude of stunned wonder that the poet feels—or would have us feel. Obviously, to rearrange MacLeish's lines would be to weaken his poem. For one thing, the closing rimed couplet in the original throws great emphasis on the final *nothing at all*. In both sound and sense, "the black pall / Of nothing, nothing, nothing—nothing at all" seems more powerful than "canceled skies / Of nothing. . . ." In this sonnet, as in Shakespeare's sonnets, the concluding couplet firmly concludes.

Charles Martin, ROUGH DRAFT, page 846

Why is this poem called "Rough Draft"? Life and art, the poet seems to say, are not the same. To the artist (whether sculptor, painter or poet), life, though charming and vital, merely provides material for a rough sketch—material that must be shaped and polished before it, in turn, becomes art. Life is imperfect. In the scene at the swimming pool, the girl wisecracks, pops gum, shrieks, and kicks "unglamorously." The young man is "too stocky and too short / For idoldom." Their passage inartistically "scribbles and erases" the waves. The scene becomes art only after the artist (in this case the poet?) has transformed it. Perhaps, too, these adolescents' puppy-love is only a rough draft for mature lovemaking.

Note the play on words in "her not quite skin- / tight pink maillol." Aristide Maillol (1861–1944) was a French sculptor of heroic female nudes. The word for a bathing suit made out of coarsely knitted, stretchable jersey fabric is *maillot*, pronounced like *Maillol* without the final *l*. The word as the poet uses it echoes his theme. What is "the cold print / Forbidding it"? Probably a poolside sign that says "No running."

Charles Martin, who lives and teaches in New York, has translated the poems of Catullus. "Rough Draft" is from his *Steal the Bacon* (Baltimore: Johns Hopkins UP, 1987).

Andrew Marvell, TO HIS COY MISTRESS, page 846

QUESTIONS FOR DISCUSSION:

1. "All this poet does is feed some woman a big line. There's no time for romance, so he says, 'Quick, let's hit the bed before we hit the dirt.'" Discuss this summary. Then try

making your own, more accurate one. Suggestion: The poem is divided into three parts, each beginning with an indented line. Take these parts one at a time, putting the speaker's main thoughts into your own words.

There's a grain of truth to this paraphrase, rude though it be. We might question, however, whether Marvell's speaker is trying to hoodwink his loved one. Perhaps he only sums up the terrible truth he knows: that time lays waste to youth, that life passes before we know it. He makes no mention of "romance," by the way—that's the paraphraser's invention. A more nearly accurate paraphrase, taking the three divisions of the poem one by one, might go like this:

Lines 1–20: If we had all the room in the world and if we were immortal, then our courtship might range across the globe. My love for you could expand till it filled the whole world and I could spend centuries in praising your every feature (saving your heart for last). After all, such treatment is only what you deserve.

•Lines 21–32: But time runs on. Soon we'll be dead and gone, all my passion and all your innocence vanished.

Lines 33–46: And so, while you're still young and willing, let's seize the day. Let's concentrate our pleasure into the present moment. Although we can't make the sun stand still (like Joshua in the Bible), we'll do the next best thing: we'll joyously make time fly.

Now, obviously, any such rewording of this matchless poem must seem a piddling thing. But if students will just work through Marvell's argument part by part, they may grasp better the whole of it.

2. In part one, how much space would be "world enough" for the lovers? Exactly how much time would be enough time?

(To point out the approximate location of the Humber and the Ganges on a globe [or a simple circle drawn on a blackboard] can drive home the fact that when the poet says *world enough*, he spells out exactly what he means. A little discussion may be needed to show that in defining "enough" time, Marvell bounds it by events [the conversion of the Jews], numbers the years, and blocks out his piecemeal adoration. Two hundred years per breast is a delectable statistic! Clearly, the lover doesn't take the notion of such slow and infinitely patient devotion seriously.)

3. What is the main idea of part two? How is this theme similar to that of Housman's "Loveliest of trees"?

(Both Marvell and Housman in "Loveliest of trees" [page 831] are concerned with the passage of time; they differ on what needs to be done about it. Marvell urges action; Housman urges filling one's youth with observed beauty. Of these two expressions of the *carpe diem* theme, Housman's seems the more calm and disinterested.)

4. Paraphrase with special care lines 37–44. Is Marvell urging violence?

(In lines 37–44, Marvell's point seems to be that time works a gradual, insidious violence. It is like a devouring beast (slow-chapped), holding us in its inexorable jaws. Some students will find the imagery odd, even offensive in a love poem: *birds of prey* (who want to eat, not be eaten), the cannonball of strength and sweetness that batters life's iron gates. Violence is not the speaker's counsel, but urgency. His harsh images lend his argument intensity and force.)

5. Considering the poem as a whole, does the speaker seem playful, or serious?

(This fifth question presents an easy dichotomy, but of course Marvell's speaker is both playful and serious. In making clear the tone of the poem, a useful poem for comparison is Marlowe's "Passionate Shepherd." What are the two speakers' attitudes toward love? Marvell's seems more down-to-earth, skeptical, and passion-driven: a lover in a fallen world, not (like Marlowe's shepherd) a lover in a pastoral Eden.

If later on, in teaching figures of speech, you want some great lines for illustrations, turn back to this inexhaustible poem. There's hyperbole in lines 7–20, understatement ["But none, I think, do there embrace"], metaphor, simile, and of course the great personification of chariot-driving time.

Telling a class that Marvell was a Puritan usually shakes up their overly neat assumption. Some may be surprised to learn that one can be a Puritan and not necessarily be puritanical.

Defending the poem against charges that its logic is fallacious, a recent critic, Richard Crider, has shown that "the speaker's appeal is not merely to the lady's passion, . . . but to a more inclusive and compelling value—completion and wholeness." A good student of Aristotle's logic as well as Aristotle's ethics, Marvell's speaker calls on his listener to exercise all her human powers, among them reason. "Although no single net will capture all the resonances of the final couplet, near the heart of the passage is the thought of living life completely, in accordance with natural law" ["Marvell's Valid Logic," *College Literature* (Spring 1985): 113–21].)

George Meredith, LUCIFER IN STARLIGHT, page 848

The name Lucifer (Latin for "light-bringing") comes from Isaiah 14:12: "How art thou fallen from heaven, O Lucifer, son of the morning!" Early interpreters of Scripture thought this line referred to the fallen archangel; later translators changed *Lucifer* to *day star*. Students may not be aware that Lucifer is a name both for the Devil and for the planet Venus as the morning star. (Venus is sometimes the evening star, too, but only as the morning star is it called Lucifer.)

Meredith, however, seems well aware of the name's duality. Although his sonnet can be read first as an account of the Miltonic fiend's taking a new flight (and meeting chagrin), it is also satisfying to read the poem as a description of Venus as a planet. We are told that Lucifer *uprose* and *sank*. Amid the stars, *his huge bulk*—a *black planet*—crosses Africa and shadows the Arctic, arriving at a *middle height* in the sky. The last two lines are puzzling, but seem to refer both to God's law and to the orderly path each star and planet follows as it circles the sun.

An Italian sonnet, "Lucifer in Starlight" illustrates a clear division in meaning between octave and sestet. In lines 1–8, the fiend appears triumphant and domineering: this is his night to howl. In lines 9–14, however, he crumples in defeat once more; or, if Lucifer is also the morning star, he rises in the octave and sets in the sestet.

James Merrill, LABORATORY POEM, page 848

If you have any students who assume that poetry deals only with the pleasant and the beautiful (and how widespread this assumption is!) then you may find Merrill's clinical love poem a useful thing to surprise them with. Try paying attention to its language in particular: the outrageous pun *taking heart* in the opening line, the *kymograph* (a device with a revolving drum that records pulsations such as muscle contractions or heartbeats), *her solutions tonic or malign* (stimulants or sedatives, preservatives or poisons?). Merrill's language is exact in the extreme, a love poem written with loving care. (After all, why should love be banned from a laboratory?)

Discussion will probably focus on the last stanza. Of the "exquisite disciplines," at least one of them is Naomi's research. The turtle's heart, violently removed, becomes a fact in a theorem. Like turtles, Charles realizes, some people give their lives for some abstract perfection, whether science, art, or any knowledge in whose service one may die. There is lovely irony in the contrast between Naomi's coldly clinical dissections (she has to "take heart," build up her nerve to perform them) and the human situation ("easy in the presence of her lover").

Charlotte Mew, THE FARMER'S BRIDE, page 849

Mew's thwarted life, marked by insanity in her family, a difficult invalid mother, frustrated lesbian desires, struggles with depression, and her suicide at 59, is documented in Penelope Fitzgerald's biography of the poet, *Charlotte Mew and Her Friends* (Reading: Addison

(text pages) 849–850

Wesley, 1988)—documented as well as it can be, considering Mew's lifelong reticence about herself and her family. Mew's career as a writer spanned literary movements: her first published story appeared in *The Yellow Book* in 1894; a few years later, as Imagism bloomed, Ezra Pound was printing her work in *The Egoist*. Her work, as Val Warner points out in his introduction to *Charlotte Mew: Collected Poems and Prose* (New York: Carcanet, 1982), "is a celebration of passion deeply felt, but always denied."

As does the speaker in the Robert Browning's "Soliloquy of the Spanish Cloister" (page 792), Mew's farmer-narrator reveals more than he realizes. Make sure that students understand what has caused the farmer's bride first to shrink from her husband, then run away. They can find abundant evidence in the poem. The farmer admits in the first stanza that she was "too young maybe" when he married her, and that the courtship was brief. Clearly the bride was unprepared for sex, and this well-meaning but unromantic man scared her off. Three years after the wedding she's still afraid—of him and of all men.

Evident throughout the poem is the strength of the farmer's love for his "little frightened fay." Untutored he may be, but he has not approached her sexually since she ran away. As lines 40–41 make clear, he longs for a child. It is also apparent that he longs for his beautiful young wife. Especially after line 30, the expression of his desire for her has an intensity that soars to heart-rending lyricism. Mew makes clear that, for the speaker, the physical nearness of his wife in the final stanza is almost unbearable. Note the skill the poet demonstrates in "The Farmer's Bride," a rimed poem that seems effortless, plain speech.

James Wright said of Mew, "The truth is that she wrote maybe six of the best poems of the 20th century. She's a superb poet, and a great woman" (*Poetry Pilot* [newsletter of the Academy of American poets] Sept. 1978). Thomas Hardy called Mew "far and away the best living woman poet—who will be read when others are forgotten," and in 1918, after reading *The Farmer's Bride*, he invited her to his home at Max Gate and struck up a friendship that endured. In 1924 Virginia Woolf declared Mew "the greatest living poetess," and in her day she was also admired by John Masefield, May Sinclair, Walter de la Mare, and Siegfried Sassoon. Yet, ironically, her poems did not circulate widely and she never received her share of acclaim. Since 1953 there have been sporadic attempts to give the poet her due. Perhaps Warner's edition of her work and Fitzgerald's biography signal a fresh appreciation.

Edna St. Vincent Millay, RECUERDO, page 850

There is probably not much to do with this delightful poem but leave it alone and let students discover it for themselves. The only thing we would find in it to discuss: Would that newspaper vender really break down into tears of gratitude? We bet she is a mere literary convention here, and in real life would probably be one tough old egg.

Would this highly musical lyric make a good song? Might it be sung?

John Milton, WHEN I CONSIDER HOW MY LIGHT IS SPENT, page 850

While this famous sonnet is usually taken to refer to the poet's lost eyesight, some critics have argued that it is not about blindness at all. The familiar title "On His Blindness" was not given by Milton, but by a printer a century later.

QUESTIONS FOR DISCUSSION:

1. If the poem is not about blindness, what might it be about? (Possible suggestions: Milton's declining powers of poetry; Milton's fame as a Puritan apologist.)

2. Is *talent* a pun referring to Milton's talent for writing poetry? What other meanings of the word seem appropriate in this poem? In the New Testament parable (Matthew 25:14–30), the hidden talent is money that should have been earning interest. That Milton is thinking primarily of work and business can be plausibly argued; other words in the

poem convey such connotations—*spent, true account, day-labor,* and perhaps *useless*, which suggests the Medieval Latin word for interest, *usuria*.

The theme of frustration in life (and reconciliation to one's lot) is dealt with differently in Shakespeare's "When, in disgrace with Fortune and men's eyes" (page 872).

Marianne Moore, THE MIND IS AN ENCHANTING THING, page 851

About this poem the poet has written, "One of the winters between 1930 and 1940, Gieseking gave at the Brooklyn Academy a program of Handel, Bach, and Scarlatti, the moral of this poem being that there is something more important than outward rightness. One doesn't get through with the fact that Herod beheaded John the Baptist, 'for his oath's sake'; as one doesn't, I feel, get through with the injustice of the deaths died in the war, and in the first world war" (note on the poem in Kimon Friar and John Malcolm Brinnin's anthology *Modern Poetry* [New York: Appleton, 1951]).

These remarks, and the poem's appearance in 1944, have led critics to call "The Mind Is an Enchanting Thing" a war poem. What to make of the poet's statement? To put it baldly, perhaps she means that nobody but a bigoted Herod would condemn a superb German musician for being German and for playing splendid German and Italian music, even in the 1930s. The drift of the poem is that it takes the mind to recognize beauty and, to do so, the mind must pierce through the heart's illusions (and prejudices?). Still, one needn't speculate about the poem's wartime relevance in order to see it as a restatement of the poet's favorite poet's theme, "Beauty is truth, truth beauty." Only the mind can apprehend things truly; like Gieseking's violin and like the apteryx's beak, it is a precision instrument.

Class discussions of this poem tend to be slow, but good and ruminative. Many students greatly admire the poem, and some instructors report getting outstanding papers about it.

What form is the poem written in? Syllabics, with lines arranged in a repeated stanza pattern—like Dylan Thomas's "Fern Hill" (page 886), which students may compare with Moore's poem. In an interview, Donald Hall asked Moore, "Do you ever experiment with shapes before you write, by drawing lines on a page?" "Never, I never 'plan' a stanza," she replied. "Words cluster like chromosomes, determining the procedure. I may influence an arrangement or thin it, then try to have successive stanzas identical with the first." Asked "How does a poem start for you?" she answered, "A felicitous phrase springs to mind—a word or two, say— . . . 'katydid-wing subdivided by *sun* / till the nettings are *legion*.' I like light rhymes, inconspicuous rhymes" (interview in *The Paris Review* [Winter 1961], reprinted in *A Marianne Moore Reader* [New York: Viking, 1961]).

Hall discusses the poem helpfully, too, in his study *Marianne Moore* (New York: Pegasus, 1970). He admires the poet's ability to weave a single word through the length of the poem, subtly changing its meaning: "The word *eye*, for example, is first the eye of the mind, then the eye of the memory, then the eye of the heart, suggesting three ways of 'seeing' that do not involve sight, but insight."

Howard Nemerov, THE SNOW GLOBE, page 852

Is the globe a symbol? If so, what does it suggest? Nemerov calls it an "emblem," one that seems to have changed its meaning for him as he has grown to adulthood. When he was a child, the toy was part of the "brightness" with which his parents "would have wanted [him] beguiled." Now he is older and sadder. When he remembers the snow globe, its coldness rather than its whiteness speaks to him, and it is death of which it speaks.

The tone of the poem is wistful, nostalgic, sorrowful. The real world cannot, like the village in the snow globe, be "frozen forever." Time does not stand still.

(text pages) 853–854

John Frederick Nims, LOVE POEM, page 853

This love poem proceeds through comedy. Nims is kidding his loved one affectionately, of course, and the mingled tone of tribute, both tender and chiding, recalls Swift's wonderful "On Stella's Birthday" (page 530). Another subject is incompetence, although Nims's lady is good with "words and people and love."

Sharon Olds, THE ONE GIRL AT THE BOYS PARTY, page 854

This poem whimsically describes a talented little girl, "her math scores unfolding in the air around her," during a pool party at which all the other guests are boys. *They* in lines 2 and 15, *their* in lines 18 and 19 seem to refer only to the boys. In lines 5, 7, and 11, the word *they* apparently includes the girl. You might ask students to note the pairs of adjectives that affirm the child's strength and composure: she is "smooth and sleek" (line 3), her body is "hard and / indivisible as a prime number" (lines 5–6), her face is "solemn and / sealed" (lines 16–17). The adjectives make clear the narrator-mother's respect for her brilliant daughter. Notable too is the metaphor of wet ponytail (itself a by-now-dead metaphor!) as pencil (line 12). That and the "narrow silk suit / with hamburgers and french fries printed on it" remind us that she is in some ways a very typical little girl.

It is the mathematical figures of speech that make this poem unique. Why not ask students to point out and discuss them? Are they apt? Do they ever appear forced? Which ones succeed best?

Wilfred Owen, ANTHEM FOR DOOMED YOUTH, page 854

Metaphorically, this sonnet draws a contrast between traditional funeral trappings and the actual conditions under which the dead lie on the field of battle: with cannon fire instead of tolling bells, rifle bursts instead of the patter of prayers, the whine of shells instead of choirs' songs, the last lights in dying eyes instead of candle-shine, pale brows (of mourning girls, at home?) instead of shrouds or palls, the tenderness of onlookers (such as the poet?) instead of flowers—an early draft of the poem reads, "Your flowers, the tenderness of comrades' minds"—and the fall of night instead of the conventional drawing down of blinds in a house where someone has died.

For another Owen war poem, see "Dulce et Decorum Est" (page 528). For other war poems, see in this manual "Poems Arranged by Subject and Theme."

The poet's revisions for this poem, in four drafts, may be studied in the appendix to C. Day Lewis's edition of Owen's *Collected Poems* (London: Chatto, 1963). In its first draft, the poem was called "Anthem for Dead [not Doomed] Youth," and it went, in our reading of the photographed manuscript:

> What minute bells for these who die so fast?
> Only the monstrous anger of our guns.
> Let the majestic insults of their iron mouths
> Be as the priest-words of their burials.
> Of choristers and holy music, none;
> Nor any voice of mourning, save the wail
> The long-drawn wail of high, far-sailing shells.
> What candles may we hold for these lost souls?
> Not in the hands of boys, but in their eyes
> Shall many candles shine, and [?] will light them.
> Women's wide-spreaded arms shall be their wreaths,
> And pallor of girls' cheeks shall be their palls.
> Their flowers, the tenderness of all men's minds,
> And every dusk, a drawing-down of blinds.

(text pages) 855-856

Linda Pastan, ETHICS, page 855

As a student, the narrator, like others in her class, found her teacher's ethical puzzler irrelevant. Now the mature woman pondering the "real Rembrandt" in the musuem finds the question still remote from her vital concerns, but for different reasons. The approach of her own old age has shown her that nothing lasts, that with the onflow of years our choices, whatever they may be, fade into insignificance.

One way of entering the poem: students may be asked to sum up its theme. Is it *carpe diem?* Is the poet saying, with Housman in "Loveliest of trees," "Life is fleeting; I'd better enjoy beauty while I can"? No, for the poet seems not to believe in day-seizing. Is it *ars longa, vita brevis est?* No, for both art and life seem pitifully brief and temporary. The point, rather, is that all things pass away despite our efforts to hold on to them. But instead of telling them what the point is, you might ask students to paraphrase the poem's conclusion that "women and painting and season are almost one / and all beyond saving by children."

To discuss: In what ways does "Ethics" differ from prose? Pastan's language seems far more musical. She makes beautiful music of alliteration and assonance. Read the poem aloud. And central to "Ethics" is a huge metaphor: old woman, season, earth, painting, and poet become one — all caught in time's resistless fire.

In our last Instructor's Manual, we wondered: How many times did the speaker have to repeat that ethics course? To our relief, on a recent visit to the University of Arizona in Tucson, Linda Pastan supplied an answer reported to us by Ila Abernathy of the Poetry Center. Pastan went to the Ethical Culture School in New York City, a private school run by the Ethical Culture Society and serving both elementary and high school students. The schools' curriculum hits ethics hard: the poet was required to take once-a-week ethics classes for twelve years.

Linda Pastan chose "Ethics" to represent her in *The Poet's Choice,* an anthology of poets' own favorite poems edited by George E. Murphy, Jr. (Green Harbor, MA: Tendril, 1980).

Katherine Philips, ON HER SON, H. P., page 856

To discuss: Is this poem serious? Then how do you account for the poet's use of puns — generally playful figures of speech? Do they spoil the poem, or do they add anything?

To answer this question thoughtfully is to go deep into metaphysical lyrics. Compare the ingenuity and playfulness of Donne's "A Valediction: Forbidding Mourning," or later metaphysical poet Emily Dickinson's "Because I could not stop for Death." Is it not possible to be serious and playful at once?

At sixteen, the poet had married a well-to-do widower thirty-eight years her senior, who eventually became a member of Parliament. Happily, her daughter Katherine, born a year after the death of her son, lived to maturity. During her short life, Katherine Philips won fame for her much-admired translation of Corneille's drama *The Death of Pompey.* For a biographical sketch, an appreciation, and a choice of poems, see *Kissing the Rod: An Anthology of 17th Century Women's Verse,* edited by Germaine Greer and others (London: Virago Press, 1988) 186-203. Not only was Philips in her lifetime the best-known female poet in England, she was the best known for succeeding generations as well. Between 1667 and 1710, her poems went through four editions.

Robert Phillips, RUNNING ON EMPTY, page 856

Phillips, born in Delaware, now lives in Katonah, New York. He writes fiction and criticism asd well as poetry, and has edited the letters and the late and uncollected poems of Delmore Schwartz. He sent us the following comment about "Running on Empty":

"Running on Empty" is fairly autobiographical. I was stunned at how grudgingly my father let me use the family car once I'd obtained my driver's license at age 16. It seemed to me he withheld this symbol of my new freedom and attainment just as he withheld his affection. So when I finally had use of the car, I went hog-wild in celebration and release.

The landscape is Sussex County, Delaware—extremely flat country bisected by Route 13 (nicknamed "The Dual" because it is composed of twin lanes dually parallel in an inexorable straight line). I was pushing my luck speeding and refusing to refuel in an act of rebellion against my father's strictness (which may explain why the 12th line reads "defying chemistry" rather than the more accurate "defying physics"—my father taught high school chemistry, and even in the classroom I was subject to his discipline).

I'm rather pleased with the way the poem picks up rhythm and begins to speed when the car does (5th–7th stanzas). And I hope students relate to the central images of car and boy, one of which can be mechanically refueled and replenished, one of which cannot.

To us the word *chemistry* in line 12 carries additional meaning, as in "behavior or functioning, as of a complex of emotions" *(The American Heritage Dictionary).* In this sense, too, the narrator was surely in defiance of his father's chemistry.

Sylvia Plath, DADDY, page 857

There are worse ways to begin teaching this astonishing poem than to ask students to recall what they know of Dachau, Auschwitz, Belsen (line 33), and other Nazi atrocities. "Every woman adores a Fascist"—what does Plath mean? Is she sympathizing with the machismo ideal of the domineering male, lashing his whip upon subjugated womankind? (No way.) For an exchange of letters about the rightness or wrongness of Plath's identifying with Jewish victims of World War II, see *Commentary* (July and October 1974). Irving Howe accuses Plath of "a failure in judgment" in using genocide as an emblem of her personal traumas.

Incredible as it seems, some students possess an alarming fund of ignorance about Nazis, and some might not even recognize the cloven foot of Satan (line 53); so be prepared, sadly, to supply glosses. They will be familiar with the story of Dracula, however, and probably won't need much help with lines 71–79. Plath may be thinking of *Nosferatu,* F. W. Murnau's silent screen adaptation of Bram Stoker's novel *Dracula,* filmed in Germany in 1922. Hitler's propagandists seized on the Nosferatu theme and claimed that the old democratic order was drinking the country's blood. Plath sees Daddy as doing the same to his daughter.

Sylvia Plath, MORNING SONG, page 860

Plath's image of the cloud being effaced seems almost a premonition of her own death. But "Morning Song" is rich in humor: the wonderful self-portrait of the *cow-heavy* mother, the baby's cries coming out in *balloons* (as in a comic strip?). The title refers to the child's crying (music to wake up by!), as well as to the poem itself.

Ezra Pound, THE RIVER-MERCHANT'S WIFE: A LETTER, page 861

After the death of Ernest Fenollosa, a scholar devoted to Chinese language and literature, Pound inherited Fenollosa's manuscripts containing rough prose versions of many Chinese poems. From one such draft, Pound finished his own version of "The River-Merchant's Wife." Fenollosa's wording of the first line went:

My hair was at first covering my brows (child's method of wearing hair)

Arthur Waley, apparently contemptuous of Pound for ignoring dictionary meanings of some of the words of the poem, made a translation that began:

Soon after I wore my hair covering my forehead . . .

Pound's version begins:

While my hair was still cut straight across my forehead . . .

Pound, says the recent critic Waj-lim Yip, has understood Chinese culture, while Waley has not, even though he understands his dictionary. "The characters for 'hair/first/cover/forehead' conjure up in the mind of a Chinese reader exactly this picture. All little Chinese girls normally have their hair cut straight across the forehead." Yip goes on to show that Pound, ignorant of Chinese as he was, comes close in sense and feeling to the Li Po original. (*Ezra Pound's Cathay* [Princeton: Princeton UP, 1969] 88–92.)

What is the tone of the poem? What details make it seem moving and true, even for a reader who knows nothing of Chinese culture?

Dudley Randall, OLD WITHERINGTON, page 862

QUESTIONS FOR DISCUSSION:

1. What suggestions do you get from the name Old Witherington?
2. What is meant by "I'll baptize these bricks with bloody kindling" (line 7)?
3. Describe the impact of *tenderly* as it is used in line 19.
4. What possible significance do you see in Old Witherington's *not* cursing "his grinning adversary" (line 18) when he has cursed everyone else?
5. What is the tone of Randall's poem?

Full of withering hate, habitually intent upon cutting people down ("withering" them), Old Witherington himself withers, too. A veteran drunkard, he claims to have died before, a "million times"—presumably of loneliness, neglect, and self-destructive behavior. His neighbors jeer him. The only person who takes him seriously, it seems, is the "crazy bastard" he is about to fight. In a curious way, Witherington's adversary, simply by engaging him in battle, endows the old man with some importance in both men's eyes. Because of this opponent Witherington exists; he somehow matters. For the man with the "jag-toothed bottle," Witherington is the sun.

Complex in its feelings, Randall's poem is about much more than hate. In it we sense compassion for the old man, if not from his neighbors then surely from the poet. In his book (*A Litany of Friends* [Detroit: Lotus, 1981]) Randall has placed "Old Witherington" in the section entitled "Friends."

The scene in Randall's poem is vividly rendered, the imagery exact and alive. The "bloody kindling" in line 7 presumably refers to what the hatchet will look like when the fight is over. *Tenderly* in line 19 suggests an appropriate gingerliness on the part of the adversary. It also suggests love—a connotation equally appropriate in light of the adversary's role in affirming Witherington's worth.

John Crowe Ransom, BELLS FOR JOHN WHITESIDE'S DAUGHTER, page 863

This is an excellent poem with which to set off a discussion of sentimentality—a quality of much bad poetry that Ransom beautifully avoids. What conventional sentiments might be expected in response to a small child's death? Usual expressions of grief. But how does the poet respond? With disbelief and astonishment, delight in remembering the little girl alive, with vexation—even outrage—that she is dead.

(text pages) 863–864

Between what Ransom says and what he might have said gapes the canyon of an ironic discrepancy. "It is not a poem," Robert Penn Warren has said, "whose aim is unvarnished pathos of recollection. . . . The resolution of the grief is not on a compensatory basis, as is common in the elegy formula. It is something more modest. The word *vexed* indicates its nature: the astonishment, the pathos, are absorbed into the total body of the mourner's experiences and given perspective so that the manly understatement is all that is to be allowed. We are shaken, but not as a leaf" (see "John Crowe Ransom: A Study in Irony," *Virginia Quarterly Review* 11 [1935]: 93–112).

Vivienne Koch has underscored the colloquial phrase *brown study* ("a state of serious absorption or abstraction"—*Webster's New Collegiate*). A child in a brown study seems alien, for her nature includes "speed" and "lightness." "The repetition in the last stanza of 'brown study' in conjunction with the key word *vexed* clinches the unwillingness of the narrator to accept the 'little lady' as departed" ("The Achievement of John Crowe Ransom," *Sewanee Review* 58 [1950]: 227–61). Warren's essay, Koch's, and other good ones are reprinted in *John Crowe Ransom: Critical Essays and a Bibliography*, ed. Thomas Daniel Young (Baton Rouge: Louisiana State UP, 1968).

James Wright has recalled a four-line parody of this poem that he wrote in collaboration with four or five other undergraduates, Ransom's students at Kenyon College (interview with Wright in *The Pure Clear Word*, ed. Dave Smith [Urbana: U of Illinois P, 1982] 12):

BALLS ON JOAN WHITESIDE'S STOGY

There was such smoke in our little buggy
And such a tightness in our car stall—
Is it any wonder her brown stogy
Asphyxiates us all?

Henry Reed, NAMING OF PARTS, page 863

This is one of the most teachable poems ever written. There are two voices: the voice of the riflery instructor, droning on with his spiel, and the voice of the reluctant inductee, distracted by the springtime. Two varieties of diction and imagery clash and contrast: technical terms opposed to imagery of blossoming nature. Note the fine pun in line 24, prepared for by the rapist bees in the previous line. Note also the connotations of the ambiguous phrase *point of balance* (line 27)—a kind of balance lacking from the recruits' lives?

Students need to be shown the dramatic situation of the poem: the poor inductee, sitting through a lecture he doesn't want to hear. One would think that sort of experience would be familiar to students, but a troublesome instructors have met in teaching this poem is a yearning to make out of it a vast comment about Modern Civilization.

The poet himself has recorded the poem for *An Album of Modern Poets*, 1 (Library of Congress, PL 20). Dylan Thomas reads "Naming of Parts" even more impressively in his *Reading, Vol. IV: A Visit to America and Poems* (Caedmon, TC 1061).

Adrienne Rich, AUNT JENNIFER'S TIGERS, page 864

The poet herself has made revealing mention of this poem in a *College English* essay reprinted in *Adrienne Rich's Poetry*, ed. Barbara Charlesworth Gelpi and Albert Gelpi (New York: Norton, 1975):

> Looking back at the poems I wrote before I was 21, I'm startled because beneath the conscious craft are glimpses of the split I even then experienced between the girl who wrote poems, who defined herself in writing poems, and the girl who was to define herself by her relationships with men. "Aunt Jennifer's Tigers," written while I was a student, looks with deliberate detachment at this split. In writing this poem, composed and apparently cool as it is, I thought I was creating a portrait of an imaginary woman. But the woman suffers from the opposition of her imagination, worked out

in tapestry, and her life-style, "ringed with ordeals she was mastered by." It was important to me that Aunt Jennifer was a person as distinct from myself as possible—distanced by the formalism of the poem, by its objective, observant tone—even by putting the woman in a different generation.

Rich's feminism clearly was beginning to emerge, however, as far back as 1951, when this poem was first published. It is apparent in the poem that the poet perceived something wrong with the passive role assigned to women. The pride, confidence, and fearlessness ("masculine" virtues, whatever the sex of the tigers) of Aunt Jennifer's imaginary creations contrast sharply with Aunt Jennifer herself—a frail lady with fluttering fingers, *terrified hands*. Worth comment is the poet's use of the word *ringed*—suggesting "encircled"—to refer both to the wedding ring that "sits heavily upon Aunt Jennifer's hand" and to "ordeals she was mastered by," specifically marriage and being expected to conform. Although she goes down in defeat, her tigers triumph.

Question for discussion: Is Aunt Jennifer's plight that of the woman artist in our society? (Because of the wedding ring's weight, she must struggle to ply her instrument, the ivory needle.)

Compare Aunt Jennifer with the dead woman who once embroidered fantails in Wallace Stevens's "The Emperor of Ice-Cream" (page 566). For another, in some ways comparable, contrast between a dull world of reality and the colorful life of the imagination, see Stevens's "Disillusionment of Ten O'Clock" (page 564), in which

> Only, here and there, an old sailor,
> Drunk and asleep in his boots,
> Catches tigers
> In red weather.

Mary Slowik discusses this and other early poems of Rich in "The Friction of the Mind," *Massachusetts Review* Spring 1984: 142–60.

Adrienne Rich, THE NINTH SYMPHONY OF BEETHOVEN UNDERSTOOD AT LAST AS A SEXUAL MESSAGE, page 865

What exactly is the poet saying about Beethoven and his ninth symphony? Clearly Rich doesn't favor either. She portrays the composer as "howling," presumably because he is "in terror of impotence / or infertility, not knowing the difference"; and "yelling at Joy from the tunnel of the ego." She depicts a Beethoven who writes the Ninth Symphony to vent a terrible sexual frustration: The fist that beats the table is "bloody." Line 12 hints as well at a sort of sadomasochistic frenzy ("gagged and bound and flogged with chords of Joy").

Here are a few questions that, raised in class, might elicit a lively exchange of views. How might a music historian react to Rich's assessment? Does Rich hate Beethoven because he is a man? Does she see him as consummate artist, alienated and suffering, driven to express his fear and anger? Does she identify with Beethoven? Is she trying jealously to cut him down to size? Is the poet simply venting her anger at a composer whose work she doesn't like? Is Rich's poem a feminist statement or merely a belittlement of a great composer and his work?

Some students might also be interested in tracing Rich's development as a poet from the formalism of the earlier "Aunt Jennifer's Tigers" to the openness of this poem and the one that follows. Her feminism, too, seems more militant than it was 22 years earlier.

Adrienne Rich, TRYING TO TALK WITH A MAN, page 866

QUESTIONS FOR DISCUSSION:

1. Who are the "we" in this poem? (The reference seems to change. In line 1 the "we" includes everyone in our country, our government. In line 2 it refers to the speaker and

(text pages) 866–868

her male companion, both apparently taking part in an anti-nuclear bomb protest. In line 30, we think, "we" are women, as opposed to "You" men.)

2. What do you make of lines 26–27: "Out here I feel more helpless / with you than without you"? (What the speaker has come up against is the silence referred to in lines 19–22, the silence "that came with us / and is familiar." Perhaps this silence represents the impossibility of a woman's being able to communicate with a man. The gulf that separates the sexes, even when they try to unite in a common cause, is detailed in the poem's final three stanzas.)

Like David Axelrod in "Once in a While a Protest Poem," Rich in this poem is not only protesting one of the world's evils (hunger for Axelrod, nuclear testing for Rich), she is making a protest against the protest itself. The protesters use the wrong methods. Their zeal is misdirected. They have, with their meaningless symbols, lost sight of what really matters: suffering individual human beings.

It may be, of course, that social purpose is secondary in this poem, and that it is mainly about a couple who have lost the ability to communicate. Perhaps they're spending time at some desert resort, but their being so close is more painful than being apart; the gulf between them still remains (compare Mary Elizabeth Coleridge's "We never said farewell") as, for Rich, the gulf between the sexes remains.

Theodore Roethke, ELEGY FOR JANE, page 867

By piling up figures of speech from the natural world, Roethke in "Elegy for Jane" portrays his student as a child of nature, quick, thin, and birdlike. A *wren*, a *sparrow*, a *skittery pigeon*, Jane has a *pickerel smile* and neck curls *limp and damp as tendrils*. She waits *like a fern, making a spiny shadow*. She has the power to make shade trees and (even more surprising) mold burst into song. For her, leaves change their whispers into kisses.

Then she dies. The poet acknowledges that for him there is no consolation in nature, in the "sides of wet stones" or the moss; his grief is not assuaged. Because he mourns the girl as teacher and friend, no more, he recognizes a faint awkwardness in his grief as he speaks over her grave:

> I, with no rights in this matter,
> Neither father nor lover.

Roethke, writing about this poem in *On the Poet and His Craft* (Seattle: U of Washington P, 1965) 81–83, reminds the reader that it was John Crowe Ransom (to whose "Bells for John Whiteside's Daughter" [see page 863] this poem has often been compared) who first printed "Elegy for Jane." Roethke discusses his use of enumeration, calling it "the favorite device of the more irregular poem." He calls attention to one "of the strategies for the poet writing without the support of a formal pattern," a strategy he uses in "Elegy for Jane": the "lengthening out" of the last three lines in the first stanza, balanced by the progressive shortening of the three lines at the poem's end.

Some readers have interpreted "Elegy for Jane" as the work of a man who never had children of his own; but in fact Roethke as a young man had fathered a daughter for whom he felt great affection. Although "neither father nor lover" of Jane, he at least could well imagine a father's feelings.

Gibbons Ruark, WAITING FOR YOU WITH THE SWALLOWS, page 868

What happens in this poem? Little physical action. The speaker tells his loved one what he sees and thinks while waiting for the train that will bring her to him. While he waits, a freight train passes and a flock of birds rises from a steeple-top, wheels overhead, and returns to its perch. Greeting his love as she steps from the train, he recalls the simple but cryptic message he'd received from the swallows. Perhaps it's something like "Leave if you must, but come back home again."

(text pages) 868–870

Clearly, in this poem, feeling and language matter more than theme. Subtle in its music, this deft lyric repays reading aloud. Note the rime scheme in each stanza: *abccba*. Many of the rimes are off-rimes: *wander, thunder; street, freight; again, train; quick, back; face, voice; long, sang.*

The poem comes from *Keeping Company* (Baltimore: Johns Hopkins UP, 1983). A native of North Carolina, Ruark has taught for many years at the University of Delaware.

Paul Ruffin, "HOTEL FIRE: NEW ORLEANS", page 869

A native of Mississippi who now teaches English at Sam Houston State University, Ruffin provides this comment on his poem:

> On the evening news a few years back I watched the account of a dreadful New Orleans hotel fire in which several people died. In one vivid scene the victims were forced into a corner room several stories up where they had to face our two greatest fears, burning and falling. They appeared at the smoke-filled window, looked back toward the fire, then crawled one by one onto the ledge, clung momentarily, and dropped to their deaths. With that fire pressing them, there seemed almost a kind of faith as they turned loose into the smoky air.

This memorable poem appears in Ruffin's collection *Lighting the Furnace Pilot* (Peoria: Spoon River Poetry, 1980), and we thank Cleatus Rattan of Cisco Junior College for pointing it out to us. Its metaphors and similes invite patient unraveling.

The first stanza of Ruffin's poem serves as background for the second. By devoting six lines to the inborn fear of falling, the poet heightens the ghastliness of the second stanza, in which the people *choose* to fall because they fear fire, "an old and certain death," even more.

What are "that natural window" and "the fire / that tempers us for the sun"? Both apparently refer to the experience of being born, going from the darkness of the womb to the light of the world outside. Both also suggest the New Orleans fire and the means its victims used to escape it: passage through a dreaded window, then through a holocaust. At birth, presumably, we trust in the earth beyond the womb to sustain us—but those who leap from the hotel window flail "like children / who know the earth has failed them." It takes one act of faith to be born; another, to leap to death. The last line of the poem is a serious pun: an apt play on the expression *a leap of faith*.

Anne Sexton, THE KISS, page 869

"The Kiss" is a rhapsodic love poem, as indicated by its metaphors from music in the last stanza. The speaker's lover is the composer who makes the orchestra flame into life. Old Mary is the poet's name for herself when she was glum, dull, and ordinary. (There used to be a comic strip called "Apple Mary," later rechristened "Mary Worth," whose main character was an aged paragon of virtue, but we don't know whether this fact matters.) As other words indicate, the speaker's love is a kind of religious awakening (*a resurrection!*; *She's been elected*). The third stanza compares her new life to a construction project—the building of a boat and its launching.

Robert Phillips deals with "The Kiss" briefly, and with Anne Sexton's other poetry at greater length, in *The Confessional Poets* (Carbondale: Southern Illinois UP, 1973).

William Shakespeare, NOT MARBLE NOR THE GILDED MONUMENTS, page 870

To discuss: Is Shakespeare making a wild boast, or does the claim in lines 1–8 seem at all justified? (Time has proved him right. Here we are, still reading his lines, 480 or so years after they were written! Of course, the fact that he happened to be Shakespeare helped his prediction come true.)

For teaching this poem in tandem with Robinson Jeffers's "To the Stone-cutters," see the entry on Jeffers (in the book itself, page 835).

(text pages) 871–874

William Shakespeare, THAT TIME OF YEAR THOU MAYST IN ME BEHOLD, page 871

William Shakespeare, WHEN, IN DISGRACE WITH FORTUNE AND MEN'S EYES, page 872

Shakespeare's magnificent metaphors probably will take some brief explaining. How is a body like boughs, and how are the bare boughs like a ruined choir loft? Students will get the general import, but can be helped to visualize the images. "Consumed with that which it was nourished by" will surely require some discussion. Youth, that had fed life's fire, now provides only smothering ashes. The poet's attitude toward age and approaching death stands in contrast to the attitudes of poets (or speakers) in other poems of similar theme: blithe acceptance of impending death in Frost's "Away!" (page 810); admiration for the exultant sparrows in William Carlos Williams's "To Waken an Old Lady" (page 895); defiance in Yeats's "Lamentation of the Old Pensioner" (page 728).

Figures of speech are central to "That time of year," but barely enter into "When, in disgrace" until the end, when the simile of the lark is introduced. The lark's burst of joy suggests that heaven, called *deaf* in line 3, has suddenly become keener of hearing. Critical discussion of both sonnets goes on: *Shakespeare's Sonnets*, edited with analytic commentary by Stephen Booth (New Haven: Yale UP, 1977) is especially valuable.

William Shakespeare, WHEN DAISIES PIED AND VIOLETS BLUE, page 872

William Shakespeare, WHEN ICICLES HANG BY THE WALL, page 873

Students are usually pleased to add the word *cuckold* to their vocabularies. (The origin of the word is uncertain, but it evidently refers to the cuckoo's habit of laying its eggs in other birds' nests.) Both songs take birdcalls for their refrains. How is the owl's call evocative of winter? Despite the famous harsh realism of the winter scene, discussion may show that winter isn't completely grim, nor is summer totally carefree.

Bertrand Bronson has a good discussion of these two poems in *Modern Language Notes* 63 (Jan. 1948): 35–38.

Karl Shapiro, THE DIRTY WORD, page 873

Shapiro's theme, which the class may be asked to make explicit, is that the dirty word one secretly loves in childhood is seen in maturity to be powerless. Is it a paradox that the word-bird is said to outlive man, and yet later the speaker tells us he murdered it? It would seem, rather, that at these two moments in his poem Shapiro sees the word from two different points of view. In one sense, the word will continue to live after the speaker's death because (figuratively) it is freed from the cage of his mind and also because (literally) words live longer than their speakers do. In the last paragraph the speaker means that he neutralized the bird's magic. Simply by growing up, he abolished its power over him.

Arranged in paragraphs rather than in conventional lines of poetry, "The Dirty Word" may lead some students to ask why this is poetry, not prose. It is a good opportunity to point out that poetry is a name we can apply to any language we think sufficiently out of the ordinary and that poetry is not determined merely by arranging lines in a conventional order on a page. To a much greater extent than a prose writer usually does, Shapiro expresses himself through metaphor. Besides the central metaphor of word as bird, there are the metaphors of mind as cage, brain as bird food, self as house, skull as room, secret thoughts as closet, vocabulary as zoo, feathers as language. The poem needs to be heard aloud, for it is full of unprosaic sounds: rimes *(sweet meat, bird . . . word)*; alliterations *(buried, body, bird; worn, wing; murdered, my, manhood)*; and internal alliterations *(ripping and chopping, walls of the skull)*. It is also rich in bizarre and startling imagery.

Richard Shelton, MEXICO, page 874

Let your students know, if they don't know already, that Scorpio is a constellation in the Southern Hemisphere, also known as Scorpion. Why does the speaker in the poem call

(text pages) 874–878

Scorpio "most dangerous of friends"? Perhaps because of the venomous sting at the end of the scorpion's tail. Perhaps also because each April, when the speaker sees Scorpio in the sky, a "frenzy" takes hold of him and he sets out for Mexico on what turns out to be a wild goose chase. What is he looking for? Something holding tremendous promise, something ineffable, a whole new life. Something perhaps like, in *The Great Gatsby,* the blue light on the end of Daisy's dock. In stanza 5 the speaker refers to "last chances"—perhaps to learn that "anything is possible."

Richard Shelton teaches in the creative writing program at the University of Arizona. "Mexico," originally published in *The New Yorker,* comes from his *Selected Poems* (U of Pittsburgh P, 1982).

Charles Simic, BUTCHER SHOP, page 875

"Butcher Shop" is a constellation of metaphors. Associating the everyday instruments of a butcher's trade with things we wouldn't expect—things whose connotations are emotionally powerful—Simic works a kind of nighttime transformation. The light recalls a convict struggling to escape. Knives recall church, cripple, and imbecile. Most pervasive of the metaphors in the poem is the river of blood (lines 8 and 14). In a sense, we are nourished by a river of blood when we dine on the flesh of animals. Perhaps (like convict, cripple, and imbecile) the animals too are sufferers. Perhaps all of these victims in chorus lift the mysterious voice that the poet hears in the closing line.

David R. Slavitt, TITANIC, page 875

In "The Convergence of the Twain," Hardy censures the vanity, luxury, and pride that prompted Fate to ram the *Titanic* into an iceberg. Slavitt's poem about the same tragedy takes another tack. He makes dying on the *Titanic* sound almost like fun—all aboard!

If they sold passage tomorrow for that same crossing, who would not buy?

Slavitt's point is that, since we all have to die, it's certainly more glamorous, more desirable to do it "first-class" (note the double meaning of "go" in the last line) than to die less comfortably, in more mundane ways, and soon be forgotten.

Christopher Smart, FOR I WILL CONSIDER MY CAT JEOFFRY, page 876

Telling us more about cats than Carl Sandburg and T. S. Eliot (in "Prufrock," lines 15–22, page 809) put together, Smart salutes Jeoffry in one of several passages in *Jubilate Agno* that fall for a little while into some continuity. This fascinating poem, and the whole work that contained it, have come down to us in a jumble of manuscripts retrieved from the asylum, sorted out brilliantly by W. H. Bond in his edition of Smart's work (Cambridge: Harvard UP, 1954). Some of Smart's gorgeous lines seem quite loony, such as the command to Moses concerning cats (lines 34–35) and the patriotic boast about misinformation: the ichneumon (or *Icneumon,* line 63) is not a pernicious rat, but a weasellike, rat-killing mammal.

Read aloud, Smart's self-contained poem in praise of Jeoffry can build a powerful effect. In its hypnotic, psalmlike repetition, it might be compared with the section of Whitman's "When Lilacs Last in the Dooryard Bloom'd."

Talking with Boswell of Smart's confinement, Dr. Johnson observed,

> I did not think he ought to be shut up. His infirmities were not noxious to society. He insisted on people praying with him; and I'd as lief pray with Kit Smart as with any one else. Another charge was, that he did not love clean linen; and I have no passion for it.

A possible paper topic: "Smart's Cat Jeoffry and Blake's Tyger: How Are These Poems Similar in View?"

(text pages) 878–879

Stevie Smith, I REMEMBER, page 878

QUESTIONS FOR DISCUSSION:

1. From the first three lines, you might expect a rollicking, roughly metrical ballad or song. But as this poem goes on, how does its form surprise you?
2. What besides form, by the way, is odd or surprising here? Why can't this be called a conventional love lyric?
3. Lewis Turco has proposed the name *Nashers* for a certain kind of line (or couplet) found in the verse of Ogden Nash. Nashers, according to Turco, are "usually long, of flat free verse or prose with humorous, often multisyllabic endings utilizing wrenched rhymes." What Nashers can you find in "I Remember"?
4. What does the poet achieve by ending her poem in an exact rime *(collide/bride)*? Suppose she had ended it with another long, sprawling, unrimed line; for example, "As far as I know from reading the newspapers, O my poor coughing dear." What would be lost?
5. What do you understand to be the *tone* of this poem (the poet's implied attitude toward her material)? Would you call it tender and compassionate? Sorrowful? Grim? Playful and humorous? Earnest?
6. How does noticing the form of this poem help you to understand the tone of it?

This poem keeps pulling rugs out from under us. From its opening, we might expect the poem to be a roughly metrical ballad or song, but then the short fourth line draws us up with a jolt. Any reader still hoping for a conventional ballad might say line 4 rimes at least. However, the passage introducing the bombers (lines 5–8) shatters any such anticipation. Far from songlike, these prosaic lines sprawl, end in far-out feminine rimes (*overhead* and *Hampstead, perversely* and *Germany*), and bring to mind the artful outrages of Ogden Nash. With the Bride's question, the poem unexpectedly returns to medium-length lines, and the rime that clicks it shut *(collide, bride)* is masculine and exact once more. Though full of surprises, this poem has less anarchy than order in it.

As its playful form indicates, "I Remember" is supposed to be fun, yet at the same time it seems awful and ominous. John Simon has made some provocative remarks about it in "The poems of Stevie Smith," *Canto* 1 (Spring 1977): 197–98.

> This, you might well say, is poetry of low intensity or simply, minor poetry. And yet, and yet! Why should a young girl marry an old man of 73? Because she is dying of consumption, and because it is wartime. The younger men are off to war, and may never come back. Even the old bridegroom and the tubercular bride may not live out their short remaining terms: the bombs may kill them first. The two of them may even desire such an ending; they have refused the relative safety of an air-raid shelter. . . . Are there perversely preposterous collisions in the sky? Can there be between our own so unalike selves so unlikely a collision, explosion, orgasm? . . . "I do not think it has ever happened," he said. Still, he must have hoped that it might.

William Jay Smith, AMERICAN PRIMITIVE, page 879

We might expect a painter called an American primitive to be naive, unsophisticated, and childlike in his view. So is the speaker who draws this verbal scene. Not only do the references to Daddy seem juvenile, but so does the line "the screen door bangs, and it sounds so funny." (Smith, incidentally, has written much fine verse for children in addition to his more serious poetry, and he understands the way a child thinks and speaks.) There is, of course, an ironic distance between the speaker's point of view and the poet's. Irony is enforced, too, by contrast between the grim event and the bouncy rhythm and use of feminine rimes.

Another possible way of looking at the poem is that Daddy himself is the primitive: the primal dollar-worshipping American. The capitalization of *Dollar* (as in the familiar phrase "the Almighty Dollar") may support this view. We are not told why Daddy died, an apparent suicide, but it is evident that riches did not buy him life. Besides inviting comparison with Sylvia Plath's ironic poem about the death of a terrible "Daddy" (page 857), Smith's mock-elegy may be set beside Wallace Stevens's "The Emperor of Ice-Cream" (page 566), with students asked to compare the two in tone and in subject matter.

W. D. Snodgrass, SEEING YOU HAVE . . . , page 879

This lyric seems the soliloquy of a domesticated male who feels a perverse yen to howl with the pack. So read, the *you* is the speaker addressing himself. In *W. D. Snodgrass* (Boston: Twayne: 1978 [46–47], Paul Gaston works out a similar interpretation:

> Both seven-line stanzas balance the speaker's awareness of the nature of his woman (in the first three lines of each) with his bemused and critical wonder at his contrary nature (in the last four). This woman, "whose loves grow thick as the weeds / That keep songsparrows through the year," is also firmly rooted, even if she is no one's "tree." . . . But, as the presumed beneficiary of his woman's staunch, dependable, more than ample love, the speaker finds himself drawn, seemingly without explanation, to perverse resistance. He envies "boys / Who prowl the streets all night in packs," even though he fully perceives their vanity and insecurity, even though he hears in his own voice the contentious tone of the quarrelsome bluejays. In wondering at his own contrary nature he is not being disingenuous. . . . Yet some part of his answer lies in the very similes he chooses to characterize his woman's love, ones that convey more than anything else a sense of stultification, oppression, and encumberance. Before the thick domesticity created by his woman, he becomes again the creature of nature, self-centered, predatory, and contentious.

The poem strikes us as one of the finest, if most complicated, lyrics in Snodgrass's Pulitzer-winning collection *Heart's Needle* (1959). The poet's *Selected Poems 1957–1987* (New York: Solo, 1987) offers an excellent survey of his work.

William Stafford, AT THE KLAMATH BERRY FESTIVAL, page 880

QUESTIONS FOR DISCUSSION:

1. What is ironic in the performance of traditional dances, by a scout troop on an Indian reservation, for an audience including a sociologist? What does the sociologist signify? What is the significance of the face that other Indians are gambling outside, turning their backs on the dances? Sum up the poet's theme.

2. Why is the war chief *bashful*? How do you account for his behavior ("listening and listening, he danced after the others stopped")? (Listening past the noise of the gamblers to the quiet of mountains and river, he gets caught up in the old dance in spite of himself, and thoughtfully keeps dancing, forgetful of himself and of the modern world.)

3. How would you scan "He took two steps"? The poet introduces the statement four times—what is the effect of this repetition? (It makes for a row of heavy stresses followed by a pause. The poem is approximately the rhythm of the war chief, who makes heavy footfalls, pauses, then goes on with his dance.)

Wallace Stevens, PETER QUINCE AT THE CLAVIER, page 880

QUESTIONS FOR DISCUSSION:

1. We know that it is Peter Quince who speaks to us in the opening section; what are we to make of what follows? (We are to laugh when he sits down at the clavier. The story of Susanna is to be a tale told by a clown, not a faithful and serious recital; nor are we to take the clown's anguish [lines 5–15] in total earnest.)

(text pages) 880–883

2. Point out all the onomatopoeia you can find in the poem. Why is it appropriate to a poem whose imagery is largely taken from music?

3. "Music is feeling, then, not sound" (line 4). How is the truth of this statement demonstrated in the rest of the poem? (Each character in the poem has a theme song. Whenever we hear the music of the elders, as in lines 12–15 and 39–40, it is like that of a coarse jazz band, or like show-off violinists excitedly plucking their violins instead of playing them: "pizzicati of Hosanna"—praise not of God, but of Susanna. The elders are unimaginative men, coarse sensualists bound to the physical world. All they can hear is "Death's ironic scraping," not Susanna's music of immortality. Other theme songs are audible. Susanna, as she lolls in her bath, touches "springs" of melody—an autoerotic suggestion? The simpering Byzantine maids titter like tambourines. All have their appropriate music.)

4. "Beauty is momentary in the mind—/The fitful tracing of a portrayal; / But in the flesh it is immortal." Is this statement nonsense? What sense can you make of it? (Stevens weaves a metaphor: the beauty of a woman is like music. Not that bodily beauty lasts forever; instead, it becomes a legend, and so continues to inspire works of art that live on in human memory.)

Harold Bloom has found an affinity between "Peter Quince" and Robert Browning's dramatic monologue "A Toccata of Galuppi's." In Browning's poem, a man apparently playing the music of the Venetian composer on the clavichord finds himself remembering Venetian gallants and ladies, and their long-vanished lust. But in Stevens's opening lines, says Bloom, "it is Stevens who speaks directly of his own desire." This desire "deprecates itself, by an identification with the desire of the elders for Susanna rather than with the more refined and repressed desire of Susanna herself, in section II" (*Wallace Stevens, The Poems of Our Climate* [Ithaca: Cornell UP, 1977] 36).

Ruth Stone, SECOND HAND COAT, page 883

QUESTIONS FOR DISCUSSION:

1. Who is the speaker? (A woman, poor or on a tight budget, who has bought a second-hand coat.)

2. What distinguishes "Second Hand Coat" from prose? (Starting with line 6, the poem is an extended metaphor in which both coat and buyer become the woman who previously wore the coat.)

3. What is the tone of Stone's poem? (A little awed, perhaps, but at the same time whimsical, especially in the last two lines.)

This is the title poem of Stone's *Second-Hand Coat: Poems New and Selected* (Boston: Godine, 1987). For several appreciative comments about Ruth Stone and her work by her fellow writers, see pages 323–330 of *Extended Outlooks*, edited by Jane Cooper, Gwen Head, Adalaide Morris, and Marcia Southwick (New York: Collier, 1982).

Jonathan Swift, A DESCRIPTION OF THE MORNING, page 883

This slice of eighteenth-century London life seems replete with human failings: Betty (a conventional name for a servant) sleeping with her master and trying to hide the evidence, prisoners released from jail in order to steal. Swift's couplets describe not the highborn but the common people, for whom a hackney coach heralded dawn in place of mythology's grander chariot driven across the sky by Phoebus Apollo. Although Swift crams his lines with images of city dirt and human corruption, the humor of his poem implies considerable affection for London's streets and sinners. If students see no humor in his view, let them compare this poem with another poem about eighteenth-century streets, Blake's angry "London" (page 562), or a rhapsodic, Romantic description of a London morning, Wordsworth's "Composed upon Westminster Bridge" (page 896).

(text pages) 884–888

Alfred, Lord Tennyson, DARK HOUSE, BY WHICH ONCE MORE I STAND, page 884

In Memoriam, section 7. "This is great poetry," wrote T. S. Eliot, "economical of words, a universal emotion related to a particular place; and it gives me the shudder that I fail to get from anything in *Maud*" (Introduction to *Poems of Tennyson* [London: Nelson, 1936]). The dark house was indeed a particular place—"67, Wimpole Street," as Tennyson noted— the house of Henry Hallam. The poem contains at least two allusions, whether or not we are expected to pick them up: "And then it started, like a guilty thing" (Horatio describing the ghost in *Hamlet,* I, i, 148); and "He is not here, but is risen" (Luke 24:6). In line 11 of one manuscript version, Tennyson wrote *dripping* instead of *drizzling.* Why is *drizzling* superior? The highest moment in the poem occurs in the last line in the two spondees, at least equal in their effect to Yeats's "And the *white breast* of the *dim sea*" ("Who Goes with Fergus?" page 623).

For some of these notes we are indebted to Christopher Ricks's matchless edition of *The Poems of Tennyson* (New York: Norton, 1969).

Alfred, Lord Tennyson, ULYSSES, page 885

The following inadequate precis, meant to make lovers of Tennyson's poem irate, might be quoted to students to see whether they agree with it: A hardy old futzer can't stand life in the old folks' home, and calls on his cronies to join him in an escape, even though the whole lot of them are going to break their necks.

For criticism, see Paul F. Baum, *Tennyson Sixty Years After* (Chapel Hill: U of North Carolina P, 1948) 92–94; and John Pettigrew, "Tennyson's 'Ulysses': A Reconciliation of Opposites," *Victorian Poetry* 1 (Jan. 1963): 27–45.

Dylan Thomas, FERN HILL, page 886

Fern Hill is the farm of Thomas's aunt, Ann Jones, with whom he spent boyhood holidays. In line 2 the poet cites a favorite saying of his father's, "Happy as the grass is green." The saying is echoed again in line 38. As students may notice, Thomas likes to play upon familiar phrases and transform them, as in line 7, "once *below* [not *upon*] a time."

It came as a great shock when we first realized that this poem, which XJK had thought a quite spontaneous burst of lyric energy, is shaped into a silhouette, and that the poet contrived its form by counting syllables. Such laborious working methods were customary for Thomas. John Malcolm Brinnin has recalled seeing more than 200 separate and distinct versions of "Fern Hill"—a fact worth conveying to students who think poets simply overflow.

We take the closing line to express Thomas's view of his own poetry, lyrical and rulebound at the same time: a song uttered in chains. Of course, the last line also means that the boy in the poem was held in chains by Time, the villain, who informs the whole poem (except for stanzas 3 and 4, which see childhood as Eden). Students may be asked to trace all the mentions of Time throughout the poem, then to sum up the poet's theme. William York Tindall, who offers a line-by-line commentary, makes a fine distinction: "Not how it feels to be young, the theme of 'Fern Hill' is how it feels to have been young" (*A Reader's Guide to Dylan Thomas* [New York: Noonday, 1962]). And we'd add, "how it would have felt to grow old, if the boy had realized he wouldn't live forever."

According to Tindall (in a lecture), Thomas used to grow huffy whenever asked if he were an admirer of Gerard Manley Hopkins. Still, to hear aloud both "Fern Hill" and Hopkins's "Pied Beauty" (page 576) is to notice much similarity of sound and imagery. Hopkins studied Welsh for a time, while Thomas never did learn the language; but both at least knew of ancient Welsh poetry and its ingeniously woven sound patterns.

Thomas's magnficent (or, some would say, magnificently hammy) reading of this poem can be heard on Caedmon recording TC 1002, cassette 51002, compact disk Z1002. The recording, *A Child's Christmas in Wales and other poems,* also contains "Do Not Go Gentle into That Good Night."

(text pages) 888–891

John Updike, EX-BASKETBALL PLAYER, page 888

Updike's ex-basketball player suffers the fate that Housman's athlete escapes by dying young. Flick Webb has to live on, unsung, in "fields where glory does not stay." The man whose "hands were like wild birds" now uses those hands to pump gas, check oil, and change flat tires. "Once in a while, / As a gag, he dribbles an inner tube." In his spare time, he sits in Mae's luncheonette and "just nods / Beyond her face toward bright applauding tiers / Of Necco Wafers, Nibs, / and Juju Beads." (Are today's students still familiar with those brand names?)

Updike's light tone does not obscure the pathos of Flick's situation. (Students might be asked if they know anyone like Flick Webb.) Though Updike has written notable light verse, he says of this early poem, his second to be accepted by *The New Yorker*, that it "is 'serious' and has enjoyed a healthy anthology life, though its second stanza now reads strangely to students. . . . That is, they have never seen glass-headed pumps, or gas stations with a medley of brands of gasoline, or the word *Esso*" (foreword to a new edition of Updike's first book, *The Carpentered Hen* [New York: Knopf, 1982]).

See how quickly your class can identify the poem's form as blank verse.

Edmund Waller, GO, LOVELY ROSE, page 889

In some ways quieter than Marvell's "To His Coy Mistress" (page 846) or Herrick's "To the Virgins, to Make Much of Time" (page 825), this poem has the same theme: *carpe diem*. "Go, Lovely Rose" merits admiration for its seemingly effortless grace and for the sudden, gently shocking focus on our mortality in the poem's final stanza.

Students may enjoy reading Ezra Pound's imitative tribute to Waller: the "Envoi" to *Hugh Selwyn Mauberley*, beginning "Go, dumb-born book . . . ," in *Personae*, Pound's collected shorter poems (New York: New Directions, 1949).

Walt Whitman, I SAW IN LOUISIANA A LIVE-OAK GROWING, page 890

Whitman often regards some other living thing and sees himself reflected in it. In "Live-Oak" (one of the *Calamus* poems), the tree becomes his mirror in line 4; and one might expect the poem, like "A Noiseless Patient spider" (page 730), to extend the comparison. But the poem takes a surprising twist: Whitman himself cannot abide the oak's solitude. (This poem has not been shown to refer to any particular friends or events in the poet's life.)

Pablo Neruda's tribute to Whitman may well be applied:

> There are many kinds of greatness, but let me say (though I be a poet of the Spanish tongue) that Walt Whitman has taught me more than Spain's Cervantes: in Walt Whitman's work one never finds the ignorant being humbled, nor is the human condition ever found offended (qtd. in Gay Wilson Allen in *Poetry Pilot* [Nov. 1976]).

Richard Wilbur, TRANSIT, page 891

In this skillfully metrical poem, Wilbur chooses a subject that might appear simple: the passage of a beautiful woman as she steps from her door, pulls on her gloves, and proceeds down her walk. But the poet's observations are neither simple nor simpleminded. There's a staggering hyperbole in the bit about the sun—a metaphysical conceit that recalls Marvell's "To His Coy Mistress." Why is stanza 2 a rueful question? It seems almost an attempt to mask how bowled over by her beauty the speaker is. His admiration bursts forth again in the startling image of the whip in the last line.

Richard Wilbur, THE WRITER, page 891

A searching criticism of Wilbur's work, and this poem, is offered by Andrew Hudgins in a review of recent poetry (*Hudson Review*, Winter 1989). Sometimes Wilbur implies that it is possible to master the world and its complicated problems in much the same way that a poet, in a successful poem, masters the language—but it isn't, of course. Wilbur thus places himself in a dilemma, one he is aware of. Hudgins summarizes "The Writer" and interprets it:

> ... Hearing his daughter as she types a story in her room, he compares the house to a ship and the sound of the typewriter keys to "a chain hauled over a gunwale," while the "stuff" of his daughter's life is "a great cargo—and some of it heavy." Then, rather glibly, he wishes her a "lucky passage." As soon as he's completed the metaphor, however, he rejects the "easy figure" because he remembers how difficult the life of a writer can be. The next metaphor he advances is embedded in the anecdote of a "dazed starling" that once became trapped in the same room his daughter is now working in.... Though the poem is touching and even powerful, the implied final metaphor, and the ending of the poem, while infinitely better than the rejected first metaphor of the ship, still have a bit of its premeditated neatness about them.

Whether or not the poem is autobiography, Wilbur does have a daughter, Ellen Wilbur, now a widely published fiction writer, author of *Wind & Birds & Human Voices*, a collection of short stories (Stuart Wright, 1984; NAL paperback 1985).

Nancy Willard, MARRIAGE AMULET, page 892

In this brief love poem, metaphors express the joy that marriage brings. The speaker credits her spouse with "polishing [her] like old wood." She anticipates that eventually she "will gleam like ancient wood." In her state of wedded bliss, even aging is "warm as fleece." She wears marriage as an *amulet*—a term that may need defining—a charm against evil or injury. In the last line she playfully imagines it powerful enough to perform the same function for a sage who "will keep us in his hand for peace."

Nancy Willard is known not only for her poetry and stories but for her novel *Things Invisible to See* (New York: Knopf, 1985) and several children's books including *William Blake's Inn*. She teaches at Vassar.

Miller Williams, MECANIC ON DUTY AT ALL TIMES, page 893

A simple question: What's wrong with the title? (The sign-painter's spelling.)

Elizabeth Bishop's "Filling Station" (page 785) calls attention to the filling station itself, and to the people who run it. But in Williams's poem, the filling station proprietor merits a quick two lines. The last four lines emphasize the vast empty spaces that surround the station. The trucks are "wailing through their tires." One of the signs squeaking in the wind, we assume, says "MECANIC ON DUTY AT ALL TIMES." Like Bishop, Williams uses imagery to draw a vivid scene.

Williams is concerned mainly to show us the customer and his family. A little drama is enacted. The driver's "slow hands," moving along his belt "as if they had no part / in whatever happened here," betray his anxiety. Clearly the cost of this repair is vital to him. The "mecanic" seems a good sort even though the "license plate was another state and year." Still, the customer must feel uncomfortably at his mercy. The poem is like a brief movie shot, a compassionate slice of life.

Miller Williams recently published *Living on the Surface: New and Selected Poems* (Baton Rouge: Louisiana State UP, 1989). He lives in Fayetteville, where he teaches and directs the University of Arkansas Press.

William Carlos Williams, SPRING AND ALL, page 893

QUESTIONS FOR DISCUSSION:

1. Why cannot Williams's attitude toward spring be called "poetic" and "conventional"? What *is* his attitude toward the approaching season? By what means is it indicated?

Consider especially lines 14–15 and 24–25, and the suggestion of *contagious* in the opening line. (Spring is stealing over the land as a contagious disease infects a victim. But spring is not a disease: it has a "stark dignity.")

2. An opinion: "This poem clearly draws from the poet's experience as a pediatrician who had attended hundreds of newborns, and whose work was often to describe with clinical exactness the symptoms of his patients." Discuss. (Lines 16–18 especially seem to contain a metaphor of newborn infants. The adjectives *mottled, dried, sluggish* could occur in a physician's report. In lines 9–13 also, the description of bushes, trees, and vines seems painstakingly exact in its detail.)

Recalling his life as writer and physician in an article for a popular magazine, Williams once told how poems would come to him while driving on his daily rounds. "When the phrasing of a passage suddenly hits me, knowing how quickly such things are lost, I find myself at the side of the road frantically searching in my medical bag for a prescription blank" ("Seventy Years Deep," *Holiday* Nov. 1954: 78). "By the road to the contagious hospital" was one such poem, originally recorded on prescription blanks (Roy Miki, "Driving and Writing," in *William Carlos Williams: Man and Poet*, ed. Carroll F. Terrell [Orono: National Poetry Foundation, 1983] 113).

Scholars have speculated that the brief lines of many of Williams's poems may have been decreed by the narrow width of a prescription blank, but we don't buy that guess. Had he wanted longer lines Williams would have turned the blanks sideways, or composed in smaller handwriting.

William Carlos Williams, To Waken an Old Lady, page 895

QUESTIONS FOR DISCUSSION:

1. By which words or phrases does Williams suggest the physical ravages of old age? What very different connotations do the phrases *broken / seedhusks* and *shrill / piping* carry, as well as the suggestions of feeble and broken senility? (Broken husks suggest a feast, piping suggests merriment.)

2. What is the *dark wind*? Can a wind be literally dark? (No, it can't; Williams means dark in the sense of sinister or menacing. This wind is like the passage of time that buffets or punishes.)

3. What is the dictionary definition of *tempered*? What does the word mean in this poem?

Yvor Winters, At the San Francisco Airport, page 895

Students might be asked to compare the language of this poem with that of a neoclassical poem such as Dryden's "To the Memory of Mr. Oldham" (page 803). Both poems demonstrate that it is possible for a poet to write of a subject of personal concern and yet to select a diction relatively devoid of imagery, tending to be general and abstract. (For the result to be good poetry, the abstract words have to be accurate, as in these illustrations.)

Of all the terms in Winters's poem, the most tangible things named are light, metal, planes, and air. It is not to Winters's purpose here to number the streaks on any tulip; his concerns take in knowledge and passion, being and intelligence. At the outset Winters indicates that to see perfectly with one's physical eyes may be, in a sense, to see falsely and imperfectly. The glittering metal of the planes is a menacing distraction. He restates this observation in lines 16–17: "The rain of matter upon sense / Destroys me momently." It is not until the third stanza, when the poet is able to see beyond the immediate moment, that he achieves understanding. In the last line we are back to the original paradox, stated another way: to be awake in a merely physical light is not to be awake at all.

Like Winters's criticism, the poem praises will, reason, and intelligence. We admit, though, to some reservations about it. In lines 18–20, the diction becomes so abstract as to seem grandiose. We might wish to know more about the situation. Why is the girl leaving? Where is she going? But if we accept the poet's lofty tone—in which seeing one's daughter off on an airplane becomes an event as momentous as the launching of Cleopatra's barge—it seems an impertinence to ask.

In his fine study of Winters, the English poet and critic Dick Davis has written sympathetically of this poem:

> The farewell is given the deeper implication of the father's watching his daughter move out of his immediate care and influence into the world of her own life. . . . "That which you dare not shun" suggests the child's future journey through life, on which the poet cannot accompany her. The understated stoicism of the end is very moving, particularly if we recall that this was virtually Winters's last serious poem. . . . The closing image is not only a fine evocation of the father left alone, momentarily fixed in private thought, withdrawn from the airport's public glare, but it suggests too that light of the intellect which had become Winters's chief concern, the intellect which sees and understands but knows that it is cut off from the life it loves and watches. (*Wisdom and Wilderness: The Achievement of Yvor Winters* [Athens: U of Georgia P, 1983] 146–47).

This beautiful, knotty, generation-spanning poem has lasted in the *Introduction to Poetry* ever since the book's first edition. Today, as in the 1960s, its appeal to students seems to reach deep.

William Wordsworth, COMPOSED UPON WESTMINSTER BRIDGE, page 896

Imaginary conversation:

Instructor: What do you make of the title? Is this a poem composed upon the subject of a bridge, or a poem composed while standing on a bridge's sidewalk?
Student: The latter, obviously.
Instructor: How do you know?
Student: His eye is located up on the bridge. Otherwise he couldn't see with such a wide-angle lens.
Instructor: You genius! To the head of the class!

Whose is the "mighty heart"? Wordsworth is describing the city as a sleeping beauty about to awaken. Of course, the brightness of the scene is increased by the poet's being out for his stroll before a hundred thousand chimneys have begun to smoke from coal fires preparing kippers for breakfast. Charles Lamb, in a letter to Wordsworth, had chided the poet that the urban emotions must be unknown to him, so perhaps this famous sonnet is an answer to the charge.

Compare "The World Is Too Much with Us" (page 717) for a different Wordsworth attitude toward commerce; or compare Wordsworth's London of 1807 with Blake's "London" of 1794 (page 562)—practically the same city, but a different perspective. (Wordsworth up on the bridge at dawn, letting distance lend enchantment; Blake down in the city streets by night, with the chimney sweep, the teenage whore, and the maimed veteran.)

James Wright, A BLESSING, page 897

At first, students are likely to regard "A Blessing" as "a delicate poem about the kinship between men and horses," as Ralph J. Mills sees it (*Contemporary American Poetry* [New York: Random, 1965]). They will be right, of course; but to take them a step further, they can be asked what *blessing* the poem refers to, and to ponder especially its last three lines. In a sense, the image of stepping over barbed wire into an open pasture (line 7) anticipates the idea of stepping out of one's body into—what? Any paraphrase is going to be clumsy; but Wright hints at nothing less than the loneliness of every creature alive. Although they are together, the two ponies are lonely to an extreme and are apparently overjoyed to see people. By implication, maybe the speaker and his friend are lonely together as well. In lines 15–21 the speaker, to his astonishment, finds himself falling in love with one of the ponies; he sees her beauty as that of a girl. At this point, we might expect him to recoil and cry, "Good grief! what's the matter with me?"—but he persists and becomes enlightened, at least for a moment. Only his physical body, he realizes, keeps him alone and separated. What if he were to shed it? He'd bloom.

A master of open form, Wright knows how to break off a line at a moment when a pause will throw weight upon sense: "Suddenly I realize / That if I stepped out of my body I would break / Into blossom."

Maybe the best way to teach "A Blessing" is just to read it aloud, and then say nothing at all.

James Wright, Autumn Begins in Martins Ferry, Ohio, page 898

Martins Ferry is the poet's native town. "Dreaming of heroes," the speaker sits in the high school stadium, the only place in town where heroes are likely to appear. Certainly the heroes aren't the men portrayed in lines 2–4: beery, gray-faced, ruptured, worn out by their jobs in heavy industry. These are the same "proud fathers" who, ashamed of their failures (including their failures as lovers to their wives), won't go home but prefer to hang around taverns. Without fathers to supply them with hero figures, their sons set out to become heroes themselves on the football field. The season of their "suicidal" ritual is fittingly the season of the death of the year. Will they become heroes? Most likely they'll just break their necks.

Perhaps the fathers were once football heroes themselves, as George S. Lensing and Ronald Moran point out in *Four Poets and the Emotive Imagination* (Baton Rouge: Louisiana State UP, 1976), a study that discusses nearly the whole of Wright's work. "From this there is the suggestion that the futures of the current community heroes may be as bleak as the present time assuredly is for the fathers."

Did Wright mean to protest the violence of football—at least, football of the Martins Ferry kind? Not according to the poet himself, who once played on an Ohio River Valley semipro team. Although the high school games were "ritualized, formalized violence," they had positive qualities: "the expression of physical grace," "terrific aesthetic appeal." Wright's own high school produced not just lads doomed to frustration (like their fathers), but at least one football hero: Lou Groza, placekicker for the Cleveland Browns. (Wright made his remarks in an interview reprinted in *The Pure Clear Word: Essays on the Poetry of James Wright,* ed. Dave Smith [Urbana: U of Illinois P, 1982] 3–4.)

In the same critical anthology, Robert Hass sees football in the poem as a harvest ritual, which, like all good harvest rituals, celebrates sexual potency and the fruitfulness of the earth (two positive qualities apparently not conspicuous in Martins Ferry). "Even the stanzaic structure of the poem participates in the ritual. The first two stanzas separate the bodies of the men from the bodies of the women, and the third stanza gives us the boys pounding against each other, as if they could, out of their wills, effect a merging" (210).

Sir Thomas Wyatt, They flee from me that sometime did me seke, page 898

Surely Wyatt knew what he was about. Sounding the final *e*'s helps to fulfill the expectations of iambic pentameter in lines 2, 12, 15, 17, 20, and 21, lines that otherwise would seem to fall short. In other lines, however, Wyatt appears to make the rhythm deliberately swift or hesitant in order to fit the sense. Line 7 ("Busily seeking with a continual change") seems busy with extra syllables and has to be read quickly to fit the time allotted it. Such a metrical feat seems worthy of Yeats, as does line 11, in which two spondees ("loose gown," "did fall") cast great stress upon that suddenly falling garment.

What line in English love poetry, by the way, is more engaging than "Dear heart, how like you this?" And when have a lover's extended arms ever been more nicely depicted? (This line may be thrown into the teeth of anyone who thinks that, in descriptive writing, adjectives are bad things.)

Elinor Wylie, THE EAGLE AND THE MOLE, page 899

Wylie's poetry is imbued with "an enduring irony about society" and marked by "an austere lyricism that suggests a deeply pessimistic cast of mind"—to quote Sandra M. Gilbert and Susan Gubar in *The Norton Anthology of Literature by Women* (New York: Norton, 1985). These judgments seem borne out by "The Eagle and the Mole," in which the poet advises her readers to "Avoid the reeking herd, / Shun the polluted flock," and, later, "Avoid the lathered pack, / Turn from the steaming sheep." *Reeking, polluted, lathered,* and *steaming* seem haughty insults hurled at the masses of humankind. But are we to identify the poet with these sentiments? Apparently not, for the final directive—to lie down with disembodied bones—seems repulsive enough to negate the earlier advice.

Still, as Gilbert and Gubar suggest, Wylie did have a jaundiced view of society, perhaps resulting from the scandal that swirled around the poet when in 1910 she left her first husband and her son to elope with lawyer Horace Wylie, married and seventeen years older than she. For a full account of Wylie's colorful life, see Stanley Olson's biography *Elinor Wylie* (New York: Dial, 1979).

William Butler Yeats, LONG-LEGGED FLY, page 900

Is there a finer poem in praise of the beauty of solitary thought? It seems a good antidote to any excessive yen for togetherness. Question for discussion: What is the poet's attitude toward solitude? Are you won over by it? How does he try to convince us that solitary thought is beautiful? (By the metaphor of that graceful, swooping fly, riding alone in its skill.)

If students will compare the portrait of Helen of Troy given in Yeats's lines 11–20 with the one in H.D.'s "Helen," they will meet a striking contrast. Yeats makes Helen a very human little girl—a queen and yet a playful child. He also makes his Helen a bit Irish in that she practices a "tinker shuffle," picked up, it would seem, from watching those disreputable gypsy tinsmiths who plied their trade in the Irish countryside and sometimes danced for coppers in a village square.

Yeats's thinkers are three immortals from politics, mythology, and the arts, all figures of greatness in the scheme of Yeats's *A Vision*. As John Unterecker points out, Yeats attributes a strange aphrodisiac power to Michelangelo's art in another late poem, "Under Ben Bulben," in which a globe-trotting woman, having glanced at the fresco of Adam in the Sistine Chapel, finds herself "in heat" (*A Reader's Guide to William Butler Yeats* [New York: Noonday, 1959] 283).

According to A. Norman Jeffares, Yeats is probably thinking once more of Maud Gonne as a child (*A Commentary on the Collected Poems of W. B. Yeats* [Stanford UP, 1968] 498). He loves to imagine the childhood of that great object of his unrequited love, notably in "Among School Children."

Willliam Butler Yeats, CRAZY JANE TALKS WITH THE BISHOP, page 901

Piecing together a history from this Crazy Jane poem and others, John Unterecker has identified the Bishop as a divinity student who had courted Jane in his youth. She rejected him in favor of a wild, disreputable lover: Jack the journeyman. As soon as he got enough authority, the Bishop-to-be had Jack banished, but Jane has remained faithful to her lover (at least, in spirit). (See *A Reader's Guide to William Butler Yeats* [New York: Noonday, 1959].) In this poem, the Bishop's former interest in Jane has dwindled to a concern for her soul only. Or has it? Perhaps the Bishop, no doubt a handsome figure in his surplice, may be demonstrating Yeats's contention that fair needs foul. Jane is living in lonely squalor. The grave, she says, can affirm the truth that her friends are gone, for it holds many of them; and her own empty bed can affirm that Jack is gone, too. Still, she firmly renounces the Bishop and his advice.

Each word of the poem is exact. Love has *pitched* his mansion as one would pitch a tent. The next-to-last line ends in two immense puns: *sole or whole*. The Bishop thinks that soul is all that counts, but Jane knows that both soul and hole are needed. Such puns may be why Yeats declared (in a letter) that he wanted to stop writing the Crazy Jane series: "I want to exorcise that slut, Crazy Jane, whose language has become unendurable."

What does Yeats mean by the paradoxical statement in the last two lines? Perhaps (1) that a woman cannot be fulfilled and remain a virgin—that, since fair and foul are near of kin, one cannot know Love, the platonic ideal, without going through the door of the physical body; and (2) that the universe is by nature a yin/yang combination of fair and foul (or, as Yeats would have it in *A Vision*, a pair of intertwining gyres). Crazy Jane may be crazy, but in Yeats's view she is a soothsayer.

William Butler Yeats, THE MAGI, page 902

After writing a lesser poem than this—"The Dolls," in which dolls hurl resentment at a "noisy and filthy thing," a human baby—Yeats had a better idea. "I looked up one day into the blue of the sky, and suddenly imagined, as if lost in the blue of the sky, stiff figures in procession" (Yeats's note at the back of his *Collected Poems*). Like dolls, the Magi seem frozen, somewhat inhuman ("rain-beaten stones"), unfulfilled. They are apparently troubled that Christ, whose birth was a miracle, died as a man. In hopes of regaining the peace of the Nativity, they pursue a second journey.

Bestial will seem to students an odd word to apply to a stable floor, unless they catch its literal sense: "belonging to beasts." But they will also need to see that its connotations of brutality fit the poem and interact with *Calvary's turbulence*. Compare "The Magi" with the rough beast in "The Second Coming" (page 718), a poem written after Yeats had more fully worked out his notion that historical events move in a cycle of endless return. ("Leda and the Swan," page 634, can be brought in, too, if there is time for it.)

In comparing Yeats's unsatisfied wise men to Eliot's in "Journey of the Magi" (page 804) good questions to ask include, Which poet writes as a Christian? How can you tell?

30 *Lives of the Poets*

Instead of being strewn throughout the poetry section, biographical notes on poets are collected in this chapter. The intent is to make them easy to find and to keep them from interrupting the poetry. Biographies are supplied for poets represented by two or more poems.

Criticism: On Poetry

This additional anthology is designed not to give you anything more to teach, but to supplement your teaching resources. While there isn't anything you'll *need* to do about it, this anthology of criticism offers additional possibilities for paper topics and some broad, general subjects for discussion. It provides, moreover, brief texts of certain famous critical statements that some instructors have said they would like to have available. These include Plato's account of Socrates' theory of inspiration and his banishment of poets from the Republic, Aristotle on imitation, Sidney on nature, Samuel Johnson on the superiority of universal truths to tulips' streaks, Wordsworth on emotion in tranquility, Coleridge on the imagination, Shelley's view of poets as "unacknowledged legislators," Emerson on the relation of thought to form, Poe on long poems, Frost on the "sound of sense," and Eliot on personality. All seem part of the permanent baggage of received critical ideas that the reader who cares deeply for poetry will tote along.

The anthology is arranged in chronological order. Most of the extracts at the more recent end of the anthology have not yet become permanent, but are (we hope) lively critical notions that may interest students, cause them to reflect and perhaps to argue. Here are a few topics (for either writing assignments or class discussion) suggested by these selections, including the classic ones.

Plato, INSPIRATION, page 934

1. According to Socrates, what is the source of poets' inspiration? How is it possible for "the worst of poets" to sing "the best of songs"?

2. Can readers be inspired as well? (Yes, and critics such as Ion. This is the point of the metaphor, the magnet that attracts iron rings and makes magnets of them, too, early in this passage.)

Plato, SOCRATES BANISHES POETS FROM HIS IDEAL STATE, page 936

1. What is the tone of this dialogue? (In banishing poets, Socrates appears to have a smile on his lips. As the passage ends, he is trying to find a way to let them return.)

2. What distinction does Socrates draw between admissible poetry and inadmissible poetry? What danger to the Republic does he find inherent in epic and lyric verse? (The danger is that feeling and emotion will be encouraged to unseat reason. Therefore, only hymns to the gods and praises of famous men, legitimate and useful expressions of feeling, are to be allowed.)

3. Compare Socrates' view of poets with Shelley's (page 940).

Aristotle, TWO CAUSES OF POETRY, page 938

In your own words, sum up what Aristotle appears to mean by imitation. Can it be charged that he reduces poetry to journalism, poems to mere descriptions of the world? (But mere descriptions don't embody harmony and rhythm—two equally essential sources of a poem.)

Samuel Johnson, 'THE BUSINESS OF A POET', page 939

1. Do poets, in Johnson's view, appear to be Muse-inspired utterers of surprising statements not their own? (Look back once more on Socrates's remarks on inspiration.)

2. What, then, is a poet's task as Johnson sees it?

3. Write a one-paragraph review of a book of imagist poetry (including, say, Ezra Pound's "In a Station of the Metro" [page 569], H. D.'s "Heat" [page 582], and Elizabeth Bishop's "The Fish" [page 572]—this last a later poem, but owing much to imagism) as Dr. Johnson might have written it. Would he condemn such poets for numbering the streaks of the tulip instead of articulating universal ideas?

William Wordsworth, 'Emotion recollected in tranquillity', page 939

1. In Wordsworth's description of the poetic process, how does a poet usually go about writing a poem?
2. Take another look at Wordsworth's poem "I Wandered Lonely as a Cloud" (page 517). What light does this statement cast upon it? What lines in the poem seem to describe the same process of poetic composition?

Samuel Taylor Coleridge, Imagination, page 940

1. This won't be easy, but try to state in your own words Coleridge's doctrine of the imagination, as you understand it from this passage. What does the mind of a poet *do* in composing a poem? (Coleridge's doctrine is more fully set forth in the *Biographia Literaria*, chapter 13. But this concise description, without special philosophical terms, of the *synthesizing* power of the imagination will serve, perhaps, to give the beginning student the essence of it.)
2. Name a poem, by any poet, in which you find what Coleridge might call "the balance or reconcilement of discordant qualities." (Suggestion: "The Love Song of J. Alfred Prufrock," in which Eliot certainly blends unlikely, conflicting matter—visions of loveliness such as the mermaids with ratty and sordid urban imagery.)

Percy Bysshe Shelley, 'Unacknowledged legislators', page 940

How is Shelley's view of the value of poetry different from that of Socrates in the statement about banishing poets from the Republic? How does Shelley's thinking resemble that of Socrates in the remarks on inspiration?

Ralph Waldo Emerson, 'Meter-making argument', page 941

Have you read any poems in this book that appear to be the work of mere musicbox heads, like Emerson's acquaintance?

Edgar Allan Poe, 'A long poem does not exist', page 942

Do you agree or disagree? If Poe is right, should we discard *The Odyssey*, *The Divine Comedy*, and "Lycidas"? If he is wrong, then how do you account for the fact that certain long poems contain patches of deadly dullness?

Robert Frost, 'The sound of sense', page 942

1. Experiment: Let someone in the class follow Frost's instructions, go out into the hall, and try speaking these or other sentences in a soft voice through a closed door. What element (if any) can the hearers still recognize?
2. Point to sentences in Frost's poems that sound as though written according to his ideas. Any Frost poem will do.

Wallace Stevens, Proverbs, page 943

1. Explain any of these remarks that seem cryptic.
2. Make up a few proverbs (about poetry) of your own.

(text pages) 947–953

William Carlos Williams, 'The rhythm persists', page 944

Williams's own poetry is often regarded as an influential model of *vers libre,* or free-form verse. What isn't free or formless about it? How do Williams's remarks help you perceive what he is doing in his poems?

Ezra Pound, Poetry and music, page 947

Do you agree? Or are there any poems you know that aren't particularly musical but are worth reading? Give examples.

T. S. Eliot, Emotion and personality, page 945

Why does Eliot think Wordsworth (in his remarks on emotion recollected in tranquillity) is wrong? Compare these two poets' statements.

Yvor Winters, 'The fallacy of expressive form', page 945

1. Winters is reacting against critics who charge that traditional metrical verse is no good because it doesn't capture the nervous, start-and-stop pace of city traffic; because it doesn't reflect the sprawling formlessness of our rapidly changing society. What is his defense of formal poetry?

2. How do you suppose Winters would react to Dr. Williams's remarks about the crab and the box?

Randall Jarrell, On the charge that modern poetry is obscure, page 946

1. To the reader who protests that modern poetry is obscure, Jarrell replies, "That's not the reason you don't read it." What does he imply the reason actually is?

2. Compare Jarrell's remarks on obscurity in poetry to Coleridge's. Isn't it possible, for reasons given by Coleridge, that certain contemporary poems may indeed be obscure?

Sylvia Plath, 'The magic mountains', page 946

1. What discovery does Plath say she has made? How does it change her attitude toward writing a poem?

2. Compare her discovery with the contention of William Carlos Williams, "No ideas but in things."

Robert Scholes, 'How do we make a poem?', page 947

1. What poems have you read that are elegies? How are they different from Merwin's "Elegy"?

2. In his last paragraph Scholes spells out some thinking that a reader might do after reading this poem. What other thoughts or insights occur to you?

   (If you care to read Scholes's discussion in greater depth, see his *Semiotics and Interpretation* [New Haven: Yale UP, 1982].)

Sandra M. Gilbert and Susan Gubar, The Freedom of Emily Dickinson, page 949

1. How do these two critics interpret the poet's retreat from the world? What was "freedom" for Emily Dickinson?

2. What do Gilbert and Gubar mean by their observation that the poet contrived "a female-centered theology"? What evidence do they give for it?

3. How did Dickinson "confront her society" (in the view of Adrienne Rich)?

DRAMA

32 Reading a Play

In many parts of the country, students rarely if ever see plays other than school or other amateur productions, and the instructor may encounter some resistance to the whole idea of studying drama. But all students are steeped in film and television drama, and it may be useful to point out that such drama begins with playscripts. One might reason somewhat like this. Movies and television, it's true, give plays hard competition in our society, and a camera *does* have advantages. In moments, film can present whole panoramas and can show details in close-up that theaters (with their cumbersome sets and machinery) cannot duplicate. Movies used to be called "photoplays," but the name implies an unnecessary limitation, for there is no point in confining the camera to recording the contents of a picture-frame stage. But a play—whether staged in a proscenium theater or in a parking lot—has its own distinct advantages. It is a medium that makes possible things a camera cannot do. Unlike movies and television, a play gives us living actors, and it involves living audiences who supply it with their presences (and who can move one another to laughter or to tears). Compared, say, to the laughter of live spectators at a comedy, the "canned" laughter often dubbed into television programs is a weak attempt to persuade television viewers that they are not alone.

A Play in its Elements

Susan Glaspell
TRIFLES, page 956

The recent comeback of *Trifles* may be due, we think, not only to Glaspell's pioneering feminist views but also to its being such a gripping, tightly structured play. Several instructors have asked for its inclusion in *Literature*. Whether or not you have much time to spend on the elements of a play, we think you will find *Trifles* worth teaching; students respond to it.

Topic for writing or discussion: What common theme or themes do you find in both *Trifles* and *Antigone*? (A conflict between the law and a woman's personal duty.)

The Provincetown Players, who performed in a theater on an abandoned wharf, had a fertile summer in 1916. Besides *Trifles*, with Glaspell herself playing Mrs. Hale, their season included the first Eugene O'Neill play to be produced, *Bound East for Cardiff*. Glaspell has said that she derived the plot of *Trifles* from a murder case she had investigated as a reporter in Des Moines.

Tragedy and Comedy

John Millington Synge
RIDERS TO THE SEA, page 972

The narrative of Synge's play is gaunt in its simplicity. At the start, old Maurya has already lost six men of her family to the sea: her husband, father-in-law, and four sons. In

(text pages) 972–983

the course of the play, suspicions are confirmed: a man reported drowned is indeed Michael, son number five. At the end, Maurya has lost Bartley, her sixth and last son, and the sea can do no more.

If you use the play to illustrate the elements of drama, you might point out the *exposition:* the early conversations of the girls and their mother, informing us that Michael is missing, feared drowned, and that Maurya dreads Bartley's going on a journey from which he will never return. While the *major dramatic question* seems plain—Will Bartley survive?— there is also a minor question: Was the missing Michael the man drowned in Donnegal? From the beginning we are given to suspect that he was, and this question is definitively answered in mid-play. The *crisis* occurs when Cathleen gives her mother the bundle and confirms that Michael's body has been found. The *climax* follows almost immediately—two blasts of emotion in a row!—with the bringing in of the dead Bartley. The *resolution,* we would say, is Maurya's rise to nobility. And the *theme?* That the sea is merciless, or perhaps more accurately (since after all, the islanders derive their living from the sea), the sea indifferently gives and takes away. Or it might be argued that Maurya sums up the theme in her memorable closing line. Does the play have a *protagonist?* Some may prefer to see Maurya instead as a central character. She is not a protagonist in the sense that she causes things to happen—she can't even prevent Bartley's journey—unless we regard what happens in this play as mainly what happens inside Maurya herself: her attaining a generous compassion for all humankind.

Riders to the Sea does not fit the mold of classic Greek tragedy, as Aristotle defined it, for its central character is a peasant, not a person of high estate, and she does not bring about her own downfall. Still, unquestionably the play has the tragic spirit. Like *Oedipus the King* and *Antigone,* it shows the central character facing and accepting the inscrutable workings of the universe, finally rising to serene dignity. Maurya may begin as quarrelsome and complaining, but in the end she becomes as noble as a queen. Beholding these events, hearing the play's wonderful language, the spectator is stirred, perhaps overwhelmed, but far from depressed. And Synge's play is undoubtedly (as Aristotle would expect of a tragedy) serious, complete in itself, of a certain magnitude, and written in a language embellished with artistry. It even has a chorus: the band of keening old women. But you probably won't want to wrestle with how it is similar or dissimilar to a Greek tragedy until your students have read Sophocles. Then, Synge's play will surely reward another glance.

The world of this play—intense, stark, informed with the language of poetry—seems to turn into a symbol virtually every ordinary object it contains. Even the rope that hangs from a nail hints of death (death by hanging). On one level, Cathleen's halting her spinning wheel merely indicates her sudden fear that Michael has drowned. On another level, her spinning, a routine of daily life, is (like that life itself) suddenly interrupted and shattered by death, or the fear of it. On still another level, a spinning wheel is associated with the famous three mythological Fates, who spin, measure, and cut off the thread of life. In abruptly ceasing to spin, perhaps, Cathleen symbolically breaks Michael's thread. Definite in its hints, the sea-wind that blows open the door, like the sea itself, is hostile, powerful, and irresistible.

Other suggestive objects invite symbol-hunting: the drowned man's clothes, for instance, tied into a sea-soaked bundle with a tight, unopenable black knot—like a terrible, impenetrable secret to be hidden from the old mother. Then, too, there's the bread that Maurya fails to convey to Bartley before his death: life-sustaining food of no use to a dead man, an undelivered good like the blessing that the old mother fails to impart. What of the fine white coffin-boards that ironically cannot be held together for lack of nails? Perhaps Maurya's forgetting them suggests her unconscious wish *not* to coffin another son. Boards are harbingers of death, for Bartley's body arrives on a plank. Suggestive, too, are the red mare that Bartley rides—red, the color of blood?—and the gray pony that knocks the young man into the sea, the very steed on which Maurya in her vision or waking dream saw the dead Michael ride. Maurya herself, it seems, has beheld riders to the sea—like Egyptians about to be drowned by the wrath of God. (For this Biblical allusion, see in the text the footnote on the title of the play.) A more obvious symbol is the empty, inverted cup that Maurya places on the table with a gesture of finality, as though to signify that all is gone, that the last drop of life (or of suffering) has been drained.

Maurya's vision of her two sons presages the play's resolution. Bartley, to be sure, might not have been a vision—actually, he may have been riding that red mare, and perhaps Maurya did try (and fail) to speak with him. When he passes her and she is unable to utter a sound and fails to give him the bread and her blessing, his doom is sealed. The apparition of Michael riding that fateful gray pony seems a clear foreshadowing—as perhaps Cathleen knows when she cries, "It's destroyed we are from this day." Bartley's death is foreshadowed from early in the play when Maurya hopes the priest will stop him from going on his errand.

Presumably the young priest isn't about to stop Bartley from taking a reasonable risk and making a needed horse sale. But the priest turns out to be a bad prophet, quite mistaken in his confidence that the Lord would not claim an old woman's last remaining son. Maurya never believed the priest's assurances: "It's little the like of him knows of the sea." (We get the impression that the characters in this play are faithful Christians living in an inscrutably pagan universe, or else one run by some cruel Manichean sub-deity.) No one but Maurya strongly feels that Bartley shouldn't take the horses to the Galway fair. The young assume that life must go on, for as Cathleen declares, "It's the life of a young man to be going on the sea." But the old mother has seen enough men claimed by the sea to fear that it is ready to claim one more.

Although Maurya has suffered the trauma of having her worst fears confirmed within minutes—learning of the death of one son and beholding the body of another—and although the sea has robbed her of a husband and six sons, she is not "broken," as Cathleen thinks, but is strong and rises to tragic serenity. In her moving speeches at the play's end, she goes beyond suffering to accept the nature of life and to take a long view of the mortality of all humankind.

For discussion: James Joyce's remark that the play suffers in that disaster is worked by a pony, not by the sea. (This seems to us a quibble: the pony may kick Bartley into the sea, but the sea conveys him out to the rocks and finishes him off.)

Riders to the Sea invites close comparison with Lady Gregory's *The Workhouse Ward*: a tragedy and a comedy, both written in rich rural Irish English, both dating from the same time and produced in the same place—those few years when Dublin's Abbey Theatre was reaching its height.

Lady Gregory
THE WORKHOUSE WARD, page 983

If you wish to use *The Workhouse Ward* to teach the elements of a play, instead of *Trifles*, it will lend itself admirably to the purpose. Its plot seems one with its characters. What happens from opening to outcome follows from the kind of person Mike McInerney is—and, to a lesser extent, from the natures of Michael Miskell and Honor Donohoe. If the play has a protagonist, it is Mike, the central character who, more than the others, has the responsibility to act, to choose, to decide. As the playwright takes pains to show, Mike is impulsive. Easily kindled to anger, he is also deeply loyal, as we can tell from his hesitation when Michael Miskell pleads with him not to leave, and from his final choice.

For a play so short, *The Workhouse Ward* devotes a large part of its time to exposition: from opening line until the entrance of Mrs. Donohoe. This exposition seems well justified, for the play requires that we understand and sympathetically accept the two leading characters. We need to know that the feud between Mike and Michael is no fleeting spat but a fight that has lasted for decades. And because we know this, we realize later in the play when Mike begs his sister to make a home for Michael too, that there is a humorous contrast between Mike's verbal abuse of his crony and his profound attachment to him—that, belying his words, Mike deeply enjoys this perpetual fight and doesn't want to be done with it.

And so, carefully and thoroughly, the playwright acquaints us with the pair and the longtime bond between them. With the unexpected arrival of Mike's sister, this exposition ends and developments move forward rapidly. Passing from the verbal battles between the two old paupers to a dynamic scene involving the three principals, the playwright introduces a subtler conflict: the moral struggle within Mike. Will he or won't he go away and leave his lifelong friend and enemy? This is the play's major dramatic question. When Mike starts trying on his new clothes and imagining an idyllic new life ("Sport and pleasure, and music on the strings!"), it would seem that his departure is imminent. But there are negative foreshadowings, hints of the outcome, in Honor Donohoe's doubts about taking her brother home with her, in her complaint that she hadn't known he was bedridden. When she flatly refuses to take Michael Miskell along, the plot gains a complication, a fresh obstacle. The climax occurs in Mike's final stand: "Bring the both of us with you or I will not stir out of this." The resolution, or conclusion, ensues swiftly as Honor departs, condemning Mike to end his days in the workhouse. This resolution doesn't end the plot, for we are still treated to the closing battle, an inspired piece of stage business.

The suit of new clothes that Honor brings her brother has symbolic hints. Trying on coat and hat, Mike decks himself in dreams of a new life, but when this hope vanishes, Honor gathers up the clothes again and makes off with them. Another symbol may be Mike's pipe, which he offers to Michael, the one valuable object he owns and the thing most dear to him. Students may need to be alerted that symbols in drama may be large and portentous, like the plague that has befallen the land in *Oedipus the King*, or small and unassuming, like those in *The Workhouse Ward* (and the dead canary in *Trifles*). In a note on the play, Lady Gregory remarked, "I sometimes think the two scolding paupers are a symbol of ourselves in Ireland" (*Seven Short Plays* [New York: Putnam, 1909]). But Elizabeth Coxhead doubts that the author of the play deliberately set out to personify Belfast and Dublin in her two paupers. "It is because they are sharply real and individualized on their own ground, that they are true on a national and indeed a universal level, and that they will go on throwing their pillows at each other to the end of time." (See Coxhead's discussion in *Lady Gregory*, 2nd ed. [London: Secker and Warburg, 1966]).

This play is laden with conflict, but as the playwright touchingly demonstrates, there is more to drama than conflict alone. Although *The Workhouse Ward* is mainly the story of the lifelong battle between Mike and Michael, the fleeting visit of Honor Donohoe reveals something more: the depth of the two men's mutual bond. What is the theme? Surely it has to do with human love, people's need for one another, fear of loneliness—perhaps too, the difficulty of changing one's whole life in old age.

A bit of background: On a visit to Gort Workhouse, Lady Gregory had heard of two old inmates whose story inspired *The Workhouse Ward*. She had planned to make the incident into a play of her own, but Yeats, her fellow director of the Abbey Theatre, decreed that at the moment the company needed a play in Gaelic. So, generously, she wrote an outline of the story and handed it over to the fluent Douglas Hyde. The result was his *Tigh na mBocht*, translated into English as *The Poorhouse* and produced in 1907, with Hyde and Lady Gregory given as coauthors. Unable to forget the story, Lady Gregory later took back her plot and with Hyde's consent pared its five characters down to three and rewrote it into Kiltartanese, the dialect of her native region, noted for its colorful extravagance. The result was tighter, livelier, and immensely funnier. At the end of the original *Poorhouse*, the two old men merely "threaten one another with their pillows." For anyone interested in seeing what else Lady Gregory's revision transformed, *The Poorhouse* is available in the excellent Coole Edition of *The Plays of Lady Gregory*, IV, ed. Ann Saddlemyer (Gerralds Cross: Colin Smythe, 1971).

Lady Gregory's reputation as a poet and playwright has long been obscured by the imposing reputation of Yeats, her most prominent associate. That she had genius in her own right is attested by her many plays, among them her wonderful rendering into Kiltartan dialect of four plays of Molière (in the same Coole Edition volume cited above). Hazard Adams offers a concise account of her life and works in *Lady Gregory* (Lewisburg: Bucknell UP, 1973). Mary Lou Kohfeldt's *Lady Gregory: The Woman Behind the Irish Renaissance* (London: Deutsch, 1986) is a detailed biography.

W. C. Fields
STOLEN BONDS, page 990

It may occur to some students to wonder how this sketch could evolve into a movie titled *The Fatal Glass of Beer* when it contains no mention at all of beer. Equally puzzling is the stage direction that calls for "one trick spoon (for Snavely)." No doubt you had to be there to find out what Fields did with that spoon. Perhaps it was a soluble joke-goods spoon that, on contact with the hot soup, instantly melted.

In translating his ten-minute sketch to the screen, Fields had to lengthen it to a twenty-minute two-reeler. He added, among other things, the fatal glass: in the opening scene, Fields, playing a dulcimer with heavy gloves on, renders a sad ballad of a young man (played by the actor who plays Chester) who goes to the big city, takes one sip of beer, and staggers out of the barroom with delirium tremens. He also added a few sight gags: using a whole loaf of French bread as a soup spoon, getting a bucket stuck on his foot, working a pump that gushes ice cubes. The original script for *Stolen Bonds* is followed with only minor changes, except that the line "I think I'll go out and lock up the cow" is greatly improved: "I think I'll go out and milk the elk." And in the end, instead of assaulting Chester with French bread and nugget, Ma and Pa smash crockery over Chester's head and then fling his inert body out-of-doors.

Stereotypes of the most obvious sort, Chief Big Spear and Little Small Blanket have no function at all unless to supply local color. Native Americans shouldn't be insulted by this: the other three characters are made of cardboard too. Fields, who despised sentimentality, mercilessly takes off on the clichés of melodrama: saving the mortgaged homestead, old parents weeping over the prodigal son's return.

Yet who cannot love this zany play, so rich in low comedy? The humor flows mostly from three sources. There are verbal gags ("that pill from Medicine Hat"); sight gags (the twin lights that respond to the wrong switches, the snow that keeps blowing in the door); and expectations set up only to be suddenly (and surprisingly) frustrated, as when, with sentimental affection, Mrs. Snavely gives her long-lost, freezing son a bowl of soup, only to have Snavely reach for it himself, saying, "S'my soup, Ma." Another example:

> Snavely: He won't take old Balto, my lead dog!
> Mrs. Snavely: Why not, Pa?
> Snavely: Because I 'et him. He was mighty good with mustard.

And then of course there's the double surprise at the end:

> Mrs. Snavely: My God! Do you want to kill him?
> Snavely: Yes, I do!
> Mrs. Snavely: [*Handing him the nugget*] Then hit him with this!

No doubt students will find other examples as well. In its closing moments, *Stolen Bonds* degenerates, or evolves, into slapstick pure and simple.

If you have access to a TV, a VCR, and a copy of *The Best of W. C. Fields* (Spotlite Video, VHS V7058), a common item in video rental shops, *The Fatal Glass of Beer* will take 20 minutes to show, plus a couple of minutes to fast-forward through two other two-

(text pages) 990–995

reelers on the tape before it (*The Golf Specialist* and *The Dentist*). You'll find enough to discuss for the rest of a class hour. Perhaps the suggested writing topic at the end of this chapter (to compare the two versions and discuss what Fields altered) might appeal to someone interested in filmmaking.

Here are a few other possible writing topics:

1. Compare and contrast the humor in the Fields sketch with that in Lady Gregory's *The Workhouse Ward*. Do you find similar elements? In what important ways do the two plays differ in their approach to comedy?

2. Watch two or three old W. C. Fields films on television or VCR. Then, in 250 words or more, try to characterize Fields's humor.

3. Write a comic sketch of your own, giving yourself the main role, and perform it for the class. Then preside over a class discussion on the problems of writing and producing a comedy.

33 The Theater of Sophocles

Sophocles
OEDIPUS THE KING, page 999

One problem in teaching this masterpiece is that students often want to see Oedipus as a pitiable fool, helplessly crushed by the gods, thus stripping him of heroism and tragic dignity. (A classic bepiddlement of the play once turned up on a freshman paper: "At the end, Oedipus goes off blinded into exile, but that's the way the cookie crumbles.") It can be argued that Oedipus showed himself to be no fool in solving the riddle of the Sphinx or in deciding to leave Corinth; that no god forced him to kill Laius or to marry Jocasta.

Another problem in teaching this play is that some students want to make Oedipus into an Everyman, an abstract figure representing all humanity. But Oedipus's circumstances are, to say the least, novel and individual. "Oedipus is not 'man,' but Oedipus," as S. M. Adams argues in *Sophocles the Playwright* (Toronto: U of Toronto P, 1957). On the other hand, Freud's reading of the play *does* suggest that Oedipus is Everyman—or, better, that every man is Oedipus and like Oedipus wishes to kill his father and marry his mother. A passage from Freud's celebrated remarks about the play is given on page 1438.

Despite Freud's views, which usually fascinate students, critical consensus appears to be that Oedipus himself did not have an Oedipus complex. Sophocles does not portray Oedipus and Jocasta as impassioned lovers; their marriage was (as Philip Wheelwright says) "a matter of civic duty: having rid the Thebans of the baleful Sphinx by answering her riddle correctly, he received the throne of Thebes and the widowed queen as his due reward" (*The Burning Fountain* [Bloomington: Indiana UP, 1954]). Wheelwright also notes, incidentally, that the title *Oedipus Tyrannus* might be translated more accurately as "Oedipus the Usurper"—a usurper being (to the Greeks) anyone who gains a throne by means other than by blood succession. Actually, of course, Oedipus had a hereditary right to the throne. (Another interpretation of the play sees Laius and Jocasta as having incurred the original guilt: by leaving a royal prince to die in the wilderness, they defied natural order and the will of the gods.)

For the nonspecialist, a convenient gathering of views will be found in *Oedipus Tyrannus*, ed. Luci Berkowitz and Theodore F. Brunner (New York: Norton, 1970). Along with a prose translation of the play by the editors, the book includes the classic comments by Aristotle, Nietzsche, and Freud, and discussions by recent critics and psychologists. Seth Bernardete offers a detailed, passage-by-passage commentary in *Sophocles: A Collection of Critical Essays*, ed. Thomas Woodard (Englewood Cliffs: Prentice, 1966). Francis Fergusson has pointed out that the play may be read (on one level) as a murder mystery: "Oedipus takes the role of District Attorney; and when he at last convicts himself, we have a twist, a *coup de théâtre*, of unparalleled excitement." But Fergusson distrusts any reading so literal, and questions attempts to make the play entirely coherent and rational. Sophocles "preserves the ultimate mystery by focusing upon [Oedipus] at a level beneath, or prior to any rationalization whatever" (*The Idea of a Theatre* [Princeton: Princeton UP, 1949]). Refreshing, after you read many myth critics, is A. J. A. Waldock's *Sophocles the Dramatist* (Cambridge: Cambridge UP, 1951; reprinted in part by Berkowitz and Brunner). According to Waldock, the play is sheer entertainment, a spectacular piece of shock, containing no message. "There is no meaning in the *Oedipus Tyrannus*. There is merely the terror of coincidence, and then, at the end of it all, our impression of man's power to suffer, and of his greatness because of this power." Pointing out how little we know of Sophocles' religion, Waldock finds the dramatist's beliefs "meagre in number and depressingly commonplace."

(text pages) 999–1042

Although the religious assumptions of the play may not be surprising to Waldock, students may want to have them stated. A good summing-up is that of E. R. Dodds, who maintains that Sophocles did not always believe that the gods are in any human sense "just"; but that he did always believe that the gods exist and that man should revere them ("On Misunderstanding the Oedipus Rex," *Greece and Rome* [Oxford: Oxford UP, 1966] Vol. 13). This comment is now included in the book ("Criticism: On Drama," page 1436).

"Possibly the best service the critic can render the *Oedipus Rex*," says Waldock, "is to leave it alone." If, however, other criticism can help, there are especially valuable discussions in H. D. F. Kitto, *Greek Tragedy*, 3rd ed. (London: Methuen, 1961), and *Poiesis* (Berkeley: U of California P, 1966); Richmond Lattimore, *The Poetry of Greek Tragedy* (Baltimore: Johns Hopkins UP, 1958); and Patrick Mullahy, *Oedipus, Myth and Complex* (New York: Grove, 1948).

In the "Suggestions for Writing" at the end of the chapter (page 1042), there is one especially challenging topic (number 3): to compare translations of the play. For any student willing to pick up the challenge, we think this topic might produce a great term paper. The differences between versions, of course, are considerable. Sheppard's rendition, or Kitto's, is more nearly literal than that of Fitts and Fitzgerald and much more so than that of Berg and Clay. In the latter team's version of 1978, the persons of the tragedy all speak like formally open lyrics in current little magazines. Lots of monosyllables. Frequent pauses. Understatement. Lush imagery. Berg and Clay perform this service brilliantly, and it might be argued: why shouldn't each generation remake the classics in its own tongue? We prefer the versions of David Grene and Elizabeth Wyckoff for clarity, speakability, and (we're told) scholarly accuracy.

Still impressive is the film *Oedipus Rex* (1957), directed by Tyrone Guthrie, a record of a performance given in Stratford, Canada. Although the theater of the play is more Stratfordian than Athenian, the actors wear splendid masks. The text is the Yeats version. The film (88 minutes long, 16 mm, in color) may be bought or rented from Contemporary/McGraw-Hill Films, 1221 Avenue of the Americas, New York, NY 10020, or from their regional distributors.

In July 1984, the Greek National Theater presented a much-discussed *Oedipus Rex* at the Kennedy Center in Washington. Bernard Knox offers an admiring account of it in *Grand Street* for Winter 1985. The director, Minos Volankis, staged the play on a "circular, dark brown plate, tilted toward the audience" and etched with a labyrinth pattern. In Volankis's version, Oedipus and Jocasta cannot see the pattern and ignore it as they move about the stage, but the chorus and Teiresias are aware of the labyrinth and respectfully trace its curves in their movements. Oedipus is a clean-shaven youth, the only young person in the play—"caught in a web spun by his elders."

34 The Theater of Shakespeare

William Shakespeare
THE TRAGEDY OF OTHELLO, page 1046

For commentary on the play, some outstanding sources of insight still include A. C. Bradley's discussion in *Shakespearean Tragedy* (1904; rpt. ed. [New York: St. Martin's, 1965]); and Harley Granville-Barker, "Preface to *Othello*," in *Prefaces to Shakespeare*, II (Princeton UP, 1947), also available separately from the same publisher (1958). See also Leo Kirschbaum, "The Modern *Othello*," *Journal of English Literary History* 2 (1944), 233–96; and Marvin Rosenberg, *The Masks of "Othello"* (Berkeley: U of California P, 1961). A convenient gathering of short studies will be found in *A Casebook on Othello*, ed. Leonard Dean (New York: Crowell, 1961). For a fresh reading of the play, see Michael Black, who in *The Literature of Fidelity* (London: Chatto, 1975) argues that the familiar view of Othello as a noble figure manipulated by the evil Iago is wrong and sentimental. According to Black, we see ourselves and our destructive impulses mirrored in both characters; hence, we are disturbed.

Lately, Lynda E. Boose has closely read the confrontation scene between Othello and Brabantio, the father of Desdemona, in front of the Duke (I, iii), and has found in it an ironic parody of the traditional giving away of the bride at a marriage ceremony. Instead of presenting his daughter to Othello as a gift, the thwarted Brabantio practically hurls her across the stage at the Moor. (The scene resembles Lear's casting away of Cordelia in *King Lear*, I, i.) In most of Shakespeare's plays, the father of the bride wants to retain and possess his daughter. Prevented by law and custom from doing so, he does the next best thing: tries to choose her husband, usually insisting on someone she does not desire. But Shakespeare, in both comedy and tragedy, always stages the old man's defeat ("The Father and the Bride in Shakespeare," *PMLA* 97 [May 1982]: 325–47).

Still another opinion that students might care to discuss: "No actress could credibly play the role of Desdemona if the character's name were changed to, say, Sally" (Frank Trippett, "The Game of the Name," *Time*, 14 Aug. 1978).

General question 3: "How essential to the play is the fact that Othello is a black man, a Moor, and not a native of Venice?" That Othello is an outsider, a stranger unfamiliar with the ways of the Venetians, makes it easier for Iago to stir up Othello's own self-doubts; and so the fact seems essential to the plot. (See especially III, iii, 201–09, 228–31, 257–67.) Venice in the Renaissance had no commerce with black Africa, but Shakespeare's many references to Othello's blackness (and Roderigo's mention of the Moor's "thick lips," I, i, 63) have suggested to some interpreters that Othello could even be a coastal African from below the Senegal. On the modern stage, Othello has been memorably played by African-American actor Paul Robeson and by Laurence Olivier, who carefully studied African-American speech and body language for his performance at the Old Vic (and in the movie version). A critic wrote of Olivier's interpretation:

> He came on smelling a rose, laughing softly with a private delight; barefooted, ankleted, black. . . . He sauntered downstage, with a loose, bare-heeled roll of the buttocks; came to rest feet splayed apart, hips lounging outward. . . . The hands hung big and graceful. The whole voice was characterized, the o's and the a's deepened, the consonants thickened with faint, guttural deliberation. "Put up yo' bright swords, or de dew will rus' dem": not quite so crude, but in that direction. It could have been caricature, an embarrassment. Instead, after the second performance, a well-known Negro actor rose in the stalls bravoing. For obviously it was done with love; with

the main purpose of substituting for the dead grandeur of the Moorish empire one modern audiences could respond to (Ronald Bryden, *The New Statesman*, 1 May 1964).

For a fascinating study of the play by a white teacher of African-American students at Howard University, see Doris Adler, "The Rhetoric of *Black* and *White* in Othello," in *Shakespeare Quarterly* 25 (Spring 1974): 248–57. Iago, Roderigo, and Brabantio hold negative and stereotyped views of black Africans which uncomfortably recall modern racial prejudices. In their view, Othello is "lascivious" (I, i, 121), an unnatural mate for a white woman (III, iii, 229–33), a practitioner of black magic (I, ii, 72–74). Under the influence of Iago's wiles, Othello so doubts himself that he almost comes to accept the stereotype forced on him, to reflect that in marrying him Desdemona has strayed from her own nature (III, iii, 227). Such, of course, is not the truth Shakespeare reveals to us, and the tragedy of Othello stems from a man's tragic inability to recognize good or evil by sight alone. "Eyes cannot see that the black Othello is not the devil," Adler observes, "or that the white and honest Iago is."

In answer to general question 4 ("Besides Desdemona and Iago, what other pairs of characters seem to strike balances?"): Alvin Kernan in his introduction to the Signet edition of *Othello* comments,

> The true and loyal soldier Cassio balances the false and traitorous soldier Iago. . . . The essential purity of Desdemona stands in contrast to the more "practical" view of chastity held by Emilia, and her view in turn is illuminated by the workaday view of sensuality held by the courtesan Bianca. . . . Iago's success in fooling Othello is but the culmination of a series of such betrayals that includes the duping of Roderigo, Brabantio, and Cassio.

On general question 5: Thomas Rymer's famous objections to the play will not be easy to refute. At least, no less a critic than T. S. Eliot once declared that he had never seen Rymer's points cogently refuted. Perhaps students will enjoy siding with the attack or coming to the play's defense.

The last general question ("Does the downfall of Othello proceed from any flaw in his nature, or is his downfall entirely the work of Iago?") is a classic (or cliché) problem, and perhaps there is no better answer than Coleridge's in his *Lectures on Shakspere*:

> Othello does not kill Desdemona in jealousy, but in a conviction forced upon him by the almost superhuman art of Iago—such a conviction as any man would and must have entertained who had believed in Iago's honesty as Othello did. We, the audience, know that Iago is a villain from the beginning; but in considering the essence of the Shaksperian Othello, we must perseveringly place ourselves in his situation, and under his circumstances. Then we shall immediately feel the fundamental difference between the solemn agony of the noble Moor, and the wretched fishing jealousies of Leontes. . . . Othello had no life but in Desdemona: the belief that she, his angel, had fallen from the heaven of her native innocence, wrought a civil war in his heart. She is his counterpart; and, like him, is almost sanctified in our eyes by her absolute unsuspiciousness, and holy entireness of love. As the curtain drops, which do we pity the most?

Was the Othello-Desdemona match a wedding of April and September? R. S. Gwynn of Lamar University writes: "Has anyone ever mentioned the age difference between Othello and Desdemona? Othello speaks of his arms as 'now some nine moons wasted.' Assuming that this metaphor means that his life is almost 9/12 spent, he would be over 50! Now if a Venetian girl would have normally married in her teens (think of the film version of *Romeo and Juliet*), that would make about 30 years' difference between him and his bride." This gulf, Othello's radically different culture, his outraged father-in-law, and Iago's sly insinuations, all throw tall obstacles before the marriage.

"If we are to read the play that Shakespeare wrote," maintains Bruce E. Miller, "we must acknowledge that Othello as well as Iago commits great evil." In *Teaching the Art of Literature* (Urbana: NCTE, 1980), Miller takes *Othello* for his illustration of teaching drama and stresses that Othello went wrong by yielding to his gross impulses. In demonstrating why the play is a classic example of tragedy, Miller takes advantage of students' previously having read Willa Cather's "Paul's Case." The latter story illustrates "the difference between sadness and tragedy. Paul's death is sad because it cuts off a life that has never been fulfilled. But it is not tragic, for Paul lives and dies in this world of human affairs." But Othello's death has the grandeur of tragedy. Realizing at last that Desdemona has been true and that in slaying her he has destroyed his own hopes of happiness, the Moor attains a final serenity of spirit, intuiting the true order of things.

35 The Modern Theater

Henrik Ibsen
A DOLL HOUSE, page 1139

Rolf Fjelde's fine translation, from his edition of Ibsen's *Complete Major Prose Plays* (New York: NAL, 1978), makes clear many lines that in previous translations had seemed vague to us. But perhaps Fjelde's most important correction occurs in the title of the play, usually called *A Doll's House*. Without the *s*, the new title seems more faithful to Ibsen's original (in Norwegian, *Et dukkehjem*). This encourages us to see Torvald, and not just Nora, as doll-like and living in an unreal world. As Fjelde writes, "It is the entire house (*hjem*, home) which is on trial, the total complex of relationships, including husband, wife, children, servants, upstairs and downstairs, that is tested by the visitors that come and go, embodying aspects of the inescapable reality outside." Students may be asked to discuss which title better fits the whole sense of the play, Fjelde's or the old one.

At the heart of the play, as its title indicates, is its metaphor of a house of make-believe. In the play's visible symbols, we see Ibsen the poet. In Act I, there is the Christmas tree that Nora orders the maid to place in the middle of the room—a gesture of defiance after Krogstad had threatened her domestic peace and happiness. In the Christmas gifts Nora has bought for the children—sword, toy horse, and trumpet for the boys, a doll and a doll's bed for the girl Emmy—Nora seems to assign boys and girls traditional emblems of masculinity and femininity and (in Fjelde's phrasing) is "unthinkingly transmitting her doll-identity to her own daughter." When the curtain goes up on Act II, we see the unfortunate Christmas tree again: stripped, burned out, and shoved back into a corner—and its ruin speaks eloquently for Nora's misery. Richly suggestive, too, is Nora's wild tarantella, to merry music played by the diseased and dying Rank. Like a victim of a tarantula bite, Nora feels a kind of poison working in her; and it is ironic that Rank has a literal poison working in him as well. (The play's imagery of poison and disease is traced in an article by John Northam included in Fjelde's *Ibsen: A Collection of Critical Essays* [Englewood Cliffs: Prentice, 1965]). Significant, too, is Nora's change of costume (page 1184): taking off her fancy dress, she divests herself of the frivolous nonsense she has believed in the past and puts on everyday street attire.

Ibsen's play was first performed in Copenhagen on December 21, 1879; no doubt many a male chauvinist found it a disquieting Christmas present. Within a few years, *A Doll House* had been translated into fourteen languages. James Gibbons Huneker has described its fame: when Nora walked out on Helmer, "that slammed door reverberated across the roofs of the world." Presumably, the play's revival on Broadway in 1974 (with Liv Ullman as Nora) and the film version starring Jane Fonda may have been due to a sense that, in the present climate of women's liberation, *A Doll House* is once again meaningful.

Ibsen, to be sure, was conscious of sexual injustices. In preliminary notes written in 1878, he declared what he wanted his play to express:

> A woman cannot be herself in contemporary society; it is an exclusively male society with laws drafted by men, and with counsel and judges who judge feminine conduct from the male point of view. She has committed a crime and she is proud of it because she did it for love of her husband and to save his life. But the husband, with his conventional views of honor, stands on the side of the law and looks at the affair with male eyes.

Clearly, that is what the finished play expresses, but perhaps it expresses much more besides. A temptation in teaching Ibsen is to want to reduce his plays to theses. As Richard

Gilman says, the very name of Henrik Ibsen calls to mind "cold light, problems, living rooms, instruction" (*The Making of Modern Drama* [New York: Farrar, 1964]).

But is the play totally concerned with the problem of the "new woman"? Ibsen didn't think so. At a banquet given in his honor by the Norwegian Society for Women's Rights in 1898, he frankly admitted,

> I have been more of a poet and less of a social philosopher than people have generally been inclined to believe. I thank you for the toast, but I must decline the honor of consciously having worked for women's rights. I am not even quite sure what women's rights really are. To me it has been a question of human rights.

Elizabeth Hardwick thinks Ibsen made this statement because he had "choler in his bloodstream" and couldn't resist making a put-down before his admirers. She finds Ibsen nevertheless admirable: alone among male writers in having pondered the fact of being born a woman—"To be female: What does it mean?" (*Seduction and Betrayal* [New York: Random, 1974]). Perhaps there is no contradiction in arguing that Ibsen's play is about both women's rights and the rights of all humanity.

Another critic, Norris Houghton, suggests a different reason for the play's timeliness. "Our generation has been much concerned with what it calls the 'identity crisis.' This play anticipates that theme: Ibsen was there ahead of us by ninety years" (*The Exploding Stage* [New York: Weybright, 1971]). Houghton's view may be supported by Nora's declared reasons for leaving Torvald: "I have to stand completely alone, if I'm ever going to discover myself and the world out there" (page 1186).

The play is structured with classic severity. Its first crisis occurs in Krogstad's initial threat to Nora, but its greatest crisis—the climax—occurs when Helmer stands with the revealing letter open in his hand (page 1182). We take the major dramatic question to be posed early in Act I, in Nora's admission to Mrs. Linde that she herself financed the trip to Italy. The question is larger than "Will Nora's husband find out her secret?"—for that question is answered at the climax, when Helmer finds out. Taking in more of the play, we might put it, "Will Nora's happy doll house existence be shattered?"—or a still larger question (answered only in the final door slam), "Will Nora's marriage be saved?"

Ibsen's magnificent door slam has influenced many a later dramatist. Have any students seen Stephen Sondheim and Hugh Wheeler's musical *Sweeney Todd, The Demon Barber of Fleet Street* (1979) on stage or on television? At the end, Todd slams a door in the faces of the audience, suggesting that he would gladly cut their throats.

For a dissenting interpretation of Ibsen's play, see Hermann J. Weigand, *The Modern Ibsen* (New York: Dutton, 1960). Weigand thinks Nora at the end unchanged and unregenerate—still a wily coquette who will probably return home the next day to make Torvald toe the line.

A topic for class debate: Is *A Doll House* a tragedy or a comedy? Much will depend on how students interpret Nora's final exit. Critics disagree: Dorothea Krook thinks the play contains all the requisite tragic ingredients (*Elements of Tragedy* [New Haven: Yale UP, 1969]). Elizabeth Hardwick (cited earlier) calls the play "a comedy, a happy ending—except for the matter of the children."

To prevent North German theater managers from rewriting the play's ending, Ibsen supplied an alternate ending of his own "for use in an emergency." In this alternate version, Nora does not leave the house; instead, Helmer makes her gaze upon their sleeping children. "This is a crime against myself, but I cannot leave them," says Nora, sinking to the floor in defeat as the curtain falls. Ibsen, however, thought such a change a "barbarous outrage" and urged that it not be used. Students might be told of this alternate ending and be asked to give reasons for its outrageousness.

Citing evidence from the play and from Ibsen's biography, Joan Templeton argues that those critics who fail to see *The Doll House* as a serious feminist statement have distorted its meaning and unintentionally diminished its worth ("The *Doll House* Backlash: Criticism, Feminism, and Ibsen," PMLA: January 1989).

(text pages) 1191–1201

For a cornucopia of stimulating ideas, see *Approaches to Teaching Ibsen's* A Doll House, edited by Yvonne Shafer (New York: Modern Language Association, 1985), one of the MLA's likeable paperback series "Approaches to Teaching Masterpieces of World Literature." Thomas F. Van Laan contributes a survey of English translations of the play, concluding that Rolf Fjelde's version, though not perfect, is "about as good a rendering of *A Doll House* as we are likely to get." June Schlueter writes on using the play as an introduction to drama, and notes that, unlike *Oedipus*, the play does not create an inexorable progress toward disaster. "At any point, we feel, justifiably, that disaster might be avoided." Irving Deer recommends approaching the play by considering "how it deals with decaying values and conventions." J. L. Styan urges instructors to have a class act out the play's opening moments, before and after discussing them, so that Ibsen's wealth of suggestive detail will emerge, which students might otherwise ignore. Other commentators supply advice for teaching the play in a freshman honors course, in a course on women's literature, and in a community college. Joanne Gray Kashdan, author of this latter essay, reports that one woman student exclaimed on reading the play: "I realized I had been married to Torvald for seven years before I divorced him!"

EXPERIMENT AND THE ABSURD

Maria Irene Fornes
A VIETNAMESE WEDDING, page 1191

Why not have students act out *A Vietnamese Wedding*, complete with props? You could assign the parts of Florence, Remy, Aileen, and Irene, and then have those four assign the other parts as instructed in the text. With any luck, such an exercise might show students how an arresting theatrical effect can be produced from even the most limited means. Fornes calls for a large array of props, but common items will do: paper flowers for the garlands, brazil nuts or dried beans for betel nuts, etc. At least a week in advance, ask for two or three volunteers to work together as prop managers. With a little ingenuity, they should be able to round up everything required.

Before any of the cast members are chosen, the class might consider the following points:

1. What should determine who is to play the parts of the four original characters? On what basis will those four choose each of the other characters from the audience?

2. Is who plays what part likely to make a difference to the impact of the performance on the audience? Explain.

3. What sort of emotional impact will you try for during the final procession? By what available means will you seek to achieve it?

After the performance, students might discuss the following questions:

1. Was the performance as theatrical as a real wedding? What made it so—or not so?

2. Of what help were the props to the total effect of the performance?

3. To what extent did the performance change or confirm your ideas about Vietnamese society?

A Vietnamese Wedding was first performed during the height of U.S. involvement in Vietnam: in February 1967 at New York's Washington Square Methodist Church, as part of a protest called Angry Arts Week.

(text pages) 1191–1201

Here is a dramatic event stark in its simplicity: the re-enactment of a traditional Vietnamese betrothal and marriage ceremony. Such a ritual might appear hard to discuss and impossible to analyze, but in fact the work raises intriguing questions and invites us to decide for ourselves what drama and the theater are all about. For one thing, it hands the instructor an occasion to remind students that Western drama originally had its origins in religious services: in the Athenian observance of feast days, when plays were performed; in the medieval tropes or brief plays that accompanied the Easter mass. You might recall the tenth-century *Quem Quaeritis* ("Whom seekest thou?") play, in which three priests represent the three Marys at the empty tomb, and a fourth plays the angel who announced, "He is not here. He is risen as He foretold." So well-received was this bit of theater that more plays like it were introduced; but English drama had to move outside to the churchyard as it added secular events. Fornes may seem to attempt a radical new kind of theater, but actually she returns to an old tradition of drama—perhaps the oldest of all.

By nature, of course, a wedding is a highly dramatic event, charged with feeling, as anyone who has attended one well knows. What weddings have students witnessed (or taken part in) that have impressed them so? (And why?) Are weddings "plays"? No, but some of them require many rehearsals. If *A Vietnamese Wedding* is not a play like most plays—it has no plot, no conflict, no conventional structure of crisis, climax, and resolution—yet to watch it is a theatrical experience. Like any play, it has action, a concentration on significant events that happen to central participants. Like a wedding ceremony, a conventional play, too, is a kind of ritual or re-enactment, depending for its effect on costumes, props, music, and other artifice. As Richard Gilman remarks in his preface to *Promenade & Other Plays* (New York: Winter House, 1971), *A Vietnamese Wedding* is the play of Fornes that "least resembles conventional drama, even of a radical kind. . . ."

> It calls upon members of the audience to participate in its rites, without having to learn any roles or indeed to "act" at all, and upon the rest of the spectators to imagine themselves present at something historical and actual. Yet from this sober summons to reality, so lacking in the superficies of drama, we experience a strange displacement; in imitating an exotic social custom and limning it as though it were an actual event, we find ourselves in the very heart of the country of the dramatic. For theater is the imagining of possible worlds, not the imitation of real ones, and what could be more unreal to us than a ceremony like this play? In enacting it we learn not how other people live but how we are able to imagine ourselves as others, which is what drama is about. If Maria Irene Fornes had given us nothing else, it would be a remarkable thing to have accomplished.

If we empathize with *A Vietnamese Wedding*, perhaps we do so in part because we bring to it outside knowledge, some predispositions in its favor. When the piece was first staged, the audience's consciousness of the sufferings of the Vietnamese people undoubtedly tinged their feelings toward the participants; even today, are not our reactions similar? To re-enact a traditional Vietnamese ceremony is to make a statement affirming the beauty and value of an unfamiliar civilization.

Fornes throws open for debate a fascinating problem: how to distinguish between art and life. Not incidentally, *A Vietnamese Wedding* appeared at a moment in the 1960s when "guerilla theater" ensembles in New York and elsewhere, for propaganda purposes, were setting up dramatic situations in public places without telling the spectators that the situation had been planned. In *The Laundromat Play*, for instance, two women argue over who owns some clothes; one throws bleach in the other's face, and the play ends with a chorus chanting anti-Vietnam-war slogans. The playwright's directions include "make sure there is an empty machine." (For details of this and other staged events, see *Guerilla Street Theater* ed. Harry Lesnick [New York: Avon Books, 1973].) It was a time in the American theater when, as Arthur Sainer has observed, everything came into question: "the place of the

261

performer in the theater; the place of the audience; the function of the playwright and the usefulness of a written script; the structure of the playhouse, and later the need for any kind of playhouse" (*The Radical Theater Notebook* [New York: Avon Books, 1975] 15). These questions may never be resolved; challengingly, Fornes reopens them.

Tom Stoppard
THE REAL INSPECTOR HOUND, page 1202

Comments on the general questions:

1. Moon is a second-string play reviewer who longs to be chief reviewer for his paper. Birdboot, senior reviewer for a rival paper, feels so guilty about two-timing his wife that his conversations tend to evolve into protests that he is upright and innocent.

2. Higgs is the first-string play reviewer whose job Moon covets; as we learn later, it is his corpse that we see lying on stage.

3. Birdboot, in his musings to himself (page 1229), realizes that Puckeridge has planned the deaths of both Higgs and Moon, critics higher up the job ladder that he wants to climb. Birdboot is slain in order to prevent his warning Moon. This is confirmed by Moon in his speech to Puckeridge just before being shot: "You killed Higgs and Birdboot tried to tell me."

4. When Birdboot is drawn on stage to take a phone call from his wife, he steps into the role of Simon. It fits him: Like Simon, he loves Cynthia and wishes to break off with Felicity. As soon as Birdboot is killed, Simon occupies the dead critic's seat in the audience and assumes his job of reviewing the play. Moon and the false Hound also trade identities: Moon becomes Hound and starts trying to solve the mystery, while Hound takes his place in the front row.

5. In the end, Magnus, the real Inspector Hound, Cynthia's vanished husband Albert, and Puckeridge all turn out to be the same man.

6. Moon falsely accuses the slain Birdboot of murdering Higgs. In truth, Birdboot was only a victim and was slain while trying to save Moon's life. Apparently Moon makes his false accusation in order to divert suspicion from himself. Having longed for Higgs's job, he will naturally be a suspect. Drawn more deeply into the Muldoon Manor play, he fails to recognize the killer stalking him in reality: Puckeridge.

7. Stoppard is kidding popular plays and films of drawing-room detection: Agatha Christie's *The Mousetrap* and its countless imitations. Students will probably have seen similar detective stories and television programs. Also, in Moon and Birdboot's critical pronouncements, Stoppard makes fun of the pompous jargon of play reviewers. If you assign students to write a play review, they might do well to read Stoppard first, and so meet a few classic clichés to avoid. More than mere literary burlesque, Stoppard's play (it might be argued) questions rigid distinctions among plays, reviews of plays, and "life."

Unravelling the twin plots of this ingenious comedy, students will be drawn (with any luck) into some lively arguments. As Moon and Birdboot watch a play on stage, we soon see that it is an inconsequential whodunit full of old-hat stage tricks (Felicity's entrance preceded by a bouncing tennis ball, the character who pretends to be an invalid in a wheelchair); and time-worn melodramatic situations (the isolated mansion, the escaped madman roaming the countryside, the drawing room full of suspects). Gradually, the two reviewers become involved in the action of the play-within-a-play. Birdboot is drawn on stage by a phone call; Moon, by the slaying of his friend. In the final irony, third-string reviewer Puckeridge wins both the top job Moon had coveted and Cynthia, the woman whom Birdboot had desired. (Puckeridge, oddly enough, isn't even listed in the play's cast of characters.)

Students may enjoy pointing out some of the bits and happenings in *The Real Inspector Hound* that are manifestly absurd: the pertinent radio messages that always blare on cue; the obvious dead body that no one notices; Puckeridge's disguises; the false Hound's pontoon boots; the comically stilted dialogue, such as Mrs. Drudge's opening speech (into the phone): "Hello, the drawing-room of Lady Muldoon's country residence one morning in early Spring?" (as though she were speaking a playwright's stage directions).

As if taking lighthearted revenge against drama critics, Stoppard satirizes not only their inflated pomposities but their haphazard and unethical methods. In his very first speech, Birdboot reveals that he and some other play reviewers have got together in the theater bar and have already decided what to say about the play, even before seeing it. In his attempts to get Moon to praise the actress playing Felicity and inflate her reputation, he is obviously trying to use his influence to win sexual favors.

Where does the play stop and the offstage world begin? Stoppard's two plots are tightly interwoven. We don't know when the play ends whether the police will enter and cart Puckeridge off to prison or whether all the actors in the play-within-a-play will simply stand up and join hands for the customary curtain calls. (Good students might be challenged to write a brief epilogue and to perform it—at least in a reading—for the class.)

The Paris Review (Winter 1988) contains a revealing interview with Tom Stoppard in which he is asked a question that bears on *The Real Inspector Hound:* What does he do on the first night of a play?

> Stoppard: The first *audience* is more interesting than the first night. We now have previews, which makes a difference. Actually, my play *The Real Inspector Hound* was the first to have previews in London, in 1968. Previews are essential. The idea of going straight from a dress-rehearsal to a first night is frightening. . . . I hate first nights. I attend out of courtesy for the actors and afterwards we all have a drink and go home.

36 *Evaluating a Play*

This chapter may be particularly useful for students to read before they tackle a play about whose greatness or inferiority you have any urgent convictions. The chapter probably doesn't deserve to be dealt with long in class, but it might lead to a writing assignment: to comment on the merits of any play in the book.

If you assign students to write a play review (see "Writing about a Play," page 1522), you might like to have them read this chapter first. To help them in forming their opinions, they may consult the list of pointers on page 1235–36.

37 Plays for Further Reading

Arthur Miller
DEATH OF A SALESMAN, page 1238

QUESTIONS

1. Miller's opening stage directions call for actors to observe imaginary walls when the action is in the present, and to step freely through walls when the scene is in the past. Do you find this technique of staging effective? Why or why not?

2. Miller has professed himself fascinated by the "agony of someone who has some driving, implacable wish in him" (*Paris Review* interview). What—as we learn in the opening scene—are Willy Loman's obsessions?

3. What case can be made for seeing Linda as the center of the play: the character around whom all events revolve? Sum up the kind of person she is.

4. Seeing his father's Boston side-girl has a profound effect on Biff. How would you sum it up? (Biff's faith in his father is shattered. From then on, Biff seeks failure, avoids success.)

5. Apparently Biff's discovery of Willy's infidelity took place before World War II, about 1939. In this respect, does *Death of a Salesman* seem at all dated? Do you think it possible, in the present day, for a son to be so greatly shocked by his father's sexual foible that the son's whole career would be ruined?

6. How is it possible to read the play as the story of Biff's eventual triumph? Why does Biff, at the funeral, give his brother a "hopeless" look?

7. How are we supposed to feel about Willy's suicide? In what way is Willy, in killing himself, self-deluded to the end?

8. What meanings do you find in the flute music? In stockings—those that Willy gives to the Boston whore and those he doesn't like to see Linda mending? In Biff's sneakers with "University of Virginia" lettered on them (which he later burns)? In seeds and gardening?

9. Of what importance to the play are Charley and his son Bernard? How is their father-son relationship different from the relationship between Willy and Biff?

10. What do you understand Bernard to mean in telling Willy, "sometimes . . . it's better for a man just to walk away" (page 1282)?

11. Explain Charley's point when he argues, "The only thing you got in this world is what you can sell. And the funny thing is that you're a salesman, and you don't know that" (page 1283). (Miller, in his introduction to the play, makes an applicable comment: "When asked what Willy was selling, what was in his bags, I could only reply, 'Himself.'")

12. What do you make of the character of Ben? Do you see him as a realistic character? As a figment of Willy's imagination?

13. Suppose Miller had told the story of Willy and Biff in chronological order. If the incident in the Boston hotel had come early in the play, instead of late, what would have been lost?

14. Another death of another salesman is mentioned in this play: on page 1274, that of Dave Singleman. How does Willy view Singleman's death? Is Willy's attitude our attitude?

15. In a famous speech in the final Requiem, Charley calls a salesman a man who "don't put a bolt to a nut"; and Charley recalls that Willy "was a happy man with a batch of cement." Sum up the theme or general truth that Charley states. At what other moments in the play does this theme emerge? Why is Willy, near death, so desperately eager to garden?

(text pages) 1238–1306

16. When the play first appeared in 1949, some reviewers thought it a bitter attack upon the capitalist system. Others found in it social criticism by a writer committed to a faith in democracy and free enterprise. What do you think? Does the play make any specific criticism of society?

17. Miller has stated his admiration for Henrik Ibsen: "One is constantly aware, in watching his plays, of process, change, development." How does this comment apply to *A Doll House*? Who or what changes or develops in the course of *Death of a Salesman*?

Directed by Elia Kazan, with Lee J. Cobb superbly cast as Willy Loman, *Death of a Salesman* was first performed on Broadway on February 10, 1949. Originally, Miller had wanted to call the play *The Inside of a Head,* and he had planned to begin it with "an enormous face the height of the proscenium arch which would appear and then open up." Fortunately, he settled upon less mechanical methods to reveal Willy's psychology. In later describing what he thought he had done, Miller said he tried to dramatize "a disintegrating personality at that terrible moment when the voice of the past is no longer distant but quite as loud as the voice of the present." *Death of a Salesman* has often been called "poetic," despite its mostly drab speech. At first, Miller had planned to make its language more obviously that of poetry and in an early draft of the play wrote much of it in verse. He then turned it into prose on deciding that American actors wouldn't feel at home in verse or wouldn't be able to speak it properly. Miller's account of the genesis of the play is given in his introduction to his *Collected Plays* (New York: Viking, 1959).

In the same introduction, Miller remarks why he thinks the play proved effective in the theater but did not make an effective film. Among other reasons, the movie version transferred Willy literally to scenes that, in the play, he had only imagined, and thus destroyed the play's dramatic tension. It seems more effective—and more disturbing—to show a man losing touch with his surroundings, holding conversations with people who still exist only in his mind. Keeping Willy fixed to the same place throughout the play, while his mind wanders, objectifies Willy's terror. "The screen," says Miller, "is time-bound and earth-bound compared to the stage, if only because its preponderant emphasis is on the visual image. . . . The movie's tendency is always to wipe out what has gone before, and it is thus in constant danger of transforming the dramatic into narrative." Film buffs may care to dispute this observation.

Miller's play is clearly indebted to naturalism. Willy's deepening failure parallels that of his environment: the house increasingly constricted by the city whose growth has killed the elms, prevented anything from thriving, and blotted out human hope—"Gotta break your neck to see a star in this yard." Heredity also works against Willy. As in a Zola novel, one generation repeats patterns of behavior established by its parent. Both Willy and Biff have been less successful than their brothers; presumably both Willy and his "wild-hearted" father were philanderers; both fathers failed their sons and left them insecure. See Willy's speech on page 1260. "Dad left when I was such a baby . . . I still feel—kind of temporary about myself."

The play derives also from expressionism. Miller has acknowledged this debt in an interview:

> I know that I was very moved in many ways by German expressionism when I was in school: . . . I learned a great deal from it. I used elements of it that were fused into *Death of a Salesman.* For instance, I purposefully would not give Ben any character, because for Willy he *has* no character—which is, psychologically, expressionist because so many memories come back with a simple tag on them: something represents a threat to you, or a promise (*Paris Review* 38 [Summer 1966]).

Ben is supposed to embody Willy's visions of success, but some students may find him a perplexing character. Some attention to Ben's speeches will show that Ben does not give a realistic account of his career, or an actual portrait of his father, but voices Willy's dream

(text pages) 1238–1306

versions. In the last scene before the Requiem, Ben keeps voicing Willy's hopes for Biff and goads Willy on to self-sacrifice. Willy dies full of illusions. Unable to recognize the truth of Biff's self-estimate ("I am not a leader of men"), Willy still believes that Biff will become a business tycoon if only he has $20,000 of insurance money behind him. One truth gets through to Willy: Biff loves him.

Class discussion will probably elicit that Willy Loman is far from being Oedipus. Compared with an ancient Greek king, Willy is unheroic, a low man, as his name suggests. In his mistaken ideals, his language of stale jokes and clichés, his petty infidelity, his deceptions, he suffers from the smallness of his mind and seems only partially to understand his situation. In killing himself for an insurance payoff that Biff doesn't need, is Willy just a pitiable fool? Pitiable, perhaps, but no mere fool: he rises to dignity through self-sacrifice. "It seems to me," notes Miller (in his introduction to his *Collected Plays*), "that there is of necessity a severe limitation of self-awareness in any character, even the most knowing . . . and more, that this very limit serves to complete the tragedy and, indeed, to make it all possible." (Miller's introduction also protests against measuring *Death of a Salesman* by the standards of classical tragedy and finding it a failure.)

In 1983 Miller directed a successful production in Peking, with Chinese actors. In 1984 the Broadway revival with Dustin Hoffman as Willy, later shown on PBS television, brought the play new currency. Miller added lines to fit the short-statured Hoffman: buyers laughing at Willy call him "a shrimp." The revival drew a provocative comment from Mimi Kramer in *The New Criterion* for June 1984: she was persuaded that Miller does not sympathize with Willy Loman and never did.

> Since 1949 certain liberal attitudes—towards aggression, ambition, and competitiveness—have moved from the periphery of our culture to its center, so that the views of the average middle class Broadway audience are now actually in harmony with what I take to have been Miller's views all along. In 1949 it might have been possible to view Willy as only the victim of a big, bad commercial system. In 1984, it is impossible not to see Miller's own distaste for all Willy's attitudes and petty bourgeois concerns, impossible not to come away from the play feeling that Miller's real judgment of his hero is that he has no soul.

For a remarkable short story inspired by the play, see George Garrett's "The Lion Hunter" in *King of the Mountain* (New York: Scribner's 1957).

A natural topic for writing and discussion, especially for students who have also read *Othello* and a play by Sophocles: How well does Miller succeed in making the decline and fall of Willy Loman into a tragedy? Is tragedy still possible today? For Miller's arguments in favor of the ordinary citizen as tragic hero, students may read his brief essay "Tragedy and the Common Man" (in "Criticism: On Drama," page 1436).

For other comments by Miller and a selection of criticism by various hands, see *Death of a Salesman: Text and Criticism*, ed. Gerald Weales (New York: Viking, 1967). Also useful is *Arthur Miller: A Collection of Critical Essays*, ed. Robert W. Corrigan (Englewood Cliffs: Prentice, 1969). In *Arthur Miller* (London: Macmillan, 1982), Neil Carson seeks to relate *Death of a Salesman* to the playwright's early life.

Sophocles
ANTIGONE, page 1306

Questions

1. Some critics argue that the main character of the play is not Antigone, but Creon. How can this view possibly be justified? Do you agree? (Antigone disappears from the play

in its last third, and we are then shown Creon in conflict with himself. Creon suffers a tragic downfall: his earlier decision has cost him his wife and his son; Eurydice has cursed him; and in the end, he is reduced to a pitiable figure praying for his own death. Still, without Antigone the play would have no conflict; surely she suffers a tragic downfall as well.)

2. Why is it so important to Antigone that the body of Polyneices be given a proper burial? (She must "honor what the gods have honored" [line 80]. See also the footnote on line 2.)

3. Modern critics often see the play as centering around a theme: the authority of the state conflicts with the religious duty of an individual. Try this interpretation on the play and decide how well it fits. Does the playwright seem to favor one side or the other? (The pious Sophocles clearly favors Antigone and sees divine law taking precedence over human law; but Creon's principles [most fully articulated in lines 187–199 and 269–288] are given fair hearing.)

4. Comment from a student paper: "Antigone is a stubborn fool, bent on her own destruction. Her insistence on giving a corpse burial causes nothing but harm to herself, to Haimon, Eurydice, and all Thebes. She does not accomplish anything that Creon wouldn't eventually have agreed to do." Discuss this view.

5. Explain the idea of good government implied in the exchange between Creon and Haimon: "Am I to rule by other mind than mine?"—"No city is property of a single man" (694–95).

6. What doubts rack Creon? For what reasons does he waver in his resolve to punish Antigone and deny burial to the body of Polyneices? In changing his mind, does he seem to you weak and indecisive? (Not at all; he has good reason to pull down his vanity and to listen to the wise Haimon and his counselors.)

7. In not giving us a love scene in the tomb between Antigone and Haimon, does Sophocles miss a golden opportunity? Or would you argue that, as a playwright, he knows his craft?

David Grene has pointed out that the plots of the *Antigone* and the *Oedipus* have close similarities. In both, we meet a king whose unknowing violation of divine law results in his own destruction. In both, the ruler has an encounter with Teiresias, whom he refuses to heed. Creon relents and belatedly tries to take the priest's advice; Oedipus, however, defies all wise counsel (introduction to "The Theban Plays" in *Complete Greek Tragedies*, ed. David Grene and Richard Lattimore [U Chicago P, 1959] II: 2–3).

Students with experience in play production might be asked to suggest strategies for staging *Antigone* today. In their commentary on the play, translators Dudley Fitts and Robert Fitzgerald make interesting suggestions. The Chorus had better not chant the Odes in unison, or the words will probably be unintelligible; let single voices in the Chorus take turns speaking lines. The solemn parados might be spoken to the accompaniment of a slow drumbeat. No dancing should be included; no attempt should be made to use larger-than-life Greek masks with megaphone mouths. More effective might be lifelike Benda type masks, closely fitting the face. "If masks are used at all, they might be well allotted only to those characters who are somewhat depersonalized by official position or discipline: Creon, Teiresias, the Chorus and Choragos, possibly the Messenger. By this rule, Antigone has no mask; neither has Ismene, Haimon, nor Eurydice" (*The Oedipus Cycle: An English Version* [New York: Harcourt, 1949] 242–43).

Entertaining scraps from our sparse knowledge of the life of Sophocles are gathered by Moses Hadas in *An Ancilla to Classical Reading* (Columbia UP, 1954). The immense popular success of *Antigone* led to the playwright's being elected a general, although as he himself admitted, he was incompetent in battle. Many reports attest to his piety, his fondness for courtesans and boys, and his defeat of an attempt by his sons to have him declared an imbecile. Plutarch relates that when Sophocles, then past ninety, read to the jury from his latest work, *Oedipus at Colonus*, he "was escorted from the lawcourt as

(text pages) 1336–1341

from a theater, amid the applause and shouts of all." One account of the playwright's death is that he strangled while reading aloud a long, breathless sentence from *Antigone*.

Suggestions for Writing. Compare and contrast the character of Creon in the two plays. (In *Oedipus the King,* he is the reasonable man, the foil for the headstrong Oedipus.)

How important to the play is Haimon? Ismene? Eurydice?

How visibly does the family curse that brought down Oedipus operate in *Antigone*? Does fate seem a motivating force behind Antigone's story? (In *Antigone* fate plays a much less prominent part; the main characters—Creon, Antigone, and Haimon— seem to decide for themselves their courses of action.)

Wendy Wasserstein
THE MAN IN A CASE, page 1336

QUESTIONS

1. At what point in Wasserstein's play does Byelinkov most nearly escape from his "case"? What causes him to depart from his characteristic ways? (Varinka's openness and infectious good cheer begin melting Byelinkov's straitlaced reserve when she insists that he dance with her in the garden and he actually complies. He goes so far as to place a lilac in her hair and then playfully, so as not to forget, to write down in his notebook that he will repeat this gesture every year on this day.)

2. Do you think Varinka so cheerfully puts up with her fiancé's quirks because she is desperate to be married, or does she really love him? (While it's true that Varinka wants to be married—see her speech that begins "Byelinkov, I am a pretty girl of thirty"—it seems clear that she does fondly admire her fiancé and relishes the thought of being "the master of Greek and Latin's wife." Feigned affection would be out of character for such an openhearted, uninhibited person as Varinka.)

3. Does Byelinkov's retreat at the end of the play seem too extreme, or does the playwright adequately prepare us for it? If so, how? (All through the play, except briefly during Byelinkov's dance with Varinka, Wasserstein's dialogue calls attention to the teacher's timidity, fastidiousness, and rigidity. The moment when he puts a lilac in Varinka's hair is a startling departure from his usual behavior. For a few moments we can hope that, given time, Varinka might soften his hard edges. But the terrible fact that Varinka rides a bicycle is too much for him. The only way he can deal with such a scandal is by snuffing out the beam of light that has briefly penetrated his self-imposed prison.)

Students who set out to compare Wasserstein's play with the Chekhov story on which it is based will probably remark that the most obvious differences between the two versions stem from the genre in which each author works. In Chekhov's version, the engagement between Byelinkov and Varinka is a story within a story, a story wrapped in a case, as it were. Because Wasserstein tells the same story in a one-act play, she has to compress and simplify. She cuts the number of characters to two, eliminating even Afanasi, Byelinkov's cook. And whatever we learn about the two characters, we have to learn from their dialogue and their actions.

Wasserstein humanizes Chekhov's Byelinkov—most notably in the scene in which he places lilacs in his fiancée's hair. Wasserstein borrows from Chekhov the galoshes and the bed curtains, but she invents other details that convey with admirable economy the kind of person Byelinkov is: his allergy to apricots, his unvarying breakfast menu, his ways of "preserving those things which are left over," to name a few. With the same economy she conveys through dialogue the boisterousness with which Chekhov has endowed Varinka, her appealing tendencies to laugh, dance, and sing.

269

(text pages) 1342–1388

In both Chekhov and Wasserstein it is the bicycle that causes the break between Byelinkov and Varinka, but the specific turning point is different in the two versions. In Chekhov's story, Varinka laughs at Byelinkov when she finds him in an undignified heap at the bottom of the stairs. In the play she laughs at him because his galosh gets caught in the bicycle. In both versions, of course, the schoolmaster's pride is fatally wounded.

Of the two writers, Chekhov is more specific about the case, making clear for instance that everything Byelinkov wears and the languages he teaches are ways to hide himself from the reality around him. Along with much of Chekhov's plot and two of his characters, Wasserstein borrows Chekhov's title but lays much less stress on its symbolism.

Tennessee Williams
THE GLASS MENAGERIE, page 1342

QUESTIONS

1. How do Amanda's dreams for her daughter contrast with the realities of the Wingfields' day-to-day existence?

2. What suggestions do you find in Laura's glass menagerie? In the glass unicorn?

3. In the cast of characters, Jim O'Connor is listed as "a nice, ordinary, young man." Why does his coming to dinner have such earthshaking implications for Amanda? For Laura?

4. Try to describe Jim's feelings toward Laura during their long conversation in Scene VII. After he kisses her, how do his feelings seem to change?

5. Near the end of the play, Amanda tells Tom, "You live in a dream; you manufacture illusions!" What is ironic about her speech? Is there any truth in it?

6. Who is the main character in *The Glass Menagerie*? Tom? Laura? Amanda? (It may be helpful to review the definition of a *protagonist*.)

7. Has Tom, at the conclusion of the play, successfully made his escape from home? Does he appear to have fulfilled his dreams?

8. How effective is the device of accompanying the action by projecting slides on a screen, bearing titles and images? Do you think most producers of the play are wise to leave it out?

For Williams's instructions for using the slide projector, see "How to Stage *The Glass Menagerie*" in "Criticism: On Drama," page 1446. Personally, we think the slide projector a mistake. In trying to justify it, Williams underestimates the quality of his play's spoken lines—but what do your students think?

The gracious world of the old South lives on in Amanda's memories. No doubt its glories shine brighter as the years go by, but all three members of the Wingfield family, in their drab little apartment, live at several removes from the real world. Laura is so shy that she cannot face strangers, yet her mother enrolls her in a business school where she is, of course, doomed to failure. Next, quite ignoring the fact that Laura has no contact with anyone outside her own family, Amanda decides that her daughter ought to marry, and cheerfully sets about finding her a gentleman caller. Some students will want to see Amanda as a silly biddy and nothing more, so that it may help to ask: In what ways is she admirable? (See Williams's initial, partially admiring description of her in the cast of characters.)

A kindly, well-intentioned young man, Jim O'Connor is a self-styled go-getter, a pop psychologist. Like Biff Loman in *Death of a Salesman*, Jim is a high school hero whose early promise hasn't materialized. He was acquainted with Laura in school but now remembers her only when prompted. Laura's wide-eyed admiration for him flatters Jim's vanity, and in her presence he grows expansive. Gradually, Laura awakens in him feelings of warmth and protectiveness, as well as a sense that her fragility bespeaks something as precious and rare as her glass unicorn. It is with genuine regret that he shatters her tremulous, newly risen hopes with the revelation that he is engaged to be married to Betty, a young woman as unremarkable as himself.

Laura's collection of glass animals objectifies her fragility, her differentness, her removal from active life. Significantly, the unicorn is her favorite. "Unicorns, aren't they extinct in the modern world?" asks Jim; and he adds, a few lines later, "I'm not made out of glass." When Jim dances with Laura and accidentally breaks off the unicorn's horn, the mythical creature becomes more like the common horses that surround him, just as Laura, by the very act of dancing, comes a few steps closer to being like everyone else. Although Jim can accept the broken unicorn from Laura as a souvenir, he cannot make room in his life for her. Her fleeting brush with reality does not in the end alter her uniqueness or release her from her imprisonment.

Amanda's charge that Tom manufactures illusions seems a case of the pot calling the kettle black. As we know from Amanda's flighty talk and far-fetched plans for Laura, the mother herself lives in a dream world; but she is right about Tom. A would-be poet, a romantic whose imagination has been fired by Hollywood adventure movies, Tom pays dues to the Merchant Seamen's union instead of paying the light bill. So desperate is he to make his dreams come true, he finally runs away to distant places, like his father before him. In truth, each character in the play has some illusions—even Jim, who dreams of stepping from his warehouse job into a future as a millionaire television executive. And as Tom's commentaries point out, at the time of the play's action all Americans seem dazzled by illusions, ignoring the gathering threat of World War II. "In Spain, there was Guernica! But here there was only hot swing music and liquor, dance halls, bars, and movies, and sex that hung in the gloom like a chandelier and flooded the world with brief, deceptive rainbows" (page 1361).

For a challenging study of the play, see Roger B. Stein, "The Glass Menagerie Revisited: Catastrophe without Violence," *Western Humanities Review* 18 (Spring 1964): 141–53. (It is also available in *Tennessee Williams: A Collection of Critical Essays*, ed. Stephen S. Stanton [Englewood Cliffs: Prentice, 1977].) Stein finds in the play themes of both social and spiritual catastrophe: the failure of both Christianity and the American dream. Although some of the play's abundant Christian symbolism and imagery would seem just decoration, students may enjoy looking for it. Scene V, in which Tom tells his mother that Laura will have a gentleman caller, is titled on the screen "Annunciation." Laura says she has dreaded to confess she has left business school, because her mother, when disappointed, wears a look "like the picture of Jesus' mother." Amanda is also identified with the music of "Ave Maria." When Tom comes home drunk, he tells Laura of seeing the stage magician Malvolio, an Antichrist who can escape from a nailed coffin and can transform water to wine (also to beer and whiskey). Jim O'Connor is another unsatisfactory Savior: he comes to supper on a Friday night and (symbolically?) is given fish, but unlike the Christ whose initials he shares, he can work no deliverance. Laura is described as if she were a saint, or at least a contemplative. When she learns that Jim is engaged to Betty, "the holy candles in the altar of Laura's face have been snuffed out" (page 1384). Compare Williams's instructions to lighting technicians in his production notes:

> Shafts of light are focused on selected areas or actors, sometimes in contradistinction to what is the apparent center. For instance, in the quarrel scene between Tom and Amanda, in which Laura has no active part, the clearest pool of light is on her figure. This is also true of the supper scene. The light upon Laura should be distinct from the others, having a peculiar pristine clarity such as light used in early religious portraits of female saints or madonnas.

Most suggestive of all, Williams keeps associating candles with lighting. Amanda's candelabrum, from the altar of the Church of the Heavenly Rest, had been warped when the church was struck by lightning. And when Tom, in his final speech, calls on Laura to blow her candles out, he declares that "nowadays the world is lit by lightning." The playwright suggests, according to Stein, that a hard, antireligious materialism now prevails. (At least, this line of reasoning may be worth an argument.)

The character of Laura, apparently, contains traits of Williams's sister Rose. For some of the play's autobiographical background, see "Writing a Play," page 1533. In a recent memoir, William Jay Smith, who knew Williams in St. Louis as a fellow college student at Washington University, remarks on the background of the play.

> I am frequently amused by those who take Tom's autobiographical projection of his family in *The Glass Menagerie* literally and picture him as having inhabited a rundown, seedy old house, if not a downright hovel. The house on Arundel Place, with its Oriental rugs, silver, and comfortable, if not luxurious furniture, was located in an affluent neighborhood. . . . Our entire bungalow on Telegraph Road would have fitted comfortably into one or two of its rooms. Mrs. Williams presided over it as if it were an antebellum mansion (*Army Brat* [New York: Persea, 1980] 190).

August Wilson
JOE TURNER'S COME AND GONE, page 1388

QUESTIONS

1. What does Bynum's "shining man" represent? What is the significance of his telling Bynum to rub blood on himself? What action of Loomis's in Act III does this ritual foreshadow?

2. What is implied when, in Act 2, Scene 2, Bynum asks Loomis whether he has ever been in Johnstown?

3. Who is Joe Turner? What does he represent?

4. At what moment does the crisis occur in *Joe Turner's Come and Gone*?

5. After reading Wilson's play, what would you say is the Secret of Life that Bynum learns from the shining man?

6. What, if anything, does Wilson's play have to say about religion?

7. For discussion: Do you think the play would have been more effective had Herald Loomis and his wife decided to stay together once they had been reunited?

8. Comment on the spelling of Loomis's name. Is he a herald? If so, what message does he impart?

9. What is the theme of *Joe Turner's Come and Gone*? Is it stated in the play, or only implied?

> David Savran remarks about Wilson's play:
>
> *Joe Turner's Come and Gone*, which takes place in 1911, performs a ritual of purification, setting African religious tradition against American Christianity. It documents the liberation of the spiritually bound Herald Loomis, who years before had been pressed into illegal servitude by the bounty hunter named in the play's title. In the course of the play the details of everyday life in a Pittsburgh boarding house give way to the patterns of African religion and ritual. With the help of Bynum, an African healer, a "Binder of What Clings," Loomis effects his own liberation. He recognizes that his enslavement has been self-imposed; this man "who done forgot his song" finds it again. Bynum explains to him: "You bound on to your song. All you got to do is stand up

and sing it, Herald Loomis. It's right there kicking at your throat. All you got to do is sing it. Then you be free" (*In Their Own Words: Contemporary American Playwrights* [New York: Theatre Communications Group, 1989] 289).

The lines Savran quotes, appearing in the play's final scene, are perhaps as good a statement of the play's theme as there is.

The troubled Loomis is a herald, it seems—Bynum's "One Who Goes Before and Shows the Way" (see Bynum's long speech in Act I, Scene 1). What Loomis learns, and shows the others, is that African-Americans, if they search, can find within themselves and their African traditions the power to be free. This is apparently the Secret of Life that Bynum has learned from his "shining man." Everyone has to find his own song. Only then can he make his mark on life.

Bynum, the conjure man, is a pivotal character. At the start of the play, he is the only one who has found his song (though Bertha and Jeremy seem closer to having found theirs than do some of the other characters). Loomis, when he first appears, is still under the influence of Joe Turner, the cruel bounty hunter who personifies all the evils of slavery. Thus Loomis frightens the others, seems to them crazy and unpredictable. Though Seth knows where Martha Pentecost is, he refuses to tell Loomis. In fact, Loomis is a man searching for his song.

Bynum also realizes that, after their long period of slavery and separation, black people have to seek and find one another. That's why his magic is aimed at bringing people together. That's also why he likes the People Finder, and why he encourages Jeremy to go "down to Seefus" to play his guitar, even though Bertha warns him the place might get raided. "That's where the music at," Bynum says. "That's where the people at. The people down there making music and enjoying themselves. Some things is worth taking the chance going to jail about." Several of the characters in Wilson's play are in search of the right person to connect with. What makes this play so life-affirming is that some of them, by reaching out, find what they're looking for.

The turning point in the play seems to come at the end of Act I, when Loomis has his vision, making such a commotion that Seth tells him he'll have to leave the boarding house. Only Bynum realizes how crucial that vision is to Loomis's spiritual health. By this time he clearly believes that Loomis is a shining man. That's why he asks, on the following day, whether Loomis has ever been in Johnstown. It was in Johnstown that Bynum had the experience with the shining man that he tells Selig about in Act I, Scene 1.

By singing the Joe Turner song in Act II, Scene 2, Bynum gets Loomis to unburden himself, to reconnect with his own African roots. As if by instinct, Loomis seems to know he can do so by rubbing himself with his own blood, thus acting out the ritual Bynum has described to Selig in Act I, Scene 1. Wilson is clearly aware that blood functions as a Christian symbol of purification. Martha urges Loomis to be washed in the blood of the Lamb. But Loomis rejects the Christianity that sustains his wife. Purification comes for Loomis and Bynum not through Christianity but through the powerful African rituals of their forefathers. "I don't need nobody to bleed for me!" Herald says. "I can bleed for myself." Loomis's song is "the song of self-sufficiency." When he learns to sing it, he becomes a shining man.

Students need to pay attention to Wilson's prologue ("The Play," preceding Act I, Scene 1). The world of *Joe Turner's Come and Gone* is a now-vanished corner of American society—a world of poor drifters, of migrants from the cotton fields to the booming Pittsburgh steel mills of 1911. But although times have changed, the situation of the characters ("foreigners in a strange land" seeking "a new identity") may recall that of any new settlers in a big city, whether Africans, Hispanics, or Asians. For class discussion: How does the play remind us of the lives (and problems) of minority people today, who find themselves transplanted to an American city from a very different culture?

(text pages) 1388–1435

For another topic for discussion, have students read Wilson's comments to his interviewer, Bill Moyers, in "Black experience in America" (the last selection in Chapter 38, "Criticism: On Drama"). How does the play reflect Wilson's views? Particularly interesting may be Wilson's opinion that African-Americans were ill advised to leave the South.

Here are three suggested writing topics:

1. The importance of magic in *Joe Turner's Come and Gone*.

2. Is Wilson a symbolist? (That he imparts a message and that he portrays real and recognizable people does not prevent his play from being richly suggestive. Joe Turner, Bynum's "shining man," and the blood rituals that are part of the quest for an individual song all hint at larger meanings. Students may find others as well.)

3. For a long paper, one entailing some research in a library: In what ways has life for most African-Americans changed since the 1911 of which Wilson writes? In what ways have problems and conditions of life stayed the same?

Wilson's play was first performed in a staged reading in 1984 at a playwright's conference at the Eugene O'Neill Theater Center. Later, in 1986 and 1987, it had two productions by the Yale Repertory Theater. Shortly thereafter, in 1988, a successful Broadway production received great acclaim: "haunting, profound, indescribably moving" (Frank Rich in *The New York Times*); "Wilson's best play" (William A. Henry III in *Time*).

Students might be asked to comment on this summing up by Henry in *Time* for April 11, 1988, and if necessary argue with it:

> At the end, when Loomis seems pathetically shorn of his consuming purpose . . . the most spiritual boarder perceives in him instead the "shiny man" of a folkloric religious vision. In that moment, spectators too find themselves transported from pity to admiration: Loomis has transformed his pointless suffering into an ennobling search for life's meaning.

Shortly after Wilson received a second Pulitzer Prize for *The Piano Player* in April 1990, he made a few revealing comments to an interviewer. Nothing in his work is autobiographical, he declared; nothing he has written has been taken from his own experience. He has successfully avoided studying other playwrights. He claims to have read nothing by Shakespeare except *The Merchant of Venice* (in high school), nothing by Ibsen, Miller, or Tennessee Williams. The only other playwrights whose work he admits to knowing are Amiri Baraka, Ed Bullins, and Athol Fugard. He doesn't go to the theater himself, hasn't been to a movie in ten years. "Part of this creative isolation is self-protective fear," explains the interviewer, Kevin Kelly. "Wilson is afraid of tampering with those chaotically rich and whimsically independent forces in his head, terrified of confusing their voices and stories with the voices and stories of other writers" ("August Wilson's True Stories," *Boston Globe*, April 29, 1990).

38 Criticism: On Drama

Again, as in the earlier "Criticism" chapters on fiction and poetry, the purpose of this little anthology of critical passages is to supply you with further resources, but only if you wish any. Aristotle's celebrated discussion of the nature of tragedy is here given at greater length than in the chapter on "Tragedy" (where it is only briefly quoted), while the comments of Tennessee Williams and Arthur Miller (both complete essays) may prove useful to assign for reading along with their plays.

These selections raise many questions for more general discussion or for writing topics. Here are a few of these questions.

Aristotle, Tragedy, page 1436

1. According to Aristotle, what sort of man is the most satisfactory subject for a tragedy? (A man neither entirely virtuous nor completely vicious: "a man of great reputation and great prosperity" who comes to grief because of some great error.)

2. Try this description of a man on the character of Oedipus. How well does it fit? (Like a tailor-made sweater. Oedipus may be Aristotle's finest illustration of a tragic hero of middling virtue. Far from perfect, Oedipus is impulsive and imperious.)

3. What do you make of Aristotle's views of women in his sixth paragraph? (Ask this one, and set off some fireworks! In noting that women are an inferior class, Aristotle isn't condemning them but observing his own society: the rulers of Athens made certain that women were subjugated. Aristotle's claim that women lack certain varieties of courage and wisdom might be a more interesting statement to argue over.)

4. Consider the advice that any extravagant incidents should be kept outside a tragedy. In *Oedipus the King,* what "extravagant" events in the story are not shown on stage? (The most fantastic is probably the meeting of Oedipus with the Sphinx; we are merely told, in the closing speech of the Chorus, that Oedipus "knew the famous riddles.")

Sigmund Freud, The destiny of Oedipus, page 1438

1. Explain Freud's remark that most of us have succeeded in repressing our Oedipal desires "insofar as we have not become psychoneurotics."

2. Suggest possible explanations—other than Freud's explanation—for the powerful impact of *Oedipus the King* upon a modern audience.

E. R. Dodds, Sophocles and divine justice, page 1439

1. What is Dodds's central point? Does it make sense to you? Could you love (or at least worship) an unjust God?

2. What evidence can you find in *Oedipus the King* to justify Dodds's view? (Notice that in the closing scene of the play, Oedipus continues to pray, to call on God to be kinder to his children than to him. Creon denies that God hates Oedipus.)

Charles Paul Segal, Antigone's womanly nature, page 1440

1. After thinking about Segal's points, discuss: How is Creon a kind of Archie Bunker? (As Segal shows, Creon likes simple dichotomies: man-woman, superior-inferior.)

(text pages) 1440–1447

2. Creon sees in Antigone, according to Segal, "a challenge to his whole way of living and his basic attitudes toward the world. And of course he is right." Explain in your words what Segal is driving at. Confirm or deny his insight by referring to evidence in the play.

Thomas Rymer, THE FAULTS OF OTHELLO, page 1441

1. What chief complaints against *Othello* does this critic lodge?
2. How would you answer them?

Samuel Johnson, "SHAKESPEARE HAS NO HEROES", page 1442

1. "Shakespeare has no heroes"—what do you understand from this remark? What is a hero, in Johnson's sense of the word?
2. We might expect that Dr. Johnson, a follower of classical "rules of criticism" (as he says), might condemn Shakespeare for mingling tragedy and comedy in one play. But why doesn't he? How does he justify—even praise—Shakespeare's practice?
3. Both Johnson and Bernard Shaw (page 1443), find that Shakespeare places his characters in extreme situations, even impossible ones. What does each critic think of Shakespeare for doing so? (Johnson admires the Bard for imagining human nature "in trials to which it cannot be exposed." In Shaw's opinion, the fact that our uncles seldom murder our fathers, as in *Hamlet,* renders Shakespeare's plays of less consequence for us.)

Bernard Shaw, IBSEN AND THE FAMILIAR SITUATION, page 1443

1. How, according to Shaw, does *A Doll House* reflect Ibsen's originality? In the play's time, what was so new about it?
2. How does Shaw explain the origin of drama?
3. Which playwright does Shaw prefer: Ibsen or Shakespeare? Why?
4. (See Question 3 for Samuel Johnson, above.)

Virginia Woolf, WHAT IF SHAKESPEARE HAD HAD A SISTER?, page 1444

1. What are the obstacles, as Woolf presents them, that would have prevented a gifted Englishwoman of the seventeenth century from rising to the heights that Shakespeare attained?
2. What positive point is Woolf driving at? (That women, like men, deserve a climate in which their gifts will be free to germinate.)
3. How—and why—has the situation of a gifted woman changed in our own day?

Edward Albee, THE THEATER OF THE ABSURD, page 1445

1. What, according to Albee, should a play do besides entertain?
2. What paradox does he find in the nature of "the supposed Realistic theater"?

Tennessee Williams, HOW TO STAGE THE GLASS MENAGERIE, page 1446

1. How do both Albee (above) and Williams feel about theatrical "realism"?
2. How does Williams argue for his use of the slide projector? If you were producing *The Glass Menagerie,* would you follow the playwright's instructions and use the projector, or leave it out?
3. What other antirealistic devices would Williams employ? Would you expect them to be effective?

Arthur Miller, TRAGEDY AND THE COMMON MAN, page 1448

1. In arguing that a tragedy can portray an ordinary man, how does Miller find an ally in Sigmund Freud? (See Miller's second paragraph.)

2. According to Miller, what evokes in us "the tragic feeling"? Compare his view with Aristotle's. (Unlike the Greek theorist, Miller finds the sense of tragedy arising not from pity and fear but from our contemplating a character who would give his life for personal dignity.)

3. In Miller's view, why is not tragedy an expression of pessimism? What outlook *does* a tragedy express?

4. Consider what Miller says about pathos (page 1450), and try to apply it to *Death of a Salesman*. Does the play persuade you that Willy Loman could have won his battle? That he isn't witless and insensitive? Or is the play (in Miller's terms) not tragic, but only pathetic?

Maria Irene Fornes, FINDING A WAY OUT, page 1450

1. What dividing line does Fornes draw between a play and real life? ("Characters are not real people. . . .")

2. Consider Fornes's declaration that a play teaches us something, changes our understanding. What do we learn from *A Vietnamese Wedding*?

August Wilson, BLACK EXPERIENCE IN AMERICA, page 1451

1. What does Wilson see as a part of his purpose in writing plays? ("To see some of the choices that we as blacks in America have made.")

2 . How does *Joe Turner's Come and Gone* illuminate any such choice? (Wilson has respect and affection for black folk culture, with its elements of myth and magic. Perhaps, he hints, it risks disappearance when transplanted to the urban North.)

3. What light do Wilson's remarks throw on the line in the play, "Everyone has to find his own song"? (These characters, as the playwright sees them, are looking for their African identities. That they are embarked on any such a quest won't be obvious to most readers. Students will need to do some thinking and discussing in order to make sense of Wilson's claim. What is an African identity? How does Herald Loomis end up with one at the end? Carefully reread the playwright's explanatory stage directions in the last moments of the play.)

4. In describing the six black men having breakfast, what does Wilson apparently like and admire? (Their spontaneous, easygoing, natural playfulness and humor; their fun-loving sense of brotherhood. To Wilson's mind, are these traits part of "an African identity"? Maybe so.)

SUPPLEMENT: WRITING

Writing about Literature

This Supplement is no more than a brief guide and a small work of reference. If you find it useful, you might ask students to read it before they plan their papers. As succinctly as it can, the section "Writing about Literature" sets forth a few general critical approaches to a story, poem, or play; then it escorts the student through the various procedures of finding a topic and organizing, drafting, revising, and finishing a paper.

Such matters need not divert much classroom time from the livelier task of reading literature. Perhaps it would be sufficient, after students have read "Writing about Literature," to take a few minutes to question them on this material, to make sure at least that they are aware of it. At the same time, you might invite their own questions about essay writing (to which, no doubt, you will have your own answers).

If this Supplement fulfills its purpose, it will save you some breath and spare you from reading many innocent almost-plagiarisms, floating unidentified quotatons (of the kind that suddenly interrupt the student's prose with Harold Bloom's prose in quotation marks), and displays of ill-fitting critical terminology.

Some instructors assign few papers and exact from their students rather highly polished prose; others prefer to keep students scribbling away constantly, on the assumption that the practice in writing is valuable (whether or not the instructor reads all their output word by word). Instructors who favor the latter approach report that they simply assign selections in the book that have questions after them, and ask students to answer some questions in writing outside of class. These papers are collected, and the instructor later skims through them and selects a few of the livelier points to quote and discuss at the next class. (The papers are not returned.)

Once—at the end of a class in which argument had waxed over the question "Is 'Naming of Parts' an antiwar poem or isn't it?"—XJK made the mistake of cutting off the discussion and telling students to go home and write their opinions down on paper. The result was to cool future class discussions: students were afraid that if they talked animatedly, they would be told to write. A different approach is that of an instructor who would halt a class discussion that had grown driveling or bad-tempered or without heart and cry, "For God's sake, let's all stop talking! Now get out your pencils and write me a paragraph. . . ." He claimed that in the next class the discussion improved markedly.

Robert Wallace, THE GIRL WRITING HER ENGLISH PAPER, page 1468

This fine poem has no purpose here but to interrupt the editorial prose and help put students in a writing mood. But it is worth reading with an eye to its metaphors: Crumpled up drafts are like the wreckage of Eden (lovely hyperbole!); writing about a poem is like plowing furrows.

In the previous edition, this poem had a different last line, so that it ended, "stars would be overhead, / their light come in." Since then, the poet has rewritten the ending, for reasons he explains in his textbook *Writing Poems, Second Edition* (Boston: Little, Brown, 1987):

> A vague dissatisfaction with the last line persisted. I didn't know what the problem was until—again reading the poem to an audience, in Kansas—the *right* line suddenly came to me. I had been aware, I think, that "their light come in" wasn't really

necessary, only emphasizing what line 11 already does: "stars would be overhead." But stopping there would leave the rhythm unfinished, and the stanza pattern uncompleted.

In the instant before I said the line, I knew how it should really go—and changed it. At once the weakness of "their light come in" was also plain. It left the *room* vaguely in place, with the starlight coming in an unmentioned window, rather than simply overhead. Instead of implying such a limit, the new line extends mysteriously the little world of the farm.

Writing about a Story

If your students complain, "I've never written about *stories* before—what am I supposed to do?" you can have them read this section. We can't imagine spending whole class hours with this material; it is supplied here mainly to provide students with illustrations of acceptable papers written by each of three usual methods, and a few pointers on format and mechanics. If you like, you can assign this section for outside reading when you first make a writing assignment.

XJK comments: The card report (page 1478) may well be God's gift to the instructor overwhelmed with papers to grade. At least, I can't take credit for its creation. This demanding exercise first impressed me as a student in the one course I took at Teachers College, Columbia. The professor, Lennox Grey, assigned us aspiring literature teachers to pick ten great novels we hadn't read and to write card reports on them. Among the novels were *War and Peace* and *Les Misérables,* and although Grey allowed us as many as two cards to encompass them, I must have spoiled a pack of cards for every novel I encompassed. But the task was an agreeable challenge, and I felt it obliged me to look more closely at fiction than I ever had. Later, as a graduate student in Ann Arbor, I found the same device heavily worked by Kenneth Rowe, Arthur Miller's teacher of playwriting, in a popular course in modern drama. Every week, students were expected to read two full-length plays and to turn in two card reports. Nearly a hundred students swelled the course, and Rowe employed two teaching assistants to do the grading. As one of them, I soon realized the beauty of the method. Even a novice like me could do a decent job of grading a hundred card reports each week without being crushed under the toil, either. For an hour a week Rowe met with the other assistant and me and superintended our labors a little, and we thrashed out any problem cases.

If you care to give such an assignment a try, don't feel obliged to write a card report of your own as a Platonic ideal to hold your students' reports up to. When you gather in the sheaves, you can compare a few of them (looking hard at the reports of any students whom you know to be intelligent and conscientious) and probably will quickly see what a better-than-average report on a story might encompass. In grading, it isn't necessary to read every item on every card: you may read the plot summaries with intermittent attention and concentrate on the subtler elements: symbol, theme, evaluation. Because extensive remarks by the instructor don't seem called for (and anyway, wouldn't fit on the card), your comments may be short and pointed. If a student reporting on "The Tell-Tale Heart" has omitted a crucial symbol, you may simply query, "The eye?" or "The heartbeat?" One can probably grade thirty card reports in less than an hour and do an honest job; whereas a set of thirty essays, even brief ones, takes at least four hours.

By asking students to produce so few words, you need not feel that their writing skills are being slighted. To get the report to fit the card, a good student has to do several drafts and revisions, none of which the instructor has to read. A shoddy job by a student who hasn't thoroughly read the story is painfully obvious. Once in a while, after a surfeit of expansive essays, I have asked a class for a card report just to rest my eyes and to remind them of the virtues of concision. Some students inevitably grumble, but most are in for a reward; some will even be delighted that the assignment is so clearly defined and limited! (Although, as reproduced in the text, the sample card is smaller than 5 by 8 inches, it illustrates exactly the amount of wordage that can be crammed on one card, even if typed with a pica typewriter.)

Warn your students to allow plenty of time to do this job right. Stephen Marcus, of the University of California, Santa Barbara, tells us that some of his students were

(text pages) 1470–1482

appalled to find it took them two or three hours to write one card report. That sounds about par for the assignment; if you want to abbreviate it, you can omit some of the required elements.

To read nothing but card reports all term is, however, a mind crusher. Essays, of course, develop different skills of thinking and organization and will probably still seem necessary.

In choosing an essay topic, if given a choice, many students have trouble deciding how large a topic to attempt in an assigned word length, and many are tempted to choose a topic larger than they can handle. Some want to make sure they'll have enough ideas to fill the word length. Even if you don't care to assign any of the topics suggested in the text, having students read the list on pages 1480–82 may give them a clearer notion of the right amplitude of topics for papers of various lengths.

If this list, and the "Suggestions for Writing" at the end of most chapters, and your own inspiration don't suffice, still more topics for writing may be quarried from the questions after the stories. (For "Stories for Further Reading," the questions are in this manual.)

Writing a Story

This section, we trust, will prove useful in at least two ways. For the instructor who uses the book in a creative writing course, it will provide support. (The instructor who, in teaching composition, makes at least one creative writing assignment will also find uses for it.) Second, those interested in the process of composing (whether or not interested in writing stories) will find in this section many glimpses into the workshops of fine writers.

For a textbook wholly devoted to the problems and techniques of writing stories, see Janet Burroway's *Writing Fiction: A Guide to Narrative Craft*, 2nd ed. (Boston: Little, Brown, 1987).

Besides the "Suggestions for Writing" at the end of this chapter, here are a few ideas offered by Elaine Evanson in *Exchange: A Newsletter for Teachers of Writing*, a publication of the Academic Achievement Center, University of Wisconsin, Stevens Point (vol. 13, no. 1, 1988):

Continue the story. Take the plot further. Add an ironic ending.
Write the diary of one of the characters.
Produce a newspaper that might have originated in the setting of the story.

Writing about a Poem

Here are notes on the two poems contained in "Writing about a Poem."

Robert Frost, DESIGN, page 1493

"Design" is fruitful to compare in theme with Walt Whitman's "A Noiseless Patient Spider" (page 730). One could begin by comparing the early versions of the two poems, Whitman's "The Soul, reaching" and Frost's "In White." What are the themes of these versions? It is more difficult to tell from these vaguer, more general statements. In rewriting, both poets seem not only to have made their details more specific, but also to have defined their central ideas.

If you wish to deal with this section in class, you might have students read "Design" and then the two student papers about it (pages 1493 and 1497). What did these writers notice about the poem that you didn't notice? What did you notice about it that they left out?

Besides Jarrell's classic explication, many other good discussions of the poem may be consulted. Elizabeth Drew has a succinct explication in *Poetry: A Modern Guide to Its Understanding and Enjoyment* (New York: Norton, 1959), and there is a more detailed reading by Richard Ohmann in *College English* 28 (Feb. 1967): 359–67.

Abbie Huston Evans, WING-SPREAD, page 1499

The student's evaluation seems just to us. While "Wing-Spread" is not so vivid a cameo as "Design," nor so troubling in its theme, and while it contains trite rimes (except for *beryl/peril*), we think it a decent poem and admirably terse.

Insufficiently recognized (like most poets), Evans (1881–1983) had a long, productive life. Her *Collected Poems* was published in 1970 by the University of Pittsburgh Press. There are dozens of poems better than "Wing-Spread" in it.

SUGGESTIONS FOR WRITING

Here are a few more topics for paper assignments to supplement the list on page 1504 in the book.

TOPICS FOR BRIEF PAPERS (250–500 WORDS)

1. A *précis* (French, from Latin: "to cut short") is a short abstract or condensation of a literary work that tries to sum up the work's most essential elements. Although a précis, like a paraphrase, states the poet's thought in the writer's own words, a paraphrase is sometimes as long as the original poem, if not longer. A précis, while it tends to be much briefer than a poem, also takes in essentials: theme, subject, tone, character, events (in a narrative poem), and anything else that strikes the writer as important. A précis might range in length from one ample sentence to a few hundred words (if, say, it were condensing a long play or novel, or a complex longer poem). Here, for instance, is an acceptable précis of Robert Browning's "Soliloquy of the Spanish Cloister":

> The speaker, a monk in a religious community, voices to himself while gardening the bitter grudge he has against Brother Lawrence, one of his fellow monks. He charges Lawrence with boring him with dull talk at mealtime, sporting monogrammed tableware, ogling women, drinking greedily, ignoring rituals (unlike the speaker, who

after a meal lays knife and fork in a cross—which seems overly scrupulous). Having vented his grudge by slyly scissoring Lawrence's favorite flowering shrubs, the speaker is now determined to go further, and plots to work Lawrence's damnation. Perhaps he will lure Lawrence into misinterpreting a text in Scripture, or plant a pornographic volume on him. So far gone is the speaker in his hatred that he is even willing to sell his soul to the devil if the devil will carry off Lawrence's; and so proud is the speaker in his own wiles that he thinks he can cheat the devil in the bargain. Vespers ring, ending the meditation, but his terrible grudge seems sure to go on.

As the detailed précis makes clear, Browning's poem contains a chronicle of events and a study in character. The précis also indicates the tone of the poem and (another essential) its point of view.

Students might be supplied with a copy of the above material to guide them and be asked to write précis of four or five poems, chosen from a list the instructor compiles of six or eight poems in the section "Poems for Further Reading."

2. Find a poem that you like, one not in the book so that it may be unfamiliar to other members of the class. Insert into it a passage of five or six lines that you yourself write in imitation of it. Your object is to lengthen the poem by a bit of forgery that will go undetected. Type out the whole poem afresh, inserted lines and all, and have copies duplicated for the others in the class. Then let them try to tell your forged lines from those of the original. A successful forgery will be hard to detect, since you will have imitated the poet's language, handling of form, and imagery—indeed, the poet's voice.

Topics for More Extensive Papers (600–1,000 words)

1. Relate a personal experience of poetry: a brief history of your attempts to read it or to write it; a memoir of your experience in reading poetry aloud; a report of a poetry reading you attended; an account of how reading a poem brought a realization that affected you personally (no instructor-pleasing pieties!); or an account of an effort to foist a favorite poem upon your friends or to introduce young children to poetry. Don't make up any fabulous experiences or lay claim to profound emotions you haven't had; the result could be blatantly artificial ("How I Read Housman's 'Loveliest of trees' and Found the Meaning of Life"). But if you honestly can sum up what you learned from your experience, then do so, by all means.

2. Write an imitation or a parody. This and the following topic may result in a paper of fewer words than the essay topics, but the amount of work required is likely to be slightly more.

(Note: This assignment will be too much of a challenge for some students and not all ought to be required to do it. But those who possess the necessary skill may find themselves viewing the poet's work as if they were insiders.) The instructor has to insist that the student observe the minimal formal requirements of a good imitation. A convincing imitation of, say, Thomas Hardy can hardly be written in Whitmanic free verse. Students may be urged to read whole collections of work in order to soak up a better sense of the poet. This assignment asks much but the quality of the results is often surprising. Honestly attempted, such an exercise requires far more effort from students than the writing of most critical essays, and probably teaches them more.

3. After you have read several ballads (both folk ballads and literary ballads), write a ballad of your own, at least twenty lines long. If you need a subject, consider some event recently in the news: an act of bravery, a wedding that took place despite obstacles, a murder or a catastrophe, a report of spooky or mysterious happenings. Then in a prose paragraph, state what you learned from your reading of traditional or literary ballads that proved useful to you as a ballad composer yourself.

(text pages) 1491–1506

TOPICS FOR LONG PAPERS (1,500 WORDS OR MORE)

1. Leslie Fiedler, the critic and novelist, once wrote an essay in which he pretended to be a critic of the last century ("A Review of *Leaves of Grass* and *Hiawatha* as of 1855," *American Poetry Review* 2 [Mar.–Apr. 1973]). Writing as if he subscribed to the tastes of that age, Fiedler declared Whitman's book shaggy and shocking and awarded Professor Longfellow all the praise. If you can steep yourself in the literature of a former age (or recent past year) deeply enough to feel confident, such an essay might be fun to write (and to read). Write about some poem once fashionabe, now forgotten; or about some poem once spurned, now esteemed. Your instructor might have some suggestions.

2. For a month (or some other assigned period of time), keep a personal journal of your reading of poetry and your thinking about it. To give direction to your journal, you might confine it to the work of, say, half a dozen poets who interest you; or you might concentrate on a theme common to a few poems by various poets.

Writing a Poem
(Some notes by XJK)

These notes are provided mainly for the instructor who employs *An Introduction to Poetry* in a creative writing course. Some may be of interest, however, to anyone who in teaching composition includes a unit on writing poems. Such an instructor will probably have firm persuasions about poetry and about the teaching of poets. Instead of trying to trumpet any persuasions of my own, let me just set down some *hunches* that, from teaching poetry workshops, I have come to feel are mostly true.

In reading a student's poem, you have to look at it with your mind a blank, reserving judgment for as long as possible. Try to see what the student is doing, being slow to compare a fledgling effort to the classics. There's no use in merely reading the poem and spotting any influences you find in it—"Ha, I see you've been reading Williams!" You can, however, praise any virtues you discover and you can tell the student firmly, kindly, and honestly any adverse reactions you feel. Point to anything in the poem that causes you to respond toward it, or against it. Instead of coldly damning the poem's faults, you can inquire why the writer said something in such-and-such a way, rather than in some other. You can ask to have anything you don't understand explained. If a line or a passage doesn't tell you anything, you can ask the student to suggest a fresh way of wording it. Perhaps the most valuable service you can perform for a student poet is to be hard to please. Suggest that the student not settle for the first words that flash to mind, but reach deeper, go after the word or phrase or line that will be not merely adequate, but memorable.

The greatest method of teaching poetry writing I have ever heard of was that of the late John Holmes. Former students at Tufts remember that Holmes seldom made comments on a poem, but often would just lay a finger next to a suspect passage and fix the student with a look of expectancy until the silence became unendurable, and the student began explaining what the passage meant and how it could be put better. (I have never made the Holmes method succeed for me. I can't keep from talking too much.)

Most workshop courses in poetry fall into a classic ritual. Students duplicate their poems, bring them in, and show them around to the class. This method of procedure is hard to improve upon. Some instructors find that the effort of screening the work themselves first and deciding what to spend time on in class makes for more cogent class sessions, with less time squandered on boring or inferior material. In general, class sessions won't be any more lively and valuable than the poems that are on hand. (An exception was a workshop I once visited years ago at MIT. The poems were literal, boring stuff, but the quality of the students' impromptu critical analyses was sensational.) Often a great class discussion will revolve around a fine poem with deep faults in it.

The severest challenge for the instructor, incidentally, isn't a *bad* poem. A bad poem is easy to deal with; it always gives you plenty of work to do—passages to delete, purple adjectives to question. The challenge comes in dealing with a truly surprising, original, and competent poem. This is risky and sensitive work because genuine poets usually know what they are doing to a greater degree than you or any other outsider does; and you don't want to confuse them with reactions you don't trust. For such rare students, all a poetry workshop probably does is supply an audience, a little encouragement, and sometimes even an insight.

There are natural temptations, of course, to which teachers of poets fall prey. Like coin collectors, they keep wanting to overvalue the talents they have on hand, to convince themselves that a student is a Gem Mint State poet, when a less personal opinion might find the student just an average specimen, although uncirculated. It's better to be too slow than too quick to encourage a student to seek nationwide publication. It is another temptation, if you have a class with a competent poet in it, to devote most of each session to that poet's latest works, causing grumblings of discontent (sometimes) among the other paying customers. I believe that a more competent poet deserves more time, but you have to conduct a class and not a tutorial.

Poetry workshops can become hideously intimate. They are bound to produce confessional or diary poems that, sometimes behind the thinnest of fictive screens, confide in painful detail the writer's sexual, psychic, and religious hang-ups. I have known poetry workshops where, by semester's end, the participants feel toward one another like the members of a hostile therapy group. That is why I believe in stressing that a poem is not merely the poet's self-revelation. It usually helps to insist at the start of the course that poems aren't necessarily to be taken personally. (See Chapter 14, "The Person in the Poem," if you want any ammunition.) Everybody will know, of course, that some poets in the class aren't capable of detached art and that a poem about a seduction may well be blatant autobiography; but believe me, you and your students will be happier if they can blow the trump in favor of the Imagination. There is no use in circulating poems in class anonymously, pretending that nobody knows who wrote them. Somebody will know, and I think that the sooner the members of the class freely admit their identities, the more easy and relaxed and open the situation will be. To know each one personally, as soon as they can, is essential.

As the workshop goes on, I don't always stick to a faithful conference schedule. Some will need (and wish for) more of your time than others, but I like to schedule at least one conference right away, at the beginning of the course. This is a chance to meet with students in private and to get a sense of their needs. I tell them to bring in a few poems they've already written, if they've written any. But I make it clear that class sessions will deal only with brand-new poems. At the end of the course, I program another such conference (instead of a final exam), sit down with each student, and ask, "Well, where are you now?"

Some students will lean on you for guidance ("What shall I write about?"); others will spurn all your brilliant suggestions and want to roar away in their own directions. Fine. I believe in offering the widest possible latitude in making assignments—but in having *some* assignments. Even the most inner-directed poet can learn something from being expected to move in a new direction. Having a few assignments will discourage the customers who think they can get through any number of creative writing courses by using the same old yellowed sheaf of papers. Encourage revision. Now and then, suggest a revision as an assignment instead of a new poem.

In "Writing a Poem" I offer a radical suggestion: that the students memorize excellent poems. Feeling like a curmudgeon for making this recommendation, I was happy to find some support for it in the view of Robert Bly, who remarks in *Coda* (June/July 1981):

> I won't even read a single manuscript now, when I visit a university workshop, unless the poet in advance agrees to memorize fifty lines of Yeats. At the first workshop I visited last fall it cut the number of graduate-student writers who wanted to see me from 15 to 2. Next year I'm changing that to fifty lines of *Beowulf*.

Bly may seem unreasonably stern, but he and I agree on the value of memorization. I believe it helps coax the writing of poetry down out of the forebrain, helps it unite with the pulse.

Bly has sane things to say, in this same article, about the folly of thirsting for publication too early. And incidentally, here's one of his unorthodox exercises for a writing workshop (imparted in an interview in the *Boston Globe Magazine* for April 10, 1988):

> One workshop, I brought in an onion for each of the students. I asked everybody to spend 10 to 15 minutes describing the exterior of the onion, using all of their senses.

That requires every bit of observation you have, to remain looking at the onion. Then, in the second part of the exercise, I said, "Now I want you to compare the onion to your mother."

That must have rocked 'em! I wonder if it produced any good results.

For a textbook wholly devoted to the writing of poetry, quite the best thing of its kind, see Robert Wallace's *Writing Poems, Second Edition* (Boston: Little, Brown, 1987).

Another book crammed with teaching hints and lively writing exercises for student poets is *The Teachers & Writers Handbook of Poetic Forms*, edited by Ron Padgett (1987), and sold by Teachers & Writers Collaborative, 5 Union Square West, New York, NY 10003. Besides supplying unstuffy definitions and examples of many expected forms, the *Handbook* deals with blues poems, collaborations, ghazals, insult poems, light verse, pantoums, performance poems, raps, renga, and more.

Although knowing something about *any* element of poetry may benefit a poet-in-training, here is a list, chapter by chapter, of material in *An Introduction to Poetry* that may be particularly useful in a creative writing class.

Chapter 2, THE PERSON IN THE POEM, page 515
    Novice poets often think of their poems as faithful diary accounts of actual experiences. This section may be useful to suggest to them that, in the process of becoming art, the raw material of a poem may be expected to undergo change.

Chapter 3, DAVID B. AXELROD, Once in a While a Protest Poem, page 553
    Assignment: Write a protest poem of your own.

Chapter 5, ABOUT HAIKU, page 577
    Assignment: Write some haiku, either original or in imitation of classic Japanese haiku.

Chapter 5, *Experiment: Writing with Images*, page 580
    A poetry writing assignment with possible examples.

Chapter 6, HOWARD MOSS, Shall I Compare Thee to a Summer's Day? page 586
    Assignment: Choosing a different famous poem, write a Mosslike version of it. Then try to indicate what, in making your takeoff, was most painful to leave out. (See also George Starbuck's parody "Margaret Are You Drug," page 600.)

Chapter 6, JANE KENYON, The Suitor, page 600
    Assignment: Write a poem similarly constructed of similes, or metaphors.

Chapter 7, PAUL SIMON, Richard Cory, page 610
    Assignment: In somewhat the fashion of Simon's treatment of Robinson, take a well-known poem and rewrite it as a song lyric. Try singing the result to a tune.

Chapter 8, *Exercise: Listening to Meaning*, page 624
    Assignment: After reading these examples, write a brief poem of your own, heavy with sound effects.

Chapter 8, READING AND HEARING POEMS ALOUD, page 637
    Assignment: Ponder this section before reading your own poems aloud in class.

Chapter 9, METER, page 649
    Assignment: After working through this section on your own, write a poem in meter.

Chapter 10, *CLOSED FORM, OPEN FORM*
    This whole chapter may be of particular value to a poetry writing class. Not only does it analyze some traditional forms, it suggests a rationale for open form verse and tries to suggest why competent verse is seldom "free."

(text pages) 1507–1521

    Assignment: After considering the definition of *syllabic verse* given in this chapter, carefully read Dylan Thomas's "Fern Hill" (page 886). Work out the form of Thomas's poem with pencil and paper, and then try writing a syllabic poem of your own.

    Assignment: Ponder, not too seriously, Wallace Stevens's "Thirteen Ways of Looking at a Blackbird" (page 682). Then, as the spirit moves you, write a unified series of small poems.

Chapter 11, *Experiment: Do It Yourself*, page 700
    An exercise in making a concrete poem.

Chapter 14, THE POET'S REVISIONS, page 727
    This section may help drive home the fact that poets often revise.

Chapter 14, TRANSLATIONS, page 731
    Assignment: Consider the translations in this section and decide what you admire or dislike in each of them. Translate a poem of your choice, from any language with which you are familiar or can follow in a bilingual edition.

Chapter 14, PARODY, page 736
    Assignment: Read these parodists, comparing their work with the originals. Then, choosing some poet whose work you know thoroughly, write a parody yourself.

Chapter 15, EVALUATING A POEM
    Warning: Although you may care to give "Telling Good from Bad" a try, this is dangerous matter to introduce into a poetry writing class. Young poets already tend to be self-consciously worried that their work will be laughed at. Save this section for late in the course, if you use it at all.

Further Notes on Teaching Poetry

These notes are offered (by XJK) in response to the wishes of several instructors for additional practical suggestions for teaching poetry. They are, however, mere descriptions of a few strategies that have proved useful in his own teaching. For others, he can neither prescribe nor proscribe.

1. To a greater extent than in teaching prose, the instructor may find it necessary to have poems read aloud. It is best if students do this reading. Since to read a poem aloud effectively requires that the reader understand what is said in it, students will need advance warning so that they can prepare their spoken interpretations. Sometimes I assign particular poems to certain people or ask each person to take his or her choice. Some advice on how to read poetry aloud is given in Chapter 20. I usually suggest only that students beware of waxing overemotional or rhetorical, and I urge them to read aloud outside of class as often as possible. If the student or the instructor has access to a tape recorder, it may be especially helpful.

2. It is good to recall occasionally that poems may be put back together, as well as taken apart. Sometimes I call on a student to read a previously prepared poem just before opening a discussion of the poem. Then, the discussion over and the poem lying all around in intelligible shreds, I ask the same student to read it over again. It is often startling how the reading improves from the student's realizing more clearly what the poet is saying.

3. I believe in asking students to do a certain amount of memorization. Many groan that such rote learning is mindless and grade-schoolish, but it seems to me one way to defeat the intellectualizations that students (and the rest of us) tend to make of poetry. It is also a way to suggest that we do not read a poem primarily for its ideas: to learn a poem by heart is one way to engrave oneself with the sound and weight of it. I ask for twenty or thirty lines at a time, of the student's choice, and then have them write out the lines in class. Some students have reported unexpected illuminations. Some people, of course, can't memorize a poem to save their souls, and I try to encourage but not to pressure them. These written memorizations take very little of the instructor's time to check and need not be returned to the students unless there are flagrant lacunae in them.

4. The instructor has to sense when a discussion has gone on long enough. It is a matter of watching each student's face for the first sign of that fixed set of the mouth. Elizabeth Bishop once wisely declared that, while she was not opposed to all close analysis and criticism, she was against "making poetry monstrous and boring and proceeding to talk the very life out of it." I used to be afraid of classroom silences. Now, I find it helps sometimes to stop a discussion that is getting lost and say, "Let's all take three minutes and read this poem again and think about it silently." When the discussion resumes, it is usually improved.

Two of the finest, most provocative essays on teaching poetry in college I have seen are these:

Alice Bloom, "On the Experience of Unteaching Poetry," *Hudson Review* (Spring 1979): 7–30. Bloom: "I am interested in the conditions of education that would lead a student to remark, early in a term, as one of mine did, that 'I wish we didn't know these were poems. Then it seems like it would be a lot easier.'"

Alan Shapiro, "The Dead Alive and Busy," *TriQuarterly* (Fall 1984): 62–70. In the main, this is a memoir of a Stanford freshman, Patty Smith, an early victim of cancer, and the way she approached poetry. In recalling her and her insightful comments on poems, Shapiro seeks an answer to a widespread problem: "It never ceases to disappoint me that the poems I have greatest difficulty teaching are the ones I care the most about."

On Integrating Poetry and Composition

How do you teach students to read poetry and, at the same time, to write good prose? Instructors who face this task may find some useful advice in the following article, first published in *The English Record*, bulletin of the New York State English Council (Winter 1981). It is reprinted here by the kind permission of the author, Irwin Weiser, director of developmental writing, Purdue University.

The Prose Paraphrase:
Integrating Poetry and Composition
Irwin Weiser

Many of us teach composition courses which demand that we not only instruct our students in writing but that we also present literature to them as well. Such courses often frustrate us, since a quarter or a semester seems too brief to allow us to teach fundamentals of composition alone. How are we to integrate the teaching of literature with the teaching of writing? What are we to do with a fat anthology of essays, fiction, poetry, or drama and a rhetoric text and, in some cases, a separate handbook of grammar and usage?

Recently, I tried an approach which seemed to provide more integration of reading and writing than I previously had felt I attained in similar courses. The course was the third quarter of a required freshman composition sequence; the departmental course description specifies the teaching of poetry and drama, but also states "English 103 is, however, primarily a composition, not a literature, course. Major emphasis of the course should be on writing." The approach I will describe concerns the study of poetry.

Because this is a writing course, I explained to my students that we would approach poetry primarily as a study of the way writers can use language, and thus our work on denotation and connotation, tone, irony, image, and symbol should help them learn to make conscious language choices when they write. Chapters in the poetry section of X. J. Kennedy's *Literature* entitled "Words," "Saying and Suggesting," and "Listening to a Voice" fit nicely with this approach. Further, because this is a writing course, I wanted my students to have frequent opportunities to write without burying myself under an even greater number of formal, longish papers than I already required. An appropriate solution seemed to be to have my students write prose paraphrases of one or two poems from those assigned for each discussion class.

During the first week of the course, we discussed and practiced the paraphrase technique, looking first at Kennedy's explanation of paraphrasing and then at his paraphrase of Housman's "Loveliest of trees, the cherry now." By reading my own paraphrase, not among the ablest in the class, I was able to place myself in the position of coinquirer into these poems, most of which I had not previously taught. This helped establish a classroom atmosphere similar to that of a creative writing workshop, one conducive to the discusson of both the poetry in the text and the writing of the students. In fact, while the primary purpose of assigning the paraphrases was to give my students extra writing practice, an important additional result was that throughout the quarter their paraphrases, not the teacher's opinions and interpretations, formed the basis for class discussion. There was rarely a need for the teacher to *explain* a poem or a passage: someone, and frequently several people, had an interpretation which satisfied most questions and resolved most difficulties.

At the end of this essay are examples of the prose paraphrases students wrote of Emily Dickinson's "I heard a fly buzz – when I died." Two of the paraphrases, at 90 and 112 words, are approximately as long as Dickinson's 92-word poem; the 160-word third paraphrase is over 75% longer because this student interpreted as she paraphrased, explaining, for example, that the narrator willed her earthly possessions in a futile attempt to hasten death. Such interpretation, while welcome, is not at all necessary, as the two shorter, yet also successful, paraphrases indicate. In fact, I had to remind students that paraphrases are not the same as analyses, and that while they might have to interpret a symbol—as these students variously explained what the fly or the King meant—or unweave a metaphor, their major task was to rewrite the poem as clear prose.

The first paraphrase is perhaps the most straightforward of this group. The author's voice is nearly inaudible. He has stripped the poem of its literary qualities—no "Heaves of Storm," only "the air before a storm"; no personification; the author is only present in the choice of the word "sad" to describe the final buzz of the fly. His paraphrase is a prose rendering of the poem with no obvious attempt to interpret it.

Paraphrase II seems to ignore the symbolic importance of the fly, and perhaps in the very casualness of the phrase "and the last thing I was aware of was this fly and its buzz" suggests the same insignificance of death from the perspective of the hereafter that Dickinson does. More interesting is this student's treatment of the willing of the keepsakes: the formal diction of "proper recipients," "standard fashion," and "officially ready to die" suggests death as a ritual. Unexpected interpretations like this appear frequently in the paraphrases, demonstrating the flexibility and richness of language, emphasizing the error in assuming that there is one *right* way to interpret a poem, and sometimes, when the interpretations are less plausible, leading to discussions of what constitutes valid interpretation and how one finds support for interpretations of what one reads.

The third paraphrase, as I suggested before, offers more interpretation as well as a stronger authorial voice than the previous two. The author adds a simile of her own, "as if the winds had ceased temporarily to catch their breaths," and more obviously than the other students uses the fly as a metaphor for death in her final sentence.

I will not take the space for a thorough analysis of these paraphrases, but I think that they suggest what a teacher might expect from this kind of assignment. Clearly, these three students have read this poem carefully and understand what it says, the first step towards understanding what it means. Small group and classroom discussions would allow us to consider these paraphrases individually and comparatively, to point out their merits and weaknesses, and then to return to the original verse with new perspectives.

Most heartening were the comments of several students during the quarter who told me that they felt more confident about reading poetry than they previously had. Though I doubt that my students are any more ardently devoted to poetry now than they were before the course began, they are not intimidated by verse on the page. They have an approach, a simple heuristic, for dealing with any unfamiliar writing. Ideally, my students will remember and use their ability to paraphrase and their ability to use their paraphrases to understand and evalute what they read when they come upon a particularly difficult passage in their chemistry or history texts during the next three years or in the quarterly reports or technical manuals or journals they will read when they leave the university and begin their careers.

Appendix: Sample Paraphrases

Paraphrase I

I heard death coming on. The stillness in the room was like the stillness in the air before a storm. The people around me had wiped their eyes dry, and they held their breaths waiting for that moment when death could be witnessed in the room. I wrote a will which gave away my possessions—that being the only part of me I could give away. A fly then flew between the light and me making a sad, uncertain buzz. My eyesight failed and I could not see to see.

Paraphrase II

I heard a fly buzz as I was about to die. The sound of the fly broke the quietness in the room which was like the calm before a storm. The people sitting around waiting for me to die cried until they could not cry anymore. They began to breathe uneasily in anticipation of my death when God would come down to the room to take me away. I had willed all of my valuables to the proper recipients in the standard fashion. I was officially ready to die, going through the final dramatic moments of my life, and the last thing I was aware of was this fly and its buzz.

Paraphrase III

I could feel the approach of death just as I could hear the buzz of an approaching fly. I knew death was buzzing around, but I did not know when and where it would land. The stillness of death was like the calmness that exists between storms, as if the winds had ceased temporarily to catch their breaths.

I was aware of the sorrow in the room. There were those who had cried because death was near, and they waited for death to stalk into the room like a king and claim its subject.

I willed all of my earthly possessions, all that could legally be assigned to a new owner, in an attempt to hasten death. But there was no way to control death; I was at the mercy of its timing. And then like the fly that finally lands on its choice place, death fell upon me, and shut my eyes, and I could no longer see.

* * *

Mr. Weiser reported in a letter that, once again, he had used the method of poetry paraphrase in his writing course, and remained pleased with it. "My students," he remarked, "no longer treat poems as holy scripts written in some mystical code, but attack them fearlessly." The course had proved fun both for them and for him, and he felt he was paying his dues to both writing and literature.

Writing about a Play

In an introductory literature course that saves drama for last, there seems never enough time to be fair to the plays available. That is why many instructors tell us that they like to have students read at least two or three plays on their own and write short papers about at least one or two of them.

If you decide to assign any such critical writing but find your time for paper-grading all the more limited as your course nears its end, you might consider assigning a card report (discussed and illustrated on pages 1476–79). Earlier in this manual (in "Writing about a Story"), we trumpet the virtues of card reports—which aren't every instructor's salvation, but which we have found to work especially well for teaching plays. The card report shown in the book—on Glaspell's *Trifles*—manages to cover a one-act play in two card faces. For a longer and more involved play, you might wish to limit the students' obligation to just a few of the play's elements (leaving out, say, *Symbols* and *Evaluation*). Otherwise, they'll need more than one card.

Among the "Topics for More Extended Papers" (page 1481), number 6 invites the student to imagine the problems of staging a classic play in modern dress and in a contemporary setting. If you prefer, this topic could be more general: Make recommendations for the production of any play. In getting ready to write on this topic, students might first decide whether or not the play is a work of realism. Ask them: Should sets, lighting, and costumes be closely detailed and lifelike, or perhaps be extravagant or expressionistic? Would a picture-frame stage or an arena better suit the play? What advice would you offer the actors for interpreting their roles? What exactly would you emphasize in the play if you were directing it? (For a few insights into methods of staging, they might read Tennessee Williams's "How to Stage *The Glass Menagerie*" on page 1446.)

Further topics for writing about plays will be found at the ends of Chapters 32, 33, 34, 35, and 36.

Writing a Play

If, by luck, you should have any up-and-coming playwrights among your students, they will probably find this necessarily brief discussion a mere whistle-whetter. All it can do, in four pages, is to point such students toward critically watching plays and, if possible, taking part in productions and performances.

Budding playwrights—the few we have known—tend to think that writing a play is mainly a matter of writing speeches to be spoken. It takes them a while to learn that the stage is more than a sounding-box for speech and that a playwright works not only with keyboard and paper but with people and properties. A book we would like to place into any budding playwright's hands is Peter Arnott's *The Theater in Its Time* (Boston: Little, Brown, 1981). A veteran director, actor, and professor of drama, Arnott writes the most readable history of world drama we know and keeps ever in mind the practicality of the stage. His chapters on the problems that actors, directors, and designers have to deal with are excellent, and he appends a day-by-day diary of a typical play rehearsal.

Like "Writing a Story" and "Writing a Poem" in this Supplement, "Writing a Play" tries to offer a few insights into the writing process. There are small revelations by Albee, James Thurber (as a playwright), and Harold Pinter, and a little information on how *The Glass Menagerie* apparently grew out of Williams's life.

Among the writing topics, we recommend number 3 in particular. To turn a short story into a play may seem an impossibly difficult challenge, but you just might find a student glad to respond to it. Such a student will learn a good deal about both fiction and drama from the inside. His or her finished script may be produced very simply—at least read by volunteer actors—in front of the class. (XJK notes: I still remember a slow-moving American literature class that leaped to life one day when a student directed a few classmates in the play he had made out of Hawthorne's "Rappaccini's Daughter." The story lent itself to melodramatic hamming, which both the cast and the audience hugely enjoyed.) For inspiration, the student might read Wendy Wasserstein's one-act play based on Chekhov's "The Man in a Case."

If any student cares to try adapting a short story for television, model scripts (and the short stories that inspired them) are available in two paperback volumes based on the PBS Television film series, *The American Short Story* edited by Calvin Skaggs (New York: Dell), at last report still in print. Volume 2 (1980) contains scenes from the television scripts for "Barn Burning" and "Paul's Case," and the entire TV script for "The Jilting of Granny Weatherall," with an interview with the scriptwriter Corinne Jacker, who discussses the problems of adapting Porter's story.